SACRAMENTO PUBLIC LIBRARY
D0051029

flowers

More Praise for *Flowers*

"*Flowers* is a delightful rendering of the natural history of plants. At its essence, it is truly botany made interesting and accessible. Anyone who ever thought they would like to learn more about flowers and their mysterious habits should read this wonderful book. Dr. Burger has done us all a service by crafting a volume that explains, in clear terms, how plants literally shaped the world we live in and why their contribution to life on this planet has profoundly impacted the rise of humankind."

David Lentz, vice president, Chicago Botanic Garden

"Our love of flowers has deep biological roots. Burger discusses complicated and important facts in a simple, forthright, and charming way. Anyone reading this book will enhance his or her love for nature by understanding the evolutionary intricacies behind the beauty. You'll never look at a flower the same way after reading Burger's enchanting and informing book."

Hugh H. Iltis, emeritus director of the Herbarium,
Department of Botany, University of Wisconsin–Madison

"Fascinating and a joy to read, William Burger's *Flowers* offers delightful nature stories as well as profound insights into the complex interrelations between plants and animals—including us—and how these webs of life came to be."

Gilbert Waldbauer, author of *Insights from Insects* and
Millions of Monarchs, Bunches of Beetles

"William Burger, with experience of a lifetime of outstanding research in flowering plant systematics and evolution, is a very qualified author on this subject. He takes the reader through the world of flowers, beginning with what they are, on to how they originated, and through to their importance in human affairs. One important message: without flowering plants, the human species would doubtless never have existed. . . . Try a taste—your view of flowers in the back garden or on the dining room table will never again be the same."

Tod Stuessy, professor and head, Department of Higher Plant Systematics and Evolution, and director, Botanical Garden, University of Vienna

william c. burger

flowers
how they changed the world

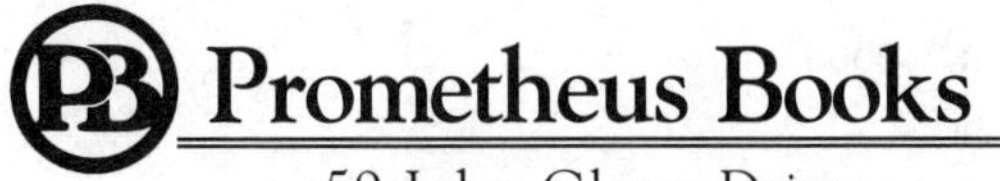

59 John Glenn Drive
Amherst, New York 14228-2197

Published 2006 by Prometheus Books

Inquiries should be addressed to
Prometheus Books
59 John Glenn Drive
Amherst, New York 14228–2197
VOICE: 716–691–0133, ext. 207
FAX: 716–564–2711
WWW.PROMETHEUSBOOKS.COM

10 09 08 07 06 5 4 3 2 1

Library of Congress Cataloging-in-Publication Data

Burger, William C.
Flowers : how they changed the world / William C. Burger.
p. cm.
Includes bibliographical references and index.
ISBN 1–59102–407–2 (hardcover : alk. paper)
1. Botany—Miscellanes. 2. Plants—Miscellanes. 3. Flowers—Miscellanes. I. Title.

QK45.2.B87 2006
582.13—dc22

2006002739

Printed in the United States on acid-free paper

contents

acknowledgments

This book is the product of a lifetime's love of nature. That appreciation, fostered by my mother, was enhanced by summers spent at our small cabin in the Hudson Highlands, north of New York City. Though struggling with serious instabilities during the Depression, my parents managed to build and sustain that woodland retreat. For a child growing up amidst the concrete and asphalt of Manhattan, summertime among fields, flowers, and forests immersed me in a completely different world. Here I could watch bumblebees at work, explore within the forest's dark interior, or listen to katydids calling through the night.

acknowledgments

It was in ninth grade that a redheaded biology teacher convinced me he wasn't working for a living. His job, I decided, was exactly what I wanted to do, and my college education was focused on that end. But then came the army, followed by the GI Bill. Suddenly, thanks to the generosity of American taxpayers, I had the wherewithal to become a biologist. Before beginning graduate work, however, I served as a laboratory assistant under Dr. Ulrich Weiss, a chemist at the New York Botanical Garden. We were part of a project working to identify antibacterial compounds extracted from soil fungi. Dr. Weiss's profound respect and genuine enthusiasm for the scientific enterprise served as an example that has guided me ever since.

American taxpayers helped play a further role in the development of this book. This time it was through the United States Agency for International Development, supporting an agricultural college in eastern Ethiopia, administered by Oklahoma State University. Here, I spent four years within a mountainous landscape of extraordinary biological and cultural variety. Teaching undergraduates who were intently diligent in their studies, and interacting with a faculty dedicated to improving agricultural productivity, all within one of the world's oldest Christian societies, was an immensely rewarding experience.

Returning to the United States, I was fortunate to join Chicago's Field Museum in its work on botanical inventories of Central America. Here, too, federal funding—through the National Science Foundation—contributed to fieldwork in Costa Rica and research over three decades, allowing me to study one of the world's most diverse floras. However, while interacting with scholars in anthropology, geology, and zoology, both here at Field Museum and at the University of Chicago, I became convinced that flowering plants were not getting their share of the credit for the bountiful diversity of today's world. Few people seemed to understand how critical these plants have been to the success of our own species. Clearly, such deficiencies were worthy of a book-length exposition. Fred Barrie, a colleague and plant taxonomist, and Carol Ulrich, gardener and editor, suggested important revisions to the first draft of this book. Earlier, the other Burgers—Melinda, Helen, and Carolyn—had listened to many of my ideas regarding the flowering plants, and their responses helped shape many aspects of the text. Finally, Prometheus Books editors Linda Regan and Mary Read cleansed the text of repetitious phrasing and made many useful suggestions.

When all is said and done, this book is about a very important group of living things that do not command the attention and respect they deserve. Because they do not crawl, sting, bite, or carry virulent diseases they are often seen as uninteresting or unimportant. Nevertheless, these are living things that, in the words of Karl Niklas, ". . . thrive without intention, build without blood or brain, move without muscle, summon without self-awareness, and feed the world without intent."[1]

introduction

Surely, flowers are among the most endearing aspects of our environment. Be it in a carefully tended garden, a tallgrass prairie, or even a vacant lot, their bright, cheery colors help make the world a more pleasant place to be. Whether borne by a little weed at the edge of your driveway or covering the surface of a tropical tree, flowers add a significant dash of color to our natural surroundings. However, such floral exuberance is limited to the lush growing season. Prolonged dry seasons or severe winters will cast a pall of lifelessness across the landscape. But start adding warmth in our northern springtime, or bring on the rains after a long tropical dry season, and the natural world becomes transformed. Both the warmer temperatures of springtime and the renewed growth of a rainy season begin a flurry of plant activity. Many species flower quickly, though others build their greenery first and display their flowers later. All participate in a seasonal progression that is repeated year after year. Likewise, fruiting and seed production are precisely timed for each species. Unusually stressful seasons, prolonged drought, or local calamity may disrupt these patterns, but plant life soon recovers and is back into synchrony with our planet's yearly journey round the sun.

Flowering plants follow in cadence with our annual cycle of seasons; that same cadence is also central to our own daily lives—as it has been ever since we were hunters and gatherers, tens of thousands of years ago. The invention of agriculture bound us even more tightly to the seasons. Though preparing the ground for planting and caring for domesticated animals may have more severely constrained our lives, these innovations helped multiply our numbers. Thanks to our new "partners"—both plants and animals—we humans now had a more assured food supply. Agriculture became a very special symbiosis between ourselves and a few species of useful plants and animals. Based on this new relationship, and in the right locations, humans were able to build grand civilizations. Even today, in our complex technological world, it is the flowering plants that provide us with nearly all the vegetable energy that sustains us. Flowering plants also provide the feed and pasture for most of the animals that help nourish us. Clearly, flowers and the plants that bear them have played a critical role in the human saga.

introduction

With a love for flowering plants that only avid gardeners can truly appreciate, my mother imparted some of her enthusiasms to me. When I was little, I was quite convinced that plants produced flowers because they were happy. I had noticed that the woods had very few flowers, and it seemed obvious that those woodland plants weren't nearly as happy as those in my mother's carefully tended garden. Better yet, that happiness was shared with busy bees and lovely butterflies that regularly visited our garden. Each year of my childhood, I became further captivated by the colorful opulence of the blooming season. The early flowers of springtime were accompanied by the joyful songs of birds. As the weeks progressed, flowering expanded across the landscape, even as the birds were quieting down. By August flowering was reaching its peak, now accompanied by a chirping chorus of little six-legged creatures. Cooler temperatures in September witnessed a sharp decline in flowering but a bonanza of fruits and vegetables. By mid-October cold temperatures had ended flowering, but now the leaves of hardwoods exploded into a final burst of brilliant color. Then, in November, the last leaves would flutter to the ground and the woods took on the barren austerity of a long winter. Insect serenades had ended with the first hard freeze; both field and forest now fell silent—except for cold winds whistling across bare branches. Long cold winter, mostly gray and unpleasant, could, on occasion, produce moments of splendor when moist snow covered every twig and branch in ermine. Finally, windy March brought longer days, warmer temperatures, and weird choruses arising from ponds and swamps as frogs and toads announced the coming of spring. By mid-April our first wildflowers were blooming again on the forest floor, and by May the flowering spectacle was beginning in fields and prairies. The blooming cycle had resumed.

As a child I had no conception of how important flowers were in the life of the landscape, or that there might have been times, eons ago, when the world's green landscapes were devoid of flowers. Nor did I realize that grasses and sedges also had flowers, albeit little ones. And I had no idea of what flowers might actually be doing—though I had figured out that plants with different kinds of flowers produced very different kinds of fruits, ranging from blueberries and tomatoes to apples and pumpkins. It would only be much later that I came to understand how all these devices functioned in the central purpose for every form of life: reproduction.

All complex living things, in many generations over time, suffer countless tribulations and eventual death. Only those species and populations that reproduce, not just successfully but abundantly, survive such onslaughts. Bad weather, virulent diseases, and hungry predators incessantly erode the living world. Only successful reproduction can counter these destructive forces. Flowers exist because they are part of the reproductive exuberance one finds throughout the living world. Flowers are functional; they are a critical way in which one group of plants has "learned" not only to survive but also to prosper.

More significantly, flowering plants are "photosynthesizing autotrophs"—they *energize themselves* by capturing the energy of sunlight. We humans and other members of the animal world are "heterotrophs," using other organisms or organic matter to energize ourselves. Thus, even though they don't seem to be too lively, green plants are the fundamental energy resource for most of the biosphere. Because flowering plants are so dominant amid the greenery that adorns our planet's land surfaces, they provide most ecosystems with their enabling energy. The flowering plants—botanists call them the angiosperms—number about 260,000 species. With *all* land plants estimated to number about 300,000 species, it is clear that the angiosperms are the biggest game in town.

However, flowering plants are by no means the most species-rich lineage of living things. When you add together all the species of butterflies, moths, beetles, bees, wasps, grasshoppers, and flies, the number of just these insect species exceeds a million. Beetles alone number over 460,000 species, with more species being discovered every day. But all those beetles don't add up to a lot of biomass. Pack a thousand of most any beetle species in a gallon jug and you'll have space left over. Here's another reason why flowering plants are so important: they are big. Though a few, like pond scum (*Lemna* and its allies), are smaller than a penny, most flowering plants are substantial in size—at least when compared to beetles. So, if we consider size and energy-capturing photosynthesis together, flowering plants turn out to be the most significant organisms of the rain forest, the tallgrass prairie, the savanna woodland, and the agricultural landscapes of our planet.

Though commonplace, a field of flowers with all its buzzing bees and chattering birds can seem to be so complex, yet so coherent in its overall functioning, as to border on the miraculous. Whether an exquisite flower or an elaborate butterfly, these manifestations of nature have inspired the age-old question: How did such wonderful creatures come to be? For many people the answer has been to imagine an intelligent Deity, and each little marvel the product of His design (later becoming known as "Intelligent Design"). That the ordinary operations of nature—birth and death, success and failure—might have fashioned so many natural wonders is difficult to imagine. Charles Darwin challenged this notion when he suggested that the manifest perfection of nature might have been forged by the selection of those that are better adapted, over eons of time. What Darwin's idea did was to interpret biological diversity as the product of a long evolutionary history. And by inferring *only* natural processes, he opened the history of life to scientific inquiry. Darwin himself, for example, conjectured that humans originated in Africa, because that's where chimps and gorillas live, the animals most similar to us. Other scientists had postulated that Asia was mankind's original home, and orangutans more closely related to us. After more than a century of intensive investigations, fossil discoveries in Africa have sustained Darwin's original suggestion. Today, using DNA sequences, we know that, of all the animals, the genetic code of chimps is, indeed, closest to our own. And we have also discovered that the peoples of Africa have more genetic diversity among themselves than is found in all the rest of humanity around the world, further supporting our African origins. By focusing on data from nature itself, and natural processes, modern science has confirmed the claims of evolution as a powerful explanatory framework: a framework which has allowed us to understand both the workings of nature and its long history. Plants and people have genes and chromosomes which function in much the same way. There is a profound unity in the diversity of life on planet Earth.

Nevertheless, many people still feel that the grandeur of nature is too vast and too intricate to be explained in this way; such sentiments have lead to the promotion of "intelligent design" as an explanatory mechanism. Though this idea has been actively discussed in the press over the last ten years, those who promote this view have an interesting problem. *Not a single scientific paper* utilizing intelligent design theory has been published over this same time period. And the reason is simple. There is absolutely no way to test this general idea or to seek substantiating evidence in the world around us. One might say that the Intelligent Designer has left no fingerprints. I believe that trying to bring "intelligent design" into modern science is akin to bringing a basketball to the golf course. People don't bring basketballs to the golf course because there's nothing they can do with them once they get there. Similarly, because "intelligent design" has no way of being tested or evaluated in the world of nature, it cannot be a part of the scientific enterprise. And that, quite simply, is why "intelligent design" has no place in courses that claim to teach biological science. Similarly, this is why, in examining and discussing flowers in the following pages, we'll adhere to standard scientific practice and discourse. A powerful methodology, modern science has given us both deep insights and a coherent picture of the natural world. Whether it's geologists studying Earth's long history, or cardiologists trying to learn more about heart disease, science has become humankind's single most intellectually successful activity. Studying plant science has been a vital part of this grand enterprise, which you shall see in the pages that follow.

In the highly acclaimed collection *The Immense Journey*, Loren Eiseley included an essay titled "How Flowers Changed the World." What startled me about this piece was not only its deep insights and eloquence but that it had been written by an anthropologist who was also a poet. No botanist or biologist whom I'm aware of has so clearly captured the true significance of flowering plants in a similar way. Perhaps, I thought, biological scientists, eagerly studying each and every tree, had failed to note the grandeur of the forest. It took a son of the open Nebraska prairie to think deeply about the flowering plants and to realize the transformation they had wrought upon our world. Eiseley's essay rings as true today as it did when it was first published almost fifty years ago. A massive accumulation of new scientific knowledge over recent decades has only strengthened his insights. Though the fossil record of plants is quite meager (when compared to that of animals), it does indicate that today's land surfaces are embellished with a diversity of life far greater than in the distant past. Yes, indeed, the flowering plants have changed the world. The poet ended his essay as follows:

> Without the gift of flowers and the diversity of their fruits, man and bird, if they had continued to exist at all, would be today unrecognizable. Archaeopteryx, the

> lizard-bird, might still be snapping at beetles on a sequoia limb; man might still be a nocturnal insectivore gnawing a roach in the dark. The weight of a petal has changed the face of the world and made it ours.[1]

Is it possible that a single lineage of green plants, not particularly animated, could fashion major changes on the land surfaces of our planet—changes that encouraged hordes of animals to follow along in explosive variety? Is it really possible that our human ascendancy was powered, if not by petals, then at least by the plants that bear them?

In this book we will do more than simply discuss flowers and remind ourselves of their importance. Our final purpose will be to show how the flowering plants have, over more than 100 million years, transformed much of the living world. But before we do that we need to get up close and personal with flowers and the plants that bear them. After that, we'll discuss what flowers do, why they do it, what helps them, and what hinders them. Then we'll consider what distinguishes the flowering plants and makes them so special. Finally, we'll examine how they have transformed terrestrial ecosystems, supported the origin of primates, and helped us humans become the masters of our planet. So let's begin our journey by taking a close look at a few of our common, yet elegant, wildflowers.

William C. Burger
Botany Department
The Field Museum
Chicago, Illinois

chapter 1

what, exactly, is a flower?

Biology, sad to say, is not an exact science. Unlike chemistry and physics, there are few simple rules and virtually nothing that can be called a law. While some general themes are plain to see, many biological categories contain a wide array of variations. One would think that something like "a flower" might be easy to define and characterize, but such is not the case. Flowers come in a huge variety of shapes and sizes, from the large and lovely water lilies, roses, and orchids to the small greenish or colorless florets of grasses and sedges. Needless to say, these "flowers" must all possess certain features and structures that have a similar function and similar basic parts, the things that really make them flowers. To get a better idea of the basic parts, we'll begin with a common flower of early spring woodlands.[1]

A Wild Geranium

Wild geraniums are small herbs of the forest floor. Thanks to a well-developed rootstock, these plants die back at the end of summer but sprout again to flower the following spring; that is to say they are *perennials*. The common wild geranium (*Geranium maculatum*) has most of its leaves arising from near the ground, with the flowering stems growing one to two feet tall. These plants of our eastern states are found only on the forest floor and along the edge of wooded areas; they flower as the leaves of trees are beginning to expand and the forest floor becomes shaded. Borne on slender stalks (the *pedicels*) at the ends of stems, the flowers themselves are about 1–1.8 inches (2.5–4 cm) broad (see fig. 1A, p. 24). The five petals make a flat or slightly curved surface, presenting a bold lilac to rose-purple display (see color insert B). Beneath these petals are five narrow lance-shaped green *sepals*. The sepals enclosed and protected the flower bud as it was developing. As flowering commenced, the sepals split apart to allow the petals to unfurl and expand. With that expansion the flower came into full bloom.

Within the five petals, around the center of the flower, we find ten *stamens*—the "male parts" of the flower. Each stamen is made up of a slender stalk, the *filament*, holding a saclike *anther* at its apex. It is within the anther that the pollen grains develop, and it is from the opening anthers that the mature pollen grains are shed. Collectively the ten stamens are called the *androecium*, from the Greek word for "male household." Note that the number of stamens, ten, is double the number of petals; we see such patterns in many other flowers as well.

Finally, in the center of the open flower we find the female parts of the flower: the *gynoecium* or "female household." As in many flowers, we find a single structure in the geranium, which we call the *pistil*. In geranium the pistil is a long tubular structure with a narrow apex (called the *style*) that will split near the top into five style-branches, or *stigmas*. It will be on the stigmas that, if conditions are right, pollen grains from other plants may be deposited by visiting insects. The thickened, rounded tubular base of the pistil is the *ovary*, and here is where a lot of the action of plant reproduction takes place. Cutting a cross section through the ovary reveals five chambers (called *locules*). Each chamber contains two *ovules*—the organs where the egg cell resides and where the seed will form following fertilization. After flowering, pollination, and fertilization, the ovule develops into the seed within the ovary that transforms itself into the fruit. The geranium fruit has a number of unusual characteristics. Only one ovule will develop into a seed in each locule and, at fruit maturity, each of the five basal parts will split apart and curl upward, still attached to the central axis of the pistil and persisting style. It is this unusual fruit that helps us recognize the genus, and gives many of its species the common name cranesbill. All together, the genus *Geranium* includes somewhere between 250 and 300 wild species and is largely confined to northern climates and to higher mountains in tropical countries. Geraniums are mem-

Figure 1. A: *Geranium maculatum* (Wild Geranium): flower from above and two enlarged stamens (*upper left*), side view of flower with two petals removed and flower bud (*center*). **B:** *Rosa carolina* (Pasture Rose): longitudinal section of flower (*right center*) and flower bud (*lower right*). Basal scales apply to leafy stems in this and the other figures.

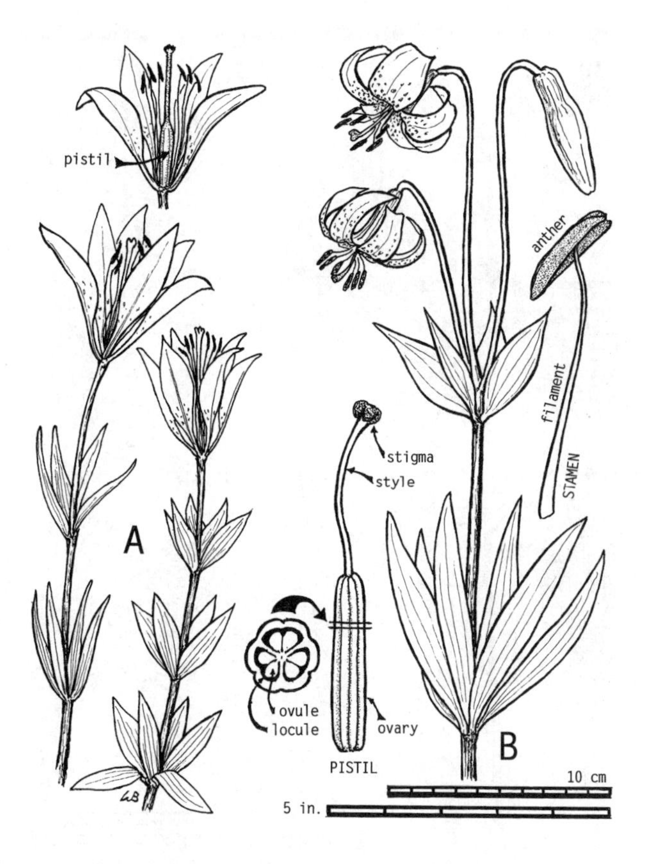

Figure 2. A: *Lilium philadelphicum* (Wood Lily): two flowering stems (*left*), flower with two tepals removed (*upper left*). **B:** *Lilium michiganense* (Michigan Lily): enlarged stamen (*upper right*), enlarged pistil with cross section of ovary (*lower center*). Note pistil composed of ovary, style, and stigma; stamen with filament and anther.

Figure 3. A: *Stachys tenuifolia* (Hedge Nettle): flowering stems (*left*), enlarged views of bilabiate flower (*top center and right*), longitudinal section of flower (*upper center*). **B:** *Campsis radicans* (Trumpet Creeper): cut-open tubular corolla viewed from beneath (*right center*).

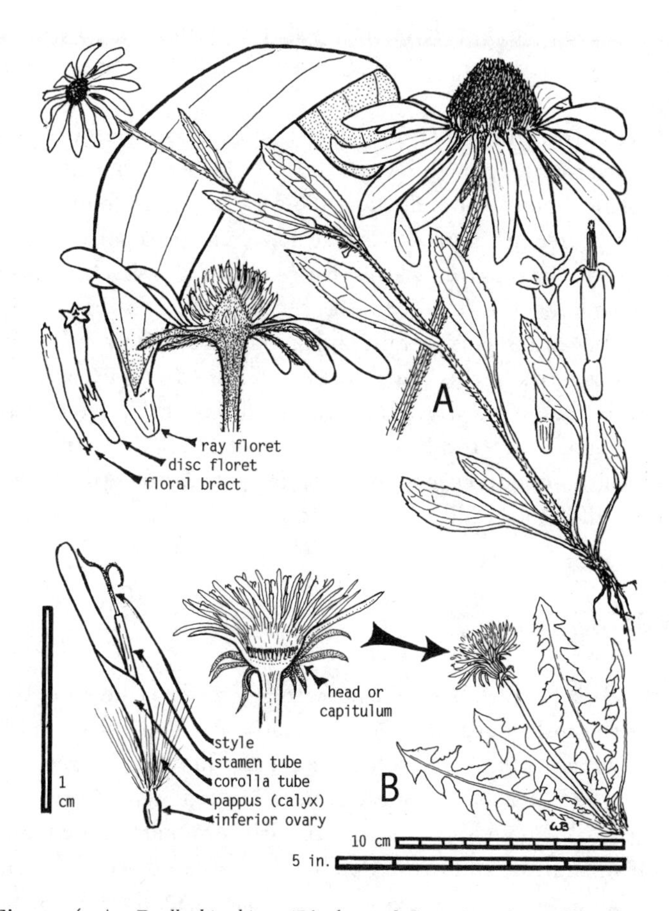

Figure 4. A: *Rudbeckia hirta* (Black-eyed Susan): enlarged head (*upper right*), cross section of head (*left center*), disk florets in female and male phases (*right center*). **B:** *Taraxacum officinale* (Common Dandelion): cross section of head (*center*), enlarged ligulate floret (*lower left*).

bers of the Geraniaceae, a family that includes the cultivated garden geraniums that actually belong to the related genus *Pelargonium*.

Finally, to get a better idea of the overall flower, we'll cut the flower longitudinally down the middle. Starting just above the pedicel on the outside of the flower, we find a circle (*whorl*) of tough greenish sepals. Next, just within the sepals, we find the five colorful *petals*. Within the base of the petals are borne ten stamens, and, finally, at the center of the flower sits the pistil. Because sepals, petals, and stamens arise from *below* the base of the ovary, this flower is said to have a *superior* ovary—an ovary that arises above the bases of the other floral parts (see fig. 1A, p. 24). To get a better idea of floral diversity, let's examine another flower with five lovely petals, but a plant from a very different family, the rose family.

A Wild Rose

A good place to start might be with a common wild rose. But since the rose genus (*Rosa*) has a great many species and cultivated varieties, let's again focus on a specific species, the pasture rose (*Rosa carolina*) of the eastern United States (see figure 1B, p. 24). These lovely flowers of early summer can be found both in prairies and open woodlands, borne on woody stems that are only one to three feet tall. Just as in the wild geranium, the most conspicuous part of these flowers is their circle of five bright pink petals. Forming a flat or saucer-shaped whorl, the petals surround a bright yellow center (see color insert C). This is very different from most cultivated roses whose many overlapping petals fill the entire flower. In contrast to ornamental roses, the center of the wild rose is open and easily accessible to visiting insects. The yellow center of the wild rose is composed of two main elements. Around the outside are between fifteen and thirty slender stamens with yellow or brownish anthers at their tips. As in the geranium, these are the "male" elements of the flower, each with a slender filament and a pouchlike anther. In the center of the flower is where we find the "female" parts of the flower, as in the geranium. But in this rose flower we do not see a pistil. Rather, the very short styles and stigmas make a rounded mound at the center of the flower. (In some roses the styles may be united to form a short column, with the stigmas at their tips.) But where in this rose flower are the ovary and its ovules?

Now, if we cut this same rose flower in half top-to-bottom, and follow the female elements down toward the stem, we'll find a thickened tubular region below the petals (see fig. 1B, p. 24). Here the thickened ovary is hidden beneath the petals and stamens. Because the sepal-lobes, petals, and stamens arise *above* the ovary, this is called an *inferior* ovary, in contrast to the geranium ovary.[2] As in the geranium, the ovary contains the ovules that, when pollinated, will grow into the seeds.

Beneath the petals and arising along the top edge of the ovary we find five triangular green structures, the sepals, or sepal-lobes. Just as in the geranium, the sepals cover and protect the flower bud in its early stages. When flowering began, the sepals opened, allowing the petals to expand, unfurl, and put on their colorful display. By picking a large flower bud just before opening, we can review its five major parts. On the outside are the tough green sepals, continuous with the outside wall of the green ovary beneath them. Removing the sepals, we find folded petals, ready to expand. Just inside of the folded petals are many slender yellowish stamens. Finally, in the center of the flower bud, we find the tops of the styles and stigmas, arising from the underlying ovary. In this configuration, partly hidden underneath petals and stamens, the more vital female parts are less likely to be damaged by insects rummaging around among the stamens and styles. Clearly, the female parts of the rose flower are protected in a way that the pistil of the geranium is not.

After pollination and fertilization the petals and stamens will wither and fall, as the ovules begin to develop into seeds. The ovary itself will enlarge to become the fruit, enclosing the seeds within. Bright red when fully ripe, we call these fruits "rose hips." The succulent wall of the rose hips is rich in vitamin C and is a tasty and nutritious snack for animals. By providing a nutritious reward, animals are enlisted in helping the rose disperse its seeds. Rose hips resemble the fruits of a number of other members of the rose family, such as apples, cherries, and peaches. You can tell that these particular fruits are derived from an inferior ovary because the persisting sepal-lobes are borne at the center of the top of the fruit—not like a geranium, or a tomato, where the persisting sepals are at the bottom, near the fruit stalk.

An interesting aspect of the rose family is that its floral elements usually number five or multiples of five (five sepals, five petals, fifteen to thirty stamens, five to twenty-five styles). We saw something similar in the geranium, though in that genus there may be only three styles on the ovary. Many families of flowering plants are very consistent in the numbers of their floral parts, but there are others where the numbers vary widely. Having looked at a geranium and a wild rose, let's next take a look at two equally colorful wildflowers in the same genus, but which represent a very different group of flowering plants.

Two Wild Lilies

We have two striking lily species in our native eastern North American vegetation. Like the pasture rose, our wild lilies also flower in early summer, and they also grow in both prairies and open woodlands. The first distinction one might note in these flowers is that they are supported on long vertical stalks (the pedicels), as was the case with the geranium. But these flowers are considerably larger. Held high, the bright reddish-orange flowers are visible from a great distance. The wood lily (*Lilium philadelphicum*, fig. 2A, p. 25) holds its flower vertically at the end of the pedicel. The six erect petals are narrowed at the base and diverge to form an open vaselike space around the sexual parts. Our other wild lily, the Turk's cap, or Michigan lily (*Lilium michiganense*), has its flowers pendant from the curved tips of the long pedicels. Here the six petals are recurved so that both stamens and styles project downward into open space (fig. 2B, p. 25; color insert A). These flowers differ from the geranium and rose not only in having six petals (rather than five) but also in that they have nothing resembling the greenish sepals. (If you look closely at the base of these six "petals," you'll find an outer whorl of three and an inner whorl of three, rather like sepals and petals.) But with showy parts all alike, they are given the technical name *tepals*. Within the colorful tepals we find a circle of six long filaments, each bearing an elongate anther. Again, filament and anther form the stamen, the basic "male" unit of the flower.

In the center of these lily flowers is the pistil, with the oblong ovary sitting above the bases of the tepals and stamens. Clearly, these are superior ovaries as in the geranium. At the top of the ovary arises a long slender style that splits into three branches or three lobes near the end. The three terminal lobes are the sticky stigmas. If we make a cross section of the lily flower's ovary, we find that it has three chambers (locules) with many little ovules attached along the ovary's central axis (see fig. 2B, p. 25).

In contrast with the five-parted flowers of the geranium and rose, these lilies exhibit lots of threes: six petal-like tepals in two series of three, six stamens also in two series of three, and a pistil with three stigmas and three locules. Lacking sepals, the flower bud is made up of the three outer tepals with their edges held tightly together. Obviously, if these get damaged in early stages of growth the flower's appearance will be marred. That's quite different from the rosebud and geranium where the pink folded petals are nicely protected by the tough sepals that aren't an important part of the show. Clearly, geraniums, roses, and lilies are very different kinds of flowers. The distinction between floral parts in fives and floral parts in threes is an important one we'll explore later in greater detail. But, for now, let's look once again at the basic parts that characterize flowers.

The Parts of a Flower

Most flowers have a stiff greenish circle or whorl of outer parts that are called the sepals and which protect the flower bud. We didn't really find these in the lilies, but that's how they are built. The next whorl is that of colorful petals. Where we can't separate petals and sepals, as in the lilies, we use the designation *tepals*. Sepals and petals or tepals together are called the *perianth* of the flower. The sepals, whether separate or partly united, are called the *calyx*, while the petals make up the *corolla*. This may seem like excess verbosity, but once we get into more complex flowers, terms like calyx, corolla, and perianth turn out to be very useful. This covers the colorful and outer protective parts of the flower (the perianth). The really sexy parts of the flower, however, get to be more complex.

The real business of flowers is sex, which we'll discuss in the next chapter, but for now we'll just look at the structures and their functions. In many animals, primitive plants, and us, there are two basic kinds of sex cells. The male sex cells are very small, but outfitted with long wiggly tails. It's their job to swim through a watery medium to the female egg cells and unite with them. The female sex cells are much plumper and usually sedentary; their large size is essential in storing energy to begin a new life. In addition, female sex cells must send out chemical signals that help the little males find their way to sexual union and reproductive success.

Flowering plants and other seed plants have gotten around the wiggly little sperm cell business by the invention of pollen. Pollen grains carry the sperm nuclei and are a "substitute" for the swimming sperm. Carried by the wind or by animals, pollen that reaches the female stigma can germinate and grow down the style toward the egg cell within the ovule. There is no longer a need for liquid water in fertilization here. This important evolutionary innovation allowed seed plants to reproduce in much drier environments than the plants that came before them. (Even today, fern, moss, and liverwort sperm must swim in a thin film of water to where the egg cell resides. That's why these plants are so rare in desert environments.)

Since we began with the outside of the flower, the sepals, and worked inward to the petals, let's continue that trajectory and first discuss the male elements of the flower, the stamens. Despite a huge range in variation among different flowers, stamens are quite similar over many different lineages. As mentioned, the stamen usually consists of a slender stalk (the filament) that bears the saclike anther at its tip. The anther usually has two parts (the *thecae*) with one or two longitudinal chambers in which the pollen is formed. When fully mature, the anther walls split open to release the pollen, to be carried by wind or animal to another flower. To withstand the rigors of travel, and especially to avoid becoming dried out over the long journey, pollen grains have a tough outer coat. However, it is only when the pollen grain lands on the stigma of the *right species* and senses the correct chemical signals that this tough coat splits open. This allows a microscopic "pollen tube" to enter the stigma and grow down the style toward the egg cell. The male parts of the flower function for only a short time as the pollen is formed, matures, and, with flowering, is shed. Again, all the stamens of a flower, taken together, are called the androecium.

In the center of the flower we encounter the female parts of the household (the gynoecium). The female elements of flowers vary much more than the male, representing many different configurations. We've already seen the difference between superior and inferior ovaries in lily, geranium, and rose flowers. And while most flowers have only a single pistil (with a single ovary) in the flower, a few, like peonies, buttercups, and the marsh marigold, can have a number of pistils in the center of their flowers. Despite this variety, there appears to be an underlying unity in the female parts. Let's start with the simplest example: a simple pistil. In this case we would have a flower with a single female element. At its base would be the thickened ovary, having a single locule with one to many ovules within. At the apex of the pistil, we find a single slender style ending in a single stigma. A more complex gynoecium is made up of a few to many separate simple pistils, bunched together in the center of the flower. As previously mentioned, buttercups and peonies are good examples of such flowers. But these form only a very small percentage of the kinds of female configurations to be found in the world of flowers.

Many flowers have a single central pistil, but with an ovary having more than one locule and several styles or stigmas. We saw this in both the geranium and the lilies. The general interpretation here is that three simple pistils may have come together in the distant past and then fused in various ways. In the case of the lily we can hypothesize that the gynoecium of the lily flower may have once been composed of three simple pistils that fused together—an earlier evolutionary union that is remembered only by the three separate locules and three stigmas. Botanists call this a three-carpellate pistil or gynoecium, assuming that it formed from three separate simple pistils or *carpels*. (The carpel is equivalent to a simple pistil.) Likewise, when a pistil has five locules and five styles or stigmas, it's assumed that these were formed from the union of five carpels in the distant past. Though a somewhat theoretical construct, the idea of carpel number in a flower is useful when trying to determine relationships between different lineages of flowering plants.

In addition to different numbers of locules within the ovary, the number of ovules in each locule may vary from one to hundreds. The ovules may be borne along the central column within the ovary (as in a chili pepper), on intruding placentae (as in a tomato), or on the outer locule wall (as in the pumpkin). Ovules may also be attached at the apex or base of the locule. All these variations illustrate how gynoecia (female households) vary so much more among flowers than do the androecia (male households). Together, the male and female flower parts are critical indicators of the family and genus to which a flower belongs.[3]

Flowers by Fusion

We have noted the hypothesized fusion of simple pistils to produce three-chambered ovaries in both geranium and the lilies. Fusion of other floral parts appears to have been a major source of evolutionary novelty among many kinds of flowers as well. A great example is the orchid flower. Fundamentally, the orchid flower isn't that different from a lily flower, having parts in threes, but orchids have some unusual innovations. For one thing, the orchid has an inferior ovary, so inferior that it may look like nothing more than the greenish stalk of the flower. As was the case in the lily, there are—or were—six petal-like tepals in the orchids. The outer three may be smaller or sepal-like, and the inner three colorful and petal-like. Usually, one of these inner petals is greatly elaborated to become the enlarged lip. In our lady slipper orchids the lip has become a very elaborate pouchlike device. And then there's the most unusual aspect of the orchid flower: stamens and parts of the pistil have become united! The upper parts of the styles and stigmas, and the stamens, have all become *completely fused* to form a single structure: the *column* (see color insert N). Both stigmas and anthers have been reduced in number and greatly transformed during this evolutionary fusion. Just as important, the pollen is no longer powdery, and it is usually held in two special structures called *pollinia*. These clever innovations have allowed orchids to do strange things no other families of flowering plants have done.[4]

The orchid flower is only one way in which rich variety has been added to the world of flowers. Fusion of the male and female parts, in fact, is extremely rare among flowering plants. Much more common is the "fusion" of petals or tepals to form a tubular or cup-shaped corolla. Let's be careful to remember that the concept of fusion-over-time is an evolutionary inference. It is based both on analysis of related flowers and by looking at the early stages of floral development. If we examine some tubular flowers in early stages of development, we find they come from three, five, or six separate minute projections, called *primordia*, on the microscopic floral apex. As the flower begins to grow, these separate little primordia come together to form a ring which then develops into a tube. In a few kinds of flowers, the petals remain separate in early stages and fuse later during development. Deep tubes and hidden recesses within the flower help protect nectar from evaporation by the desiccating wind. Also, they help orient the pollinator to enter the flower in a specific way: by their size and orientation they restrict the visitor's size and form.

Floral tubes are characteristic of many flowers. Conspicuous examples include the bright reddish corollas of the trumpet creeper (*Campsis radicans*, fig. 3B, p. 26) and the white flowers of the catalpa or umbrella tree (*Catalpa speciosa*); both are members of the Bignoniaceae family. The catalpa flowers have broad lower petal-lobes that form a landing field on which bumblebees can alight. The colorful *Tabebuia* trees of the American tropics make a big show of such flowers (color insert M). Many smaller flowers also have had their petals united to form a tube. Good examples are phlox, with a slender tube beneath the spreading corolla lobes, or the flowers of blueberries whose corolla has the shape of a little urn. In the mint family and related flowers, the corolla tubes have become strongly two-lipped (see fig. 3A, p. 26). Like the catalpa flower, there is a prominent lower lip on which the insect can land, and a smaller upper lip.

What's important about these tubular flowers is that they *constrain* the activities of the pollinators. That's quite different from rose and lily flowers, which can be approached from a variety of directions. Most tubular flowers allow the pollinator to enter the flower from only one direction. By having the insect enter in a well-defined path, stigmas can be oriented to have contact with the insect early on and to pick up pollen from a distant flower. Once that opportunity has passed, and as the insect enters deep into the floral tube to slurp up the nectar, the anthers can dump their own pollen on the insect as it leaves. In many tubular flowers, the pollen is placed on that part of the insect likely to encounter the stigma when entering another flower. These more structurally complex flowers are considered to be recent evolutionary innovations; ancient fossil flowers from the time of the dinosaurs do not show such patterns.

Floral Symmetry and Floral Multiplication

More obvious in a comparison of flowers like roses and lilies on one hand, and orchids and two-lipped mints and catalpa on the other, is their basic symmetry. Roses and lilies are radially symmetrical. Like a pie, you can slice them several ways and come up with equal halves. Orchids and two-lipped flowers are bilaterally symmetrical. You can slice them only one way to produce two similar halves. As we mentioned, bilaterally symmetrical flowers use their shapes to force their animal pollinators to enter the flower from a specific direction. This allows the flower to both deposit pollen and pick up pollen from their visitors with greater precision. However, there are many other ways in which flowers have diversified.

Another significant source of floral diversity has come from *multiplication of parts* within the flower. This may explain all those stamens in the rose flower—always in multiples of five. In many of these cases the microscopic floral bud has only five minute stamen primordia, but they divide to give multiples of five as the flower develops. Unusual floral multiplications have produced some of our most successful ornamental flowers. The garden rose, with lots of overlapping petals, is a good example. In fact, some of these flowers can no longer reproduce themselves in nature and must be propagated by gardeners. In contrast, another way of producing floral diversity is through "reduction" or the *loss of parts*. This trend has been especially common among flowers pollinated by the wind (we'll get to them shortly).

In the case of small flowers, there's yet another way in which they can become more complex. It is a simple system but it works: take all those little flowers and bring them close together. Queen Anne's lace (*Daucus carota*, color insert K) makes a fancy show by displaying its tiny little flowers close together on a little umbel and then arranging these in *another* larger compound umbel. (*Umbels* are umbrella-like, with the flowers radiating out on slender stalks from a common point of origin.) These umbels are a good example of an "inflorescence." We use the word "inflorescence" to describe the grouping or arrangement of many flowers together. Take a close look at the underside of the inflorescence of Queen Anne's lace and check out the "umbel of umbels" which positions all these many flowers over the same surface. More intriguing yet are those inflorescences that look just like *single large flowers*—with all their little flowers tightly packed together. But before we get into these, let's first look at yet another class of little flowers, flowers that often don't look like flowers.

Flowers That Don't Look Like Flowers

When we talk about flowers, most people think of the colorful entities that adorn our gardens and enliven our natural environments. However, there are many smaller flowers we rarely notice. People are sometimes surprised to hear that grasses and sedges also have flowers, but they do. These little nondescript flowers are often hidden within bracts, and they have no special coloring. (*Bracts* are small scalelike or leaflike structures that protect buds and small flowers in their early stages.) The reason they lack color is simple: they are pollinated by the wind, and the wind is blind. Thus there is no selection pressure for the flowers to become conspicuous, as is the case with animal-pollinated flowers, where visual cues play an important role in guiding the pollinator. In open habitats with plants of the same species in abundance, wind pollination can be very effective. Grasslands are a prime example; but in forests wind pollination works better if the trees are leafless, as in our early springtime.

In many plants, the young, developing flower buds are protected by enclosing bracts—in addition to the sepals. As we mentioned above, bracts are small scalelike coverings from which the flower buds emerge as they expand. In the grasses (Poaceae or Gramineae family) and sedges (Cyperaceae family), the little bracts have become important protective elements that surround one or several little florets. Because they're so small, and there is nothing resembling sepals, petals, or tepals, we call the flowers of grasses *florets*. Within the various bracts—all with special names—the grass floret possesses three or six stamens, and one pistil with two usually feathery stigmas. The little stigmas are feathery to help catch a pollen grain carried by the wind. In contrast, the anthers are a large aspect of these little florets. The reason is simple: it takes an awful lot of pollen to effect pollination by the wind (see color insert F). The ovary in grass flowers is very small and contains a solitary ovule that will produce a single seed. A few grass seeds have become humankind's most important source of food energy. These are the cereal grains, including wheat, rice, maize, barley, sorghum, oats, rye, and the millets.[5]

Sedges look like grasses but their little flowers are quite different in structure. They, too, may be enclosed by small bracts or scales, but their flowers often have three or six tiny tepals in a configuration that isn't too different from that of a lily flower. The little ovary of sedges often bears three style branches, but, like the grasses, it has only a single locule with a single ovule. Whereas grasses dominate on open dry or burned landscapes, the sedges are more common in marshes and open wet depressions. Together, grasses and sedges form a major part of open sunny savannas, grasslands, and marshes—and both are pollinated by the wind.

Another group of flowering plants we're all familiar with that have inconspicuous little flowers are the oaks (the genus *Quercus* in the family Fagaceae). The trees may be grand but their flowers are not. Not only are the flowers very small, they come in two forms: male and female. The small female flowers are rarely noticed because they're tucked away in leaf-axils (where the leafstalk meets the stem). Male flowers of oaks are borne on pendant catkins (resembling a cat's tail), easily seen dangling near the ends of twigs as the leaves are just beginning to expand in early spring. Every catkin bears many male flowers, each with a minute perianth, and several anthers on thin filaments. These male flowers have no petals or pistil, but they do produce a lot of pollen. Similarly, the female flowers have no petals and no stamens, but they do possess a short ovary with two or three prominent stigmas. Again, the reason these flowers are so small and colorless is that they are pollinated by the wind. Small flowers with reduced parts result in conserved energy; this energy can be devoted to making lots of pollen—to deal with the statistical improbability of having pollen reach a stigma of the same species through the agency of air.

If flowers range from the minute greenish flowers pollinated by the wind to pretty roses and lilies, or large water lilies pollinated by insects, what really defines the flower? The answer lies in the sexual parts: the stamens and the pistil or pistils. Most flowers have both and are therefore bisexual (or hermaphrodite); these are often referred to as "perfect flowers." If organs of only one sex are present, as in oak flowers, then the flowers are unisexual, and we call them "imperfect flowers." Plants with unisexual or imperfect flowers have several possibilities. The plant can bear both male and female flowers, as do the oaks. In this case the flowers may be unisexual but the trees are bisexual. On the other hand, the plants may be unisexual themselves, bearing flowers with the organs of only one sex. Such unisexual plants are rare in our northern temperate zone and only a bit more common in the tropics; papayas and date palms are good examples of unisexual plants. But let's get back to pretty, animal-pollinated flowers.

Flowers That Really Aren't Flowers (Technically Speaking)

There are a goodly number of flowers in our fields and gardens that aren't really flowers. Oh, they look enough like flowers to have everyone calling them flowers, but, in a technical sense, they aren't flowers. Two of our favorite woody ornamentals have pseudoflowers, where bright colorful bracts serve the function of petals and the true flowers are much less conspicuous. Our flowering dogwoods (*Cornus florida*), for example, have four large white, less often pink, bracts surrounding a small cluster of flowers. The bright bracts are the "petals" of this pseudoflower. Dogwoods put on quite a show in early spring, and because they remain small trees, they are ideal around a home or in a garden. Bougainvillea (*Bougainvillea spectabilis*) is a tropical favorite that can vine its way around a veranda or be trained to form small bushes. These pseudoflowers have three large bracts subtending three slender tubular flowers. What makes bougainvilleas so popular is that the bright pink, red, or orange bracts keep the show going for months on end, whether or not the less-conspicuous central flowers are ready for pollination.

A number of other popular ornamentals have bright colorful bracts providing a brilliant display. Of these, the poinsettia (*Euphorbia pulcherrima*) is a wintertime favorite, but here it's clearly evident that the colorful elements are modified leaves and not parts of the flower. However, we have a relative of poinsettia in our eastern North American fields and midwestern prairies that has small pseudoflowers that look to all the world like real flowers. These plants, the flowering spurge (*Euphorbia corollata*), don't resemble a poinsettia at all, though they are members of the same genus. These clever little pseudoflowers can be found flowering during July in our midwestern dunes and prairies. In this case the little flower has what clearly look like five white petals arising from a little green cup. The stamens seem normal enough but the pistil is borne on a stalk that elongates as the fruit develops (color insert I). Both our native wildflower and poinsettia (originally from Central America) are members of one of the most successful genera of flowering plants: *Euphorbia*. With about 1,600 species worldwide, *Euphorbia* also includes many large leafless cactuslike succulents in Africa. Some of these tropical euphorbia species have larger pseudoflowers (called *cyathia*) which give ample evidence of having been cobbled together from a set of bracts, several bunches of stamens, and a pistil that once was part of a separate flower. This is not so easily seen in the little "flower" of our flowering spurge. In fact these may be one of the best examples of "pseudoflowers" in nature. But there is another group of even more complex pseudoflowers in our fields and gardens.

The pseudoflowers we encounter most often are members of the sunflower family (Asteraceae). An older name, Compositae, is quite appropriate, since these pseudoflowers are composites of small flowers tightly congested together in a "head," or *capitulum*. Pull apart the bright yellow pseudoflower of a dandelion and you'll find it's made up of dozens of slender little flowers, each about a half inch (12 mm) long. These are called *ligulate flowers* because each has a single long, strap-shaped (ligulate) petal (fig. 4B, p. 27). This compact cluster of flowers is subtended by a series of overlapping green bracts at the base (called an *involucre*). Together, the tightly congested yellow flowers, with their subtending green calyx-like involucre, really do look like a single flower. And that seems to be exactly how insects see them, too. In the case of the dandelion, all the little flowers of the capitulum are the same, each with a single straplike petal. However, things get more complex when the composite capitulum consists of two different kinds of flowers.

Many genera of the sunflower family have flower heads composed of two very different kinds of flowers (or florets). In the center are the disc florets, tightly packed together. They have small, slender corolla tubes with usually five minute lobes at their tips. These corolla tubes are radially symmetrical. Together, these disk florets form the rounded center of the pseudoflower. In addition, around the outer periphery of this center, is a second kind of flower: the ray florets. These have a single long and conspicuous strap-shaped corolla, the *ray*. It is these strongly bilaterally symmetrical ray flowers that give the pseudoflower its pseudopetals. From a distance, these things surely look like flowers. Daisies, asters, and black-eyed Susans put on their impressive display by packing together these two different kinds of little flowers (fig. 4A, p. 27; color insert J).

And again, they are subtended by a stiff green involucre that often protects the many flowers while they are developing. In the case of the cultivated sunflower, the pseudoflower can be as wide as a large pancake.

Incidentally, if you're looking for saclike anthers in these flowers, you won't find them. Instead, members of the Compositae family have had their five stamens united along their lateral edges to form a long, slender tube (fig. 4B, p. 27). In most composite genera, the pollen is carried out from within this tube by the expanding style as it pushes its way up through—and out of—the staminal tube. Thus the styles "present" the pollen to potential pollinators; later, the tip of the style splits into two halves and bends backward to present the stigmatic surfaces for pollination. This complex floral morphology is what helps distinguish these plants—asters, daisies, and sunflowers—as a well-defined natural family: the Asteraceae.[6]

Another Crowd of Pseudoflowers

The aroid family (Araceae) includes many indoor ornamentals in the genera *Anthurium*, *Dieffenbachia*, and *Monstera*, as well as our woodland jack-in-the-pulpit (*Arisaema triphylla*) and skunk cabbage (*Symplocarpus foetidus*). In the aroids the flowers are very small and tightly clustered on a cylindrical central axis, called the *spadix*. The spadix, in turn, is at first enclosed by an enveloping leaflike structure called the *spathe*. The spathe may become simply strap-shaped as the flowers become mature, or the spathe may produce a fancy perianth-like enclosure for the spadix and its flowers. In the case of the jack-in-the-pulpit, the flowers are near the bottom of the spadix, nicely protected within the tubular base of the spathe, or "pulpit." Likewise, the purplish "petal" of the skunk cabbage pseudoflower is the spathe, enclosing a spherical spadix (color insert H). There are several things about the skunk cabbage that are impressive. Forget the skunk part, that's just the bad smell of the plant's tissues when they've been broken. For us nature photographers, these are the earliest large "flowers" of our northern springtime. Not only do these "flowers" serve as the harbinger of a new season, but also when we're lucky, we'll see where they have melted the ice around them in their swampy habitat. These pseudoflowers are hot stuff! Well, maybe only warm, but they certainly are among the group having the most active tissues known in plants. This ability to produce heat is, in fact, a family characteristic also found in tropical aroids, where a majority of the species live. The higher temperatures help disperse the nasty aromas of these pseudoflowers. By nasty I mean like rotting, stinking flesh. Many aroid pseudoflowers are pollinated by flies, and many flies just love the aromas of rotting flesh. Not only do the purplish-red tubular spathes lure flies into their interior where the flowers are, but also some species trap them for a while. Most often, the female flowers are receptive when the flies arrive; then the flies are held prisoner within the spathe until the male flowers shed their pollen. Soon, the barriers to escape are opened and the flies, now covered with pollen, escape—perhaps to be attracted by another spathe that has both the color and the aroma the flies find so attractive.[7]

Though not really flowers, a few aroid spathes and spadices

are spectacular. After seven years of careful tending in the University of Wisconsin's greenhouse, a specimen of a giant voodoo lily (*Amorphophallus titanum*) flowered in 2001. These plants have no stem; instead they produce a single larger leaf each year over several years. Each growing season, the leaves increase the size of the plant's underground storage organ (the *corm*) where they store the energy needed to produce inflorescence and fruit. After the last and largest leaf has finished its growing season, the leaf dies. Finally, with the beginning of the next growing season, a huge pseudoflower emerges directly from the belowground corm. In the Wisconsin greenhouse, the stalk and spathe reached six feet in height! When the spathe opened, it was three feet wide, looking rather like an upside-down purple skirt. And this monstrosity quickly produced a humongous stench, an effort that produced results. The interior of the spathe was soon mobbed by hundreds of flies. Summertime in Madison, Wisconsin, has no shortage of flies; but imagine the delight these northern flies experienced in encountering such a tropical treat. And it wasn't only flies that were attracted. Daily progress of the growth of this grand pseudoflower was communicated via the Internet and, when it finally blossomed, twenty-five thousand people came to see it. Growing wild only within the rain forests of Sumatra, the flowering of this phenomenal species has been witnessed first-hand by few people. But enough of pseudoflowers, let's get back to normal flowers.

What Does It Take to Build Flowers, Fruits, and Seeds?

An interesting aspect of virtually all flowers is that the female parts of the flower are in the *center of the flower*. The reason may be simple: that's where the plumbing is. Transforming little ovaries into larger fruits—and ovules into seeds—requires lots of nutrition. That nutrition gets pumped in via the plumbing (vascular tissue) of the stems, and this tissue is in the *center* of the stem. So it makes sense to have organs that will be doing most of the growing close to the central axis of the flower. Interestingly, this leads to another possibility. The rapid wing beats of bees and other insects generate considerable static electricity. Coming into a flower, the "charged insect" may make electrical ground with the moist stigma/style connected directly to the central plumbing system of the plant. This is grounded by water leading down to the roots. Such a transfer of charge may help propel pollen from insect to stigma—another reason for having the female parts in the center of the flower. But let's get back to the problem of building seeds.

The work of photosynthesis is done largely in the leaves, where green chlorophyll and its associated pigments capture the energy of sunlight. Acquiring the energy needed to produce flowers and fruits by means of photosynthesis takes weeks to months. Corn plants need more than two months of growth before they can begin to build flowers. Apple trees flower in early spring, but need a long summer to produce succulent apples in late September. If that's the case, how can early spring flowers like crocuses, dandelions, and daffodils put on such a charming show early in the season? The answer is simple: plants that bloom early in the spring have systems of energy storage, like bulbs, corms, and roots. Spring-flowering trees also need stored energy to produce a flush of early flowers and leaves. In the case of the apple, summertime photosynthesis is used to do two things: first, to produce the crop of sweet apples, and, at the same time, to build buds and store away enough energy to produce a flush of new flowers and leaves for the coming spring. Thus, the life of the plant must be carefully cadenced to conform to the seasons. Whether in a rain forest, northern woodland, or seasonally dry grassland, flower and seed production have to be in synchrony with the seasons. Now, before we leave this chapter, let me describe the strangest flower I have ever encountered.

The early rains had begun in eastern Ethiopia, a time when the open acacia woodlands were graced by a green filigree of delicate new foliage. To take advantage of this lovely spring-like season, a group of us decided to go camping along one of the usually dry rivers that runs southward from the Harar highlands. After finding a comfortable area and setting up our tents, we wandered about, admiring all the new growth. One of our group was watching some local Somali-speaking herders and found that they were digging something out of the ground for food. From what he saw it was a whitish mushroomlike material that they had gathered. But whatever it was, it must have been a native plant or fungus, since this was taking place in a perfectly wild, uncultivated, open acacia woodland. Well, I figured, if pastoral nomads can find these strange things, a trained botanist ought to be able to do the same. I began roaming over the flat, dark soil, newly decorated with sprouting grass and expanding little herbs, looking for anything unusual. Finally, I spotted something odd under the spreading crown of a small acacia tree. The dark brown earth was mounded up over an area about four inches in diameter, and the dirt mound was cracking open at the center. Obviously, something was beginning to emerge here. So I dug a pit around the mound, in order not to damage whatever was growing at the center. After considerable effort, I had unearthed an eight-inch long columnar specimen about three to four inches thick, weighing almost a pound. The texture of its tissue was rather dense, like that of a radish, and it was white. But this strange specimen looked like no fungus I had ever seen. Rounded in cross section, the only way to figure out what this was would be to slice it down the middle, top to bottom. After cutting it into two halves with my hunting knife, I could see the innards. Slowly, it dawned on me: this thing was a flower!

At its bottom end was a massive inferior ovary with many congested ovules borne at the top of a single locule. Above that was a mound representing, I figured, the stigmatic area. The perianth formed a thick tube, with three massive V-shaped stamens sitting on the interior walls. The top of this "perianth tube" was conical, like the end of a projectile, just what was needed for pushing up through dense soil. Back at our college, and crawling through lots of family descriptions, I discovered that this was the flower of a parasitic family, Hydnoraceae. Some many months later, after a half-specimen (pickled in alcohol) had been sent to Kew Gardens near London, it was identified as *Hydnora johannis*, a species known only from Ethiopia.

These plants—if you can call them plants—parasitize the roots of acacia trees, and the only large structure they produce is the flower itself. If I had not dug up this thick flower, its upper perianth would have emerged to a few inches above the surface of the soil. Remaining connected at the top, the perianth parts would have split and opened to allow three or four apertures for pollinators to enter the flower. This is the case with the specimen shown in color insert G. Notice the dark out-of-focus forms inside the flower; these were beetles rummaging around within the flower, and the flower's probable pollinators. I'm sure this unusual flowering extravaganza lasts for less than a week, before the aboveground parts decompose and seeds begin to develop underground. (This poses the question as to how such buried seeds might be dispersed: by wild pigs who do a lot of digging? Or is it by mole rats?) But what is especially interesting about this unusual flower is its origin from tissues that are completely parasitic on acacia roots and utterly devoid of chlorophyll. Just as noteworthy, parasite and host were both flowering at the beginning of the early rains, a very narrow window of time. More important, both host and parasite were expending a lot of energy to produce the acacia's many little pom-pom inflorescences and the one huge and hefty flower of the *Hydnora* parasite. Why, in fact, do plants expend so much energy in the construction of flowers and inflorescences? What is the fundamental purpose of flowers?

chapter 2

what are flowers for?

To get a better idea of why flowers exist, let's first take a look at a recent report regarding not a plant but an animal. Though most articles in scientific journals deal with the results of experimentation or careful observation, once in a while scientists stumble upon an unusual event in nature—a natural experiment—which becomes the basis of a scientific report. Such an article appeared in the February 2003 issue of the *Proceedings of the Royal Society of London*. The article was titled "Rescue of a Severely Bottlenecked Wolf Population by a Single Immigrant." (The term *bottlenecked* implies that the population had little genetic diversity because it had been founded by only a few individuals, having suffered a genetic *bottleneck* so to speak.)[1]

Everyone had thought that wolves (*Canis lupus*) had been exterminated in the Scandinavian peninsula by the 1960s. Shooting, poisoning, and trapping over many decades were thought to have wiped them out. But then, in 1983, a lone wolf pack was discovered in southern Sweden, separated by more than five hundred miles from other wild wolves in Finland and Russia. Genetic DNA analysis indicated that this pack had probably originated from a single pair of wolves. Averaging about ten in number, the pack had limited genetic diversity and was only barely surviving. This population was carefully monitored over the ensuing years. Then, suddenly, after 1991, the pack began expanding its numbers, though there were no apparent changes in its environment. By 2002 these wolves had increased to about one hundred animals in ten packs! What had happened?

Here was a natural "event" that modern DNA analysis could help us understand. Regular DNA sampling of the pack—a few hair follicles is all that was required—indicated that the isolated pack *had been joined by a single wolf* from afar. Here was dramatic evidence that the additional, and differing, DNA of the new immigrant wolf could change the health and the future of the pack. This example, I would like to suggest, tells us a great deal about the reason for colorful flowers.

Plants can't wander across the countryside to make new friends or bring in a batch of fresh genes. Instead, many flowering plants have developed an "advertising-and-reward" campaign, getting animals to help them out. Such animal pollinators can carry pollen over long distances. Imagine a pollen-covered bumblebee, blown off course by a gusty wind, coming upon an isolated population of the same plant species she'd been visiting. As she tanks up on additional nectar, this little bumblebee may be able to do for these isolated wildflowers exactly what the lone wolf did for the isolated pack. Just as the wolf brought in new genetic diversity to rescue a "bottlenecked" wolf population, animals carrying gene-loaded pollen from a distant source can do the same thing for plants.

However, as we mentioned earlier, there are a lot of little flowers that are not colorful and not pollinated by animals. These flowers are pollinated by the wind. Not as effective as animal pollination perhaps, but if there are enough individuals of a particular species in the landscape, and it produces a *lot* of pollen, wind pollination can also be effective. In fact, there may not be much difference in total energy expenditure between a plant that uses animal pollination versus one using wind pollination. Animal-pollinated flowers require a colorful perianth, enticing aromas, and energy-rich nectar; wind-pollinated flowers must invest in a huge production of pollen. Either way, pollination requires a major investment of energy. But why do plants of all shapes and sizes put so much energy into forming flowers? The real question is: why flowers?

What Are Flowers Really For?

The quick and simple answer to this question is: SEX! Flowers are part of the business of getting male and female sex cells together so that, once united, these cells can become the basis of a new life. In the case of flowers, the pollen and egg nuclei will fuse within the ovule to form the fertilized egg cell whose further divisions begin the formation of the embryo. Soon the embryo, encased within the energy-rich seed, will be ready for the cues that initiate its development and, when conditions are right, germination will begin the life of a new plant. But couldn't all this have been done without sex cells?

Indeed, many plants do produce sprout shoots, lateral roots, bulbils, and other forms of asexual reproduction. Some, like the dandelion, produce flowers most of which produce viable seeds—whether or not they've been pollinated. But sex is the normal form of reproduction for the huge majority of plants and animals, and this presents us with the question: why should so elaborate and wasteful a process be maintained in so many disparate living things? What's the real reason for creating sex cells (gametes) with half the ordinary number of chromosomes—cells that are useless—unless and until they unite with another sex cell?

Sex is central to the lives of all complex life forms despite its obvious expense. Think about how convoluted a process it really is. First a female has to go to all the trouble of finding a mate. And then sex means that half the genes of her offspring will not be her own. Furthermore, half of her offspring will be males who may contribute little more than sperm to the next generation. Wouldn't it be simpler just to let everybody reproduce themselves? In fact, some organisms do exactly that, but, with few exceptions, such lineages begin to decline in viability. Bad mutations accumulate, and there is no way of getting rid of them. (Mutations are inevitable changes in the genetic code, and most of these changes are deleterious.) Without sex, the genetic foundation of a lineage begins to deteriorate.

Sex constantly reshuffles the genetic material, providing both good and bad hereditary combinations. "Natural selection" is the process that will get rid of the losers, while the others will be further tested over time in an ever-changing environment. Just as important, early stages of producing sex cells—the process of meiosis—allow for chromosome repair in a way ordinary cell division—mitosis—does not. (In meiosis the divided cells end up with half the normal number of chromosomes. The resulting sex cells need to join up with another sex cell to get back to the normal adult chromosome count.) And most important, meiosis allows for "crossing-over" where sections of paired chromosomes can be exchanged; this provides new *combinations* of genes within the same chromosome. Thus, while the union of sex cells allows for all sorts of chromosomal combinations, crossing-over provides yet further rearrangements of the genes within the chromosomes themselves.

What's So Great About Sex?

As we've just noted, sex is both a repair system and a generator of new and continuing variability. In the case of humans, our forty-six chromosomes align in twenty-three homologous pairs during meiosis. (We received one set of twenty-three chromosomes from Mom, the other twenty-three from Dad.) By lining up and moving apart in two opposing directions during the process of meiosis, each daughter cell gets half of the normal complement—twenty-three in total. With the two chromosome sets (from the grandparents) being mixed up in thousands of different combinations in each parent's sex cells, there's little wonder why children of the same family can be so different. And that's not all; as we just noted, with crossing-over, there are further opportunities for new gene combinations. Finally, sex allows advantageous new genes to spread through the population much more rapidly than is possible in asexual reproduction. These advantages hold for all sexual reproduction, whether by pigweeds, penguins, or people.

All told, the fundamental reason for sex is that it produces and maintains continuing high genetic variability within populations. Thus, having varied offspring allows a species to respond to minor variations in the local environment. A genetically diverse population also has a better chance to meet the challenge of inevitable and unpredictable environmental changes. Even more significantly, the continual recombination of genes allows a species to deal with what's been called the Red Queen's admonition.

Leigh Van Valen of the University of Chicago approached the idea of evolutionary "fitness" from a rather unusual perspective. Instead of thinking about the individual plant or animal, he looked at the question in regard to all the members of a species. He suggested that "fitness" might be a measure of an entire species' use of energy in the ecosystem in which it is living. And, he reasoned, if a species is increasing its overall fitness, it must be taking a larger bite out of the energy available in that ecosystem.[2] We know that the energy of ecosystems is largely determined by the amount of sunlight falling on the system, as well as the energy stored in living and dead tissues of the system. Consequently, the amount of energy in any ecosystem is tightly constrained. Thus, Van Valen argued, as a species increases its fitness, it must affect all other species in the system *negatively*. Quoting the Red Queen in Lewis Carroll's story, he suggested that each species finds itself in a world where ". . . it takes all the running you can do, to keep in the same place." Because you are in an environment of incessantly evolving neighbors, you've got to "keep running" just to keep up with them. Ongoing new mutations, and the mixing of all our old genes through sex, help us in this never-ending race.

In fact, sex may be the only way of dealing with the environment's nastiest challenge: short-lived, rapidly mutating pathogens and parasites. Whether plant, animal, or human, the most serious challenge to a species' survival are those bacteria, viruses, fungi, and small parasites that are specifically adapted to attacking us. And, as if that weren't bad enough, these pernicious creatures are constantly evolving themselves. Thus, a dynamic genetic system is the only way to keep ahead of never-ceasing pestilence. In this regard, exchanging genes with oneself or a close relative (inbreeding) doesn't produce a whole lot of variability among the offspring. Much better is exchanging genes with more distant members of the same species (outbreeding)—and that's really what colorful flowers are for.

Inbreeding Is Bad; Outbreeding Is Good

Pollen germinates on the stigma of a flower, forms a tube that grows down the style, and brings the male nuclei close to the ovule in which the egg cell resides. The fusion of one nucleus from the pollen tube with one of the female nuclei within the ovule will effect fertilization, restore the full chromosome complement, and provide the genetic wherewithal to create another individual plant of the same species. But what happens if the pollen grain lands on the stigma of the wrong species?

The answer is simple: nothing happens. The pollen grain must land on the stigmatic surface of a member of its own species. Chemical signals from the surface of the pollen grain elicit a "go" or "no-go" response from the stigmatic surface. With the right environment on the right stigma, the pollen grain will germinate, form a pollen tube, and begin to grow down the style toward the ovule. Without that correct back-and-forth chemical signaling, fertilization doesn't have a chance. Here, as in so many other instances, good sex requires good communication.

For flowers the love story begins as wind or a pollinating animal brings pollen to the stigma in the flower of the *same* species. But what happens when pollen lands on the stigma of the plant where it originated? Since many flowers have *both* male and female elements, self-pollination is a real possibility. If the point of sex is to foster variability within the population, then it's a lot better to mix genes with others than oneself. Different lineages of plants have developed a variety of methods to reduce self-pollination—the extreme form of inbreeding. Let's first take a look at one of the most sophisticated systems.

"Self-incompatibility" is a fancy internal biochemical system that some flowers use to avoid self-pollination. Here the chemical signature of the pollen grain alerts the stigmatic surface to the fact that the pollen comes from the selfsame plant, and stops its further development! With self-incompatibility, the stigma either turns off its receptivity system or the growing pollen tube is stalled within the style. Interestingly, self-incompatibility is the converse of our own immune system. Our immune system seeks to recognize *foreign* cells or organisms and disable or destroy them. Here in the flower, the self-incompatibility system seeks to identify pollen from *the same* plant and render it nonfunctional.[3]

In a self-incompatible plant, the stigma becomes invested with active chemicals that will either accept or reject pollen before the flower opens. As the flower blooms and begins to shed its own pollen, the stigma is already prepared to discriminate between unacceptable pollen from foreign species and pollen from the same species brought in by visiting insects. But in the case of self-incompatibility, the stigmatic surface must *also* reject pollen from the same flower or the same plant, while playing active host to pollen from others of its own species. This is a very subtle chemical system, based on highly coadapted gene complexes. Very fancy indeed. But there are many lineages of flowering plants that have not developed such a sophisticated system; they avoid self-pollination in other ways.

For plants that have not developed a self-incompatibility system, there are a variety of ways to avoid self-fertilization. Surely, the simplest system is the same as the one we humans have: distinct and separate males and females. If the individual plant has flowers with organs of only one sex, there is no way it can fertilize itself. Such plants are unisexual. Flowers on these plants may have only one set of sexual organs, the stamens (male) or the pistil and ovules (female), but not both. In other cases the flower does have both stamens and ovules, but only one set of organs is really functional; the plant produces either viable pollen or functional ovules, but not both. Such flowers and plants are functionally unisexual. In these cases only those with functional female parts are capable of producing fruits and seeds.

Over four thousand years ago, the ancient Sumerians were aware of sex in plants, sort of. The very useful date palm (*Phoenix dactylifera*) comes in two forms: pollen-producing males and date-producing females. Sorry to say, the male palms cannot produce sweet and nutritious dates. Once the palms were big enough to reveal their sex, the Sumerians planted female plants into groves. However, they also learned to allow some male plants to grow to full maturity. During the flowering season priests would cut off the large male inflorescences and march through the groves of date palms waving the pollen-producing inflorescences. Date palms are pollinated largely by the wind, so this procedure worked just fine. Though these ancient peoples didn't understand the underlying details, they had figured out how to get high yields of dates by planting lots of females while maintaining enough males for the pollination festivities. Papayas (*Carica papaya*) are another important fruit crop with separate male and female plants. Such unisexual species of plants are not common, especially in more severe climates. In addition, unisexuality is a feature that's reserved mostly for woody plants; few smaller herbaceous plants are unisexual.

Other Ways to Avoid Self-Pollination

Plants with fully functional bisexual flowers, but lacking self-incompatibility, have other means of preventing self-pollination. Of these various mechanisms, precise timing is the most common. Many flowers use timing as part of a "behavior" to avoid self-pollination. One example is seen in flowers that mature and shed their pollen *well before* the pistil and ovules are ready for pollination. That is to say, these flowers have an earlier "male" phase and a later "female" phase. Such flowers are called *protandrous* (first males). Other plants use the reverse strategy; here the female parts are receptive *before* the pollen of the same plant is shed. This is called *protogyny* and the flowers *protogynous* (first females). Either way, the chances for self-pollination are greatly reduced. In some members of the avocado or laurel family (Lauraceae), stigmas may be receptive in the morning with anthers shedding pollen in the afternoon, or the stigmas may be receptive in the afternoon with anthers opening the following morning. In these cases insects searching for nectar are likely to bring pollen from other flowers *before* they encounter the flower's own pollen.

Another method of avoiding self-pollination is structural: the flower is configured in such a way that there is little likelihood of self-pollination. Two-lipped (bilabiate) flowers are good examples of this system; there are many examples of such flowers in the mint, figwort, and orchid families. These flowers are strongly bilaterally symmetrical with a narrow entry into the tubular corolla, preceded by a broader "landing field." A long style usually arches over the top of the entryway and is the first interior part of the flower for the visitor to engage. With stigmas at the end of the style, pollen from another flower is likely to be encountered first. Then, forcing themselves deeper into the flower, the insect can slurp up the nectar. However, on exiting, the pollinators are doused with pollen as they withdraw. There's not much chance for self-pollination in such structurally complex flowers (see fig. 3A, p. 26).

The flowers of the catalpa family (Bignoniaceae, color insert M) are also two-lipped, with a well-developed landing field, narrow tube, and an elongate style that holds the stigma near the entry of the flower. But some of these plants add an additional twist. The stigma is often broadly two-lipped, with the stigmatic surfaces borne on the interior face of the "lips." The lips remain spread apart until pollen from a visiting insect comes in contact with them, whereupon the lips close up; this is another mechanism for avoiding self-pollination.

Among the more than quarter million species of flowering plants, there are many kinds of pollination strategies. One of the cleverest is to keep one's flowers wide open for cross-pollination until the petals are ready to drop. This gives cross-pollination a chance. Then, if no foreign pollen has arrived, these particular flowers "fold their tent" in a very special way. These unpollinated flowers end their existence by contracting their stamens so as to bring the pollen-loaded anthers into contact with the stigma. This, of course, results in self-pollination, and the advantage of increased genetic diversity is lost. But producing self-pollinated seeds is a lot better for the survival of a species than producing no seeds at all. (Needless to say, plants haven't "decided" on these various strategies. Different plant lineages have utilized a variety of pollination systems and those that have been effective have been retained.)

All told, sex and outbreeding will maintain a dynamically variable population over time. This is what's needed to survive in an ever-changing and pathogen-filled world. Trouble is, sex requires not only good communication—as we saw in the chemistry of pollination—it also requires good timing.

Sex and the Seasons

Reproduction among animals demands not only that males and females find each other, but also that they're in the right mood when they do. There's no point in getting together if the gametes (sex cells) aren't ready to go. Since the biological purpose of lovemaking is procreation, sex cells need to be ready for productive union. Also, most environments are highly seasonal; thus it's often critical for females to give birth at an optimal time of year for optimum survival of their offspring. The same is true for plants; seeds need to be formed while growing conditions are favorable. In the case of both plants and animals, timing of sexual receptivity is usually limited to only a short period each year. For a few fish species this may mean being in the ocean surf at high tide under the light of a full moon in May. Similarly, plants need precise synchrony when getting their sex cells together. Dispersing a lot of pollen when other plants of the same species do not have receptive flowers is a huge waste. It makes no difference if your flowers are pollinated by bats, beetles, birds, or blasts of air, flowers of the same species need to be receptive *at the same time*. To make sure they flower at the right time, and produce seeds at the right time, plants need to be in tight synchrony with the seasons.[4]

Like it or not, virtually every place on planet Earth has a seasonal cycle. Earth's longitudinal axis (north pole to south pole) tilts at about 23° from the plane of our orbit around the Sun. This means that the axis "wobbles" back and forth—in respect to the Sun—as it makes its yearly trip around the Sun. In the northern summer it is the Northern Hemisphere that "leans toward" the Sun, while the Southern Hemisphere "leans away" (their wintertime). This same tilt gives us northerners the dramatic changes of seasons that provide a rhythm for our journey through the year. Cold winters, tied to short days, slowly give way to longer days and a warmer springtime. Later, the long days of summer will slowly give way to the shortening days of fall. North of the Arctic Circle, the Sun may be visible throughout the twenty-four-hour day for a few weeks in early summer, just as that same Sun remains hidden below the horizon for a few weeks in midwinter.

In the temperate latitudes, springtime warming comes only gradually, and the leaves of the forest expand slowly. For many woodland plant species, spring is the time to flower. Most of our wind-pollinated trees flower at this time, as the leaves are beginning to expand, and while the forest canopy is open. This is a great time for the windy gusts of March and April to transport pollen across the landscape and through the leafless forest.

For flowers on the forest floor, early spring is also the best time to flower. Having prepared their flower buds the summer before, and thanks to energy stored in their roots, these plants are ready to sprout as temperatures begin to warm. Their time is short; they must complete their flowering cycle before the trees have fully expanded their leaves and the forest floor turns dark. Also, by flowering early, these woodland flowers offer nourishment to newly emerging insects. Flowers of fields and prairies do not bloom so early. Vulnerable to unusual blasts of arctic air in these open sites, prairie flowers begin blooming later than those within the woodland. But once these open habitats start flowering, they'll put on a show into October. And for a really big floral spectacle, you need to see springtime in a region with a "Mediterranean flora."

Mediterranean climates have some very special qualities. Their summers are hot and dry, followed by an autumn that is cooler but also dry. In this climate the rains begin to fall in the cold of winter and on into early springtime. Constrained by this kind of annual moisture and temperature cycle, a huge majority of the vegetation will flower in grand synchrony in springtime. You'll find these kinds of floras in southern California, central Chile, southwestern Australia, South Africa, and—as the name implies—the Mediterranean region. All are located between 30 and 40 degrees of latitude (north or south), and they tend to be found on the westward side of large continental landmasses. In addition to their dramatic flowering, Mediterranean-type floras are dominated by small trees with tough little leaves, and half their plant species are short-lived annuals. The large majority of these plants flower profusely in the springtime, maturing their seeds over the long, dry summer and fall. But when we leave the so-called temperate zone and head toward the equator we encounter a very different kind of seasonality.

Tropical Seasonality

Near the equator, day length varies by less than an hour over the year, and yet there are dramatic seasonal effects. Just as the northern seasons are determined by the Sun, tropical seasons are also closely bound to the annual movement of the Sun's trajectory, south to north and back again. Only here it's not the temperature that changes so much, rather it is the pattern of rainfall. Tropical rains tend to track with the intertropical convergence zone, which stays in line with the Sun, sweeping back and forth, north and south, through the year. Thus the major seasons of the tropics—dry season and wet season—are tightly configured to the annual calendar.

As a northerner living for the first time in the tropics, I experienced firsthand the dramatic transformation of wet season to dry season while teaching in eastern Ethiopia. Having arrived in highland Ethiopia in July, I was entering a world of lush green foliage and growing crops. Afternoons were regularly punctuated by passing showers. This was obviously the rainy season. By late September many crops were mature. Bright yellow Meskel daisies were blooming all across the highlands. Then, suddenly, in early October, the winds shifted direction. Now they were blowing briskly out of the north, from across the Sahara and Arabia. Cloudless skies became a crystalline blue; the rains ceased. The Sun shone brightly day after day, week after week. Months went by without a drop of rain. Leaves were shed, the vegetation withered, and grasslands became parched and yellowed. Windy days blew about a hazy dust that seemed to penetrate everything. All but a few rivers ran dry. More months would pass, and still there was no rain. Finally, in early April, I noticed that the winds were shifting. Soon the air smelled moist and sweet; for the first time in over six months fluffy clouds were forming in the blue sky. Now the winds were coming from the southeast; the early rains were on their way! By June the rains were falling with regularity; thundershowers punctuated nearly every afternoon. Continuing through September, these rains then ceased again in early October, as suddenly as in the previous year.

Here, in northeastern Africa, we were part of the Indian Ocean's grand monsoonal weather system. The winds from the south brought moisture-laden air out of the Indian Ocean, while winds from the north brought crisp dry air. Here was a really dramatic contrast of two very different seasons: dry and wet. This is the pattern in most of Ethiopia, but farther south, in Kenya and around the equator, there are two shorter rainy periods separated by two dry periods each year. Traveling still farther south we come into another region that experiences a single rainy season; but these rains fall between October and March—during Ethiopia's dry season. Clearly, the rains are in line with the Sun, sweeping across the tropics in an annual cycle as predictable as our northern seasons. Think of that iconic photograph of planet Earth, taken from outer space, showing much of Africa and the southern part of the world brightly illuminated. This picture was taken in December, summertime in

Antarctica. Here, the intertropical convergence zone is as far south as it gets, having brought the rainy season with it, while northern Africa is in the midst of its long dry season.

The Rainy Season

In these warm tropical climates, after months without moisture, landscapes rapidly transform themselves as the rainy season begins. Thanks to continually warm temperatures—and unlike the cooler north—tropical woodlands quickly sprout their leaves. Grasslands grow anew from the dried, often burned, tufts of last year's growth. Birds begin to nest; insects emerge in large numbers ready to feast on the new growth and each other. After a long season of almost lifeless desiccation, the early rains initiate a rebirth—as in our northern springtime. Living in the seasonal tropics requires staying in synchrony with the rhythm of the year. Plants must survive the long, dry months, then quickly sprout, grow, and flower. In the northland, it is the cold that shuts down the vegetation; here in the seasonal tropics it is a lack of moisture. For northern life, both changing temperatures and the regular changes in day length signal the season's passing. Northern migratory birds sense the shortening days at the end of summer and prepare to move southward in their annual migration. In the tropics, temperatures don't change all that much; and changes in day length are measured in minutes, not hours. How can plants anticipate the rainy season? When should they renew their growth?

The early rains in Ethiopia brought colorful change.

Brown dusty thorn bush and straw-colored grasslands sprouted a new flush of greenery. Widely spreading branches of acacia trees quickly donned a gossamer of pale green young leaves, further decorated by little pom-poms of white or yellowish flowers. Big crinum lilies with stalks two feet high were brandishing blossoms seven inches long—only two weeks after the rains had begun! Neither the stalks nor the flowers were in evidence before the rains had commenced. Bugs were buzzing about everywhere. A return of moisture had brought the landscape back to life. This was the time of year when I discovered that extraordinary *Hydnora* flower emerging from the ground (color insert G).

But the story of seasonal flowering isn't really that simple. During one year, unusual rains unexpectedly fell in late December in the middle of the dry season. A number of little grasses and herbs quickly and foolishly sprouted new green shoots. But the trees, the shrubs, the big crinum lilies, and most of the rest of the flora responded *not at all*. By some means or other these plants "knew" the real rainy season was still a few long dry months away. Scientists call this a *dormancy requirement*; plants will not resume growth until a sufficient period of time has passed. These odd December rains would be short and fleeting. But when, after several more months, the early April rains finally did arrive, the entire flora responded quickly and dramatically. In eastern Africa the rains are not all that dependable. As soon as the regular rains do begin, many plants flower almost immediately. Any delay may mean that there will not be enough time to mature fruit and produce a proper crop of seeds. Since successful reproduction is "the bottom line" in the battle for survival, for higher plants that means producing a healthy seed crop.

The "efforts" some of these tropical plants will go to in order to mature their fruits and seeds was a surprise to me. *Kleinia* is a shrublike genus of African composites (Asteraceae) that has adapted to dry regions by becoming succulent-stemmed. The leaves are small and quickly lost, as the finger-thick, fleshy green stems do most of the plant's photosynthesis. Eastern Ethiopia had two easily distinguished species, one with reddish purple flowers and the other with yellow flowers. Like so many members of their family, these plants bloomed at the end of the growing season and into the early dry season. Since they flowered at the same time, I proceeded to collect stems of the two species, cut them into newspaper-length sections, and turned them into specimens. Neatly arranged in their numbered and enfolding newspaper, they were then placed in a ventilated press over moderate heat. After drying and removing them from the press, I examined my specimens.

Something was wrong; there were no flowers to be seen. Their collection numbers, relating back to the field notes, indicated the flower color; but here, in the dried material there were no flowers to be found. This puzzled me. I was sure I had collected *flowering* material, but not only were these specimens lacking any signs of flowers, they all had mature seeds ready for dispersal!

In the following year, further collections of these species produced the same results: flowers gone, seeds present. Clearly, these determined creatures had treated my drying efforts as if they were simply a continuation of the dry season. Sucking moisture from their stems, they had turned their flowers into matured seeds. (In composites the inferior ovary and its single ovule form a tight unit so that fruit and seed are essentially one and the same.) Being cut, folded, and mutilated didn't seem to bother them. These plants "knew" how to produce seeds, come what may. Another collecting incident, however, proved even more remarkable.

An Orchid in the Grass Savanna

Collecting plant specimens during the early rains of May in an open acacia grassland at the edge of Ethiopia's Rift Valley, I came upon something I hadn't expected to see in this kind of habitat: a tall orchid (*Eulophia schimperiana*). The flowers were only about an inch long, but they were borne near the top of a slender, few-branched stalk—about three and a half feet tall! This brought the flowers to a level where they stood above the dry grasses that were themselves beginning their regrowth from below. Down on the ground, among the grassy tufts, sat a cluster of the orchid's ellipsoid greenish pseudobulbs (about six inches long and an inch thick). Obviously, here was where the orchid had stored the energy needed to quickly build that tall inflorescence. Flowering in these early rains meant that there would be pollinators and, hopefully for the plant, sufficient time during the remaining rainy period to mature the seed capsules.

Carefully collecting specimens of this fascinating plant, I set aside one of the pseudobulbs and brought it back to our college greenhouse in the highlands. This orchid had been growing down in the hot Rift Valley, about 3,000 feet above sea level; now the pseudobulb would be grown in a greenhouse at 6,500 feet above sea level. In the Rift Valley the orchid had lived through many long, hot, dry seasons, but here in the cooler temperatures of the highlands the plant would be watered all year long. The pseudobulb took root, sprouted new leaves, and seemed to be very happy in its greenhouse habitat. Then it proceeded to do something remarkable. During the next three years, and despite constant watering throughout the year, this plant produced a flowering inflorescence at about the same time as I had first collected it. Each May the plant brought forth a flower stalk! At no other time did this plant give any hint of flowering—despite incessant watering. What was going on here?

When I had first encountered the plant I was sure it had responded to the early rains, after more than six months of desiccation. But here in the greenhouse there was no drought and there were no early rains; watering had been continuous. How did this plant sense that it was May? The only reasonable suggestion is that the plant was able to respond to the lengthening of the day, during the months of January to May. Here, only nine degrees north of the equator, the difference between day lengths in December and June is a bit less than an hour. Somehow that change was enough to trigger our orchid and keep it on schedule, even in its new and artificial environment. Actually, this inadvertent little experiment shouldn't have been too surprising; some varieties of rice are known to be able to respond to differences in day length of less than half an hour. Obviously, these tropical plants are keying in on the annual cycle of the sun in a far more sophisticated way than I had ever imagined.

Plants in the more temperate parts of the world, of course, have much more disparate day lengths to cue them in to seasonal timing. In comparison with a one-hour difference between the longest and shortest days at 9° north latitude, the difference between longest and shortest days at 42° latitude is about six hours. That's the latitude of the California-Oregon border, Chicago, Boston, and Rome. Exactly how plant cells stay in time with the real world and "sense" changes in day length is now being investigated by using mutants of our favorite laboratory plant: *Arabidopsis thaliana*. This is the science of *circadian rhythms*, the internal clocks that keep cells and the organisms they comprise in step with the world around them. (Incidentally, *Arabidopsis thaliana* became a laboratory favorite because it is small, is easy to grow, has a short life cycle, and has a relatively simple genome. This was also the first plant to have its complete DNA sequence identified in 2000.)

When to Fruit, When to Sprout

Offering something good to eat is a great way of having animals disperse a plant's seeds. And this may be done even more effectively if the plant provides this nourishment at a time of the year when it's really needed. Many North American trees and shrubs, such as hawthorns, viburnums, and sassafras, mature their juicy fruits in September and October, just in time for the long southward journey of so many of our migratory birds. The birds, gaining energy from the nutritious fruits, will disperse seeds over great distances. Bird-dispersed fruits are moist and usually colored red, blue, purple, black, or white. These colors stand out nicely among green vegetation and are well adapted to bird vision (which is rather similar to our own). Also, these fruits aren't very large; they're built for the size of the mouths that will swallow them.

Larger fruits evolved for larger animals with bigger mouths. Most of these larger mouths belong to furry mammals. Mammal-dispersed fruits tend to be orange, yellow, brown, or greenish, and they may have a hard outer rind or husk that needs to be broken open before the goodies can be removed (examples are pumpkins, calabash, and cacao pods). Many of these fruits begin green and astringent or acidic. As they enlarge they turn color, soon becoming sweet and succulent, just as their seeds become mature and ready for dispersal. This is a clever cadencing, transforming green, hard, and sour fruit (stay away) to something that is yellow, sweet, and juicy (come and get it). One of our favorite examples, the apple, originated in the mountains of western Asia. Here, apples become sweet, succulent, and fall to the ground at a time when bears are trying to devour as many calories as possible, just before their long winter hibernation. Many seeds may be crushed by the busily chewing bears, but a number will surely get through and be effectively dispersed.

Once dispersed, seeds may have built-in means of determining when to sprout. We already noted that many plants *did not* expand their leaves after unusual rains in the middle of the dry season in Ethiopia, probably because of dormancy requirements. Likewise, many seeds in other kinds of habitats also have dormancy requirements. Here, in our northern regions, it is important for both seeds and plants to remain dormant, even if there is an unusual warm spell in February. A cold snap in March might kill anything that had started growing earlier. In fact, many seeds of northern plants *require* a cold period of considerable length before they can respond to moisture and warmth. Such timing requirements help the seeds or the plant stay in sync with the season.

Plants and their seeds have other strategies for survival. Imagine that next summer or the next rainy season suffers a really horrendous drought; and suppose *all your seeds* had germinated in the preceding spring? Under such conditions the entire seed crop of the previous year might be lost. Bad strategy! A great many plants get around this problem by varying the time over which their seeds will sprout. Like an insurance policy, such staggered germination, over several years, means that seeds will be available in future years. In some cases, seeds require abrasion or scarification to breach the tight enclosure of a hard seed coat; this is something that is likely to happen only accidentally over a longer period of time. Other seeds require the heat of fire before the seed coat can absorb moisture and begin to germinate. Some seeds require bright sunshine while others germinate only in moist darkness. Such a variety of factors—over many species—gives us the "seed bank." Soils, both in the Torrid Zone and in the Temperate, contain many unsprouted seeds, all part of a strategy to survive over bad years. This may be one of the reasons why flowering plants suffered little during the great extinction that wiped out the dinosaurs around 65 million years ago. With lots of seeds "sleeping" in the ground, many species would sprout once conditions improved again.

Flowers and the many strategies of flowering have helped make the angiosperms one of the most successful lineages of living things on our planet. As we shall see in the next chapter, much of this success is based on the many complex relationships they have forged with their animal "friends."

chapter 3

flowers and their friends

In exploring the role of "friendship" regarding flowers, we will need to consider the whole plant as well, organisms that range in size from 200-foot-tall forest trees to little aquatics smaller than a dime. For a majority of these many different kinds of plants, the most important form of "friendship" is a relationship we can't see, and one we should appreciate more than we usually do. This is a symbiosis that occurs underground—a friendly union between the roots of plants and certain cooperative fungi. Though cryptic, this may be the most important symbiotic friendship in the living world.

First, let's recall what kind of organisms fungi happen to be. Toadstools, mushrooms, molds, yeasts, rusts, smuts, lichens, and smaller beings we rarely notice are all included in the fungal kingdom. Though once considered a subdivision of the plant world, fungi are now recognized as a kingdom of their own. Fungi (singular: fungus) are profoundly different from plants. For one thing, they are not green. None of them has chlorophyll; they cannot make their own food by photosynthesis as green plants do. Their "bodies" are mostly hair-thin filaments that wind their way through soil and organic matter. There they exude enzymes to dissolve their food and then reabsorb the predigested nourishment. (Bacteria operate in much the same fashion, but they are much smaller, not organized into complex forms, and represent a much simpler life form.) Also, the walls of fungal cells are made from chitin and not cellulose, as are the walls of plant cells. (Chitin is the stuff insect cuticles are made of.)

Fungi differ also in how they are constructed. Higher plants begin as embryos and build themselves up through the actively dividing cells in growing tips (*meristems*) at the ends of roots and twigs. Fungi have neither embryos nor meristems. They build mushrooms, lichens, and other larger structures by the interweaving of their slender filaments. How they manage this minor miracle is still a mystery. In addition, fungi have peculiar forms of sexual union—at the cellular level—unlike anything in plants. Comparisons of DNA sequences also support a separate status for the fungi and indicate that they are more closely related to microscopic animal life than they are to plants.[1]

Many scientists believe that a symbiotic relationship between fungi and land plants may go back to those times when plants were first pioneering land surfaces. For these early plants, getting water and mineral nutrients was a major problem. Roots and internal plumbing (the vascular system) were a solution to absorbing moisture from the soil and distributing it throughout the plant body. But though plant roots may be able to absorb water, they have difficulty getting essential nutrients out of soil solutions. It is here that the role of fungi is a big help. By forming a special union, called *mycorrhizae* (literally, fungus roots), land plants discovered a way of increasing their uptake of critical nutrients from the soil. What the fungus does is to pick up these vital nutrients and help *transfer them directly* into the roots. Terrific!

But why should a fungus spend energy "helping out" a green plant? The answer is that in this mycorrhizal exchange system the fungus gets energy-rich sugars in return. Here we've got a true symbiotic relationship, one in which both partners benefit. With their green foliage and the energy of sunlight—using photosynthesis—plants can produce lots of energy-rich compounds both to build themselves and to share with their fungal symbionts. In this case, the fungus-root association helps bring in nutrients that the roots themselves cannot absorb as efficiently, in exchange for sugars produced in the green leaves. Botanists estimate that over 80 percent of land plants are dependent on soil fungi for at least part of their mineral nutrition. In fact, the mycorrhizal exchange system may have been a key innovation in allowing larger plants to expand their colonization over the land 400 million years ago.

The rich greenery of the rain forest, amber waves of grain, and the floral diversity of a prairie rely on more than just sunshine and good weather. Without critical elements like nitrogen, phosphorous, potassium, and sulfur, which are available primarily from soil water, land vegetation could not develop or survive. These critical elements in the soil—and the fungi that help make them available to higher plants—are an essential part of our biosphere. For life that lives on land, this may be the world's most significant cooperative biological relationship.

And that's not all. Fungi are also important in the rotting and degradation of dead plant materials. Fungi contain enzymes capable of digesting cellulose. (As we noted, their walls are made of chitin, so they don't have the problem of digesting themselves.) Without fungi and bacteria, forests would become piled high and choked with dead plant remains. In both mineral uptake and nutrient recycling, the fungi are essential participants in ongoing vegetation dynamics. But in addition to fungi, flowers have many complex friendships.

Pollinators, a Big-time Friendship

When discussing flowers that have friends, we're mostly talking about flowers whose pigments, aromas, and symmetry have a single purpose: *advertising*. These are signals for animals: "Come and visit us." For the vast majority of flowers this is no empty proposition: there really is a reward. Nectar, a solution of water, sugars, and other nutrients, is the primary reward for the great majority of animal visitors. These pollinators rely on sugary nectar to energize their daily life activities. In addition, there are some highly specialized flowers providing oils or aromas for their very choosy pollinator clientele.[2] But more about them later; let's start with the bee family.

As a boy, I recall being fascinated by bumblebees and how busy they always seemed to be when visiting flowers. Having been told that they were gathering nectar, I imagined nectar to be as tasty for the bumblebees as ice cream was for me. Watching bumblebees still impresses me; they really do hustle. But why do they work so hard? The question is worth considering, and though the answer may be simple, it is also profound. Busy bumblebees are able to feed a greater number of offspring than do the lazy ones. Those with the genetic wherewithal to work hard will produce more progeny, but those lacking such a genetic program will produce fewer progeny and eventually fall by the wayside. Over the years and across generations the simple consequence of hard work will result in a very busy bunch of bees. Here's another example of how natural selection really works.

Similarly, flowers that give little nectar will be less visited, produce fewer seeds, and they, too, will soon be gone. In fact, a field of flowers and their pollinators has to work as an effective economic marketplace for the simple reason that if it doesn't, it's gone. The ultimate selection pressure is long-term survival for both the flowers and their pollinators, and the system's basic currency is energy. Here's where Adam Smith's rules of the free market become intertwined with Darwin's natural selection to produce a complex economy of nature. Of course, there's a lot of play in the system; there has to be. A bad year for bumblebees may be a bad year for seed production, but things ought to recover soon enough. However, if a few really bad years do come along, local populations *will* go extinct. If luck has it, those populations that survive in nearby areas will be able to disperse their seeds and offspring into the depopulated area, and the local population will flourish once again—until the next disaster. Not a pretty playbill, but one that has tested plants and animals over millions of years and thousands of disasters. Unfortunately, today, with wild vegetation confined to ever smaller areas, the chances of being repopulated after a local extirpation have been greatly reduced, and the loss of species continues inexorably throughout the world—a serious problem to which we will return. For now, let's get back on a more positive track.

Boldly symmetrical flowers, colorful perianth, alluring aromas, and sweet sugars are all major expenditures for a plant.[3] Throwing a party takes energy. Special tissues, such as nectaries (where the nectar is formed), require energy for their construction and energy in the production of the sugary nectar itself. This is the primary reward for the animals that pollinate flowers. Members of the bee family, both solitary and social bees, utilize both pollen and nectar as *the only food source* for their larvae. This makes the bee family—with over 25,000 species—especially important pollinators throughout the world. And though we humans do not use nectar ourselves, we do love honey and have tended honeybees over thousands of years. In addition, honeybees are essential pollinators for some of our important food and forage crops, ranging from apples to alfalfa. Thanks to this long and economically important history, honeybees have been the focus of intense scientific scrutiny. This research has given us unique insights into the world of the honeybee and her remarkable talents. (I use "her" since honeybee workers are infertile females.)

The Amazing Honeybee

Our knowledge of honeybees was grandly expanded by the pioneering work of Austrian zoologist Karl von Frisch and his associates. By clever experiments using very simple apparatus, von Frisch discovered that bees could see a variety of colors. They can't distinguish as many colors as we do, and they can't see as far into the red end of the spectrum as we do, but they see further into the ultraviolet. In fact, many flowers that look uniformly colored to us actually have bold patterns in the ultraviolet. His experiments also showed that bees can "see" polarized light, which allows them to determine the Sun's position from a patch of blue sky in a way we cannot. However, the most celebrated of von Frisch's findings was not their color vision. More amazingly, he discovered that bees were able to *communicate* with each other!

Bees communicate both the nature of a new nectar source *and* its direction by means of a "dance language." For shorter distances they use a circular dance, but for distances over one hundred yards they do a "waggle dance" in the form of a figure eight. These dances are performed on the vertical face of the comb within the darkness of the hive. The distance of the dance, the frequency of bees' "waggle," and the angle of the dance from the vertical indicate distance and direction. Many repetitions of the dance, and concordant dances by other newly returning bees, communicate how valuable the new nectar source might be. Apparently, bees do not come home to do a "waggle dance" unless the new source is rich in energy, significant in amount, and unencumbered by too many other nectar feeders. Though such interpretations of bee behavior have been challenged from time to time, continuing research has corroborated von Frisch's original conclusions.[4] In fact, it's been shown that a bee dancing for thirty minutes will change the angle of her dance to compensate for the movement of the Sun across the sky during that time! In addition, radar tracking has recently confirmed the efficacy of the dance language by actually tracking the flights of newly instructed bees. Clearly, when bees discover a new source of nectar, they can share that information with their hive-mates by returning with the aroma of the flowers and doing a clever little dance.

After bringing nectar to the hive, it is regurgitated to other bees over several cycles to create honey. This process transforms the sucrose of nectar into the fructose and glucose of honey. A honeybee can return with about 50 milligrams of nectar on each trip; this means it will take *two hundred thousand* trips to gather a kilogram (2.2 pounds) of nectar. Because it is so resistant to spoilage or decay, honey serves the bees as a long-term energy source. Thanks to stored honey, bees can maintain themselves and their hives over winter. As much as 15 kilograms (33 pounds) of honey may be necessary to carry a hive of twenty thousand bees through a long, cold winter. Speaking of the problem of getting through a long winter, I was startled by a scientific report I came across recently. "Bees' heads weigh less in winter than in summer" claimed the title.[5] Sure enough, researchers in Finland found that bees can reduce both their brain activity and brain size in winter, building larger brains when the flowers bloom in spring. Here's a pretty cool way to save energy in wintertime: cut down the size of your internal computer. Despite this wintertime "brain reduction," beekeepers have noted that bees can remember the directions to flower sources *over the winter*. Not bad for a brain the size of the head of a pin. (Most other pollinating insects survive the winter by hibernating or by being cold-resistant eggs or pupae.)

Worldwide, bumblebees are among the most important pollinators. As with other bees, bumblebees gather nectar and pollen to nourish their young. Adapted to colder climates, colonies of bumblebees often have few adult members and they also forage individually (having no dance to communicate information). In foraging for pollen, however, bumblebees have learned a trick that honeybees haven't. It's called "buzz pollination." If you're quiet and patient near flowering tomato or potato plants, you can even hear it. (Both plants are members of the genus *Solanum* in the Solanaceae family.) The buzzing bumblebee will fly to the flower, land, and, once properly positioned on the flower, *change the pitch* of her buzzing. Suddenly, there's a higher-pitched buzz. These potato and tomato flowers do not give nectar, and their anthers do not have slits down their sides. Rather, they have small rounded pores at the top of the anther. What the bumblebee is doing is using a higher-pitched "buzz" to *vibrate* pollen grains out of the anthers. Pretty neat trick, and you can hear them doing it! The blueberry family (Ericaceae) and the melastome family (Melastomataceae) include many species with anthers that open by small pores and are pollinated by small bees in a similar way. All told, it's estimated that about 8 percent of colorful flowers are buzz-pollinated.

Like honeybees, busy bumblebees have been the subjects of scientific study. Not only do bumblebees have to find food for their developing young, they've got to build a nest, keep their young warm, make sure they don't waste too much time flying too far, or visit flowers with too little pollen while not running out of fuel themselves. All this activity is powered by the sugars in nectar, with pollen a source of protein for the growing larvae. It's a hugely amazing enterprise when you get to look into it, as Bernd Heinrich's classic (but technical) *Bumblebee Economics* does.[6] Though a number of other insects, especially beetles, act as pollinators by feeding on pollen, it is only the bee family that exclusively feeds its young pollen and nectar. For other pollinators, sugary nectar is the primary energy source for the adults—not the larvae.

Have you ever seen a mosquito drinking nectar at a tiny flower? You've got to get really close to some very small flowers to spy this kind of activity, but it reminds us of one of the vital roles of flowers. Lady mosquitoes don't suck our blood for their own amusement or our discomfort; they need a blood meal to create a raft of eggs. Male mosquitoes visit little flowers for nectar; the reason the males don't bother us is because they don't need a blood meal to chase female mosquitoes; sugary nectar in their gas tanks does the job. Likewise, the many wasps we see at flowers are also busy sipping nectar. Once energized, many of the female wasps will go hunting for spiders and other prey to provision their larvae. Butterflies, moths, hummingbirds, sunbirds, and nectar-feeding bats are other kinds of flower visitors; these animals are especially important pollinators in the tropics. And if we're going to mention unusual pollinators, such as nectar-slurping bats, we should point out that such special friends need special accommodations.

Unusual Flowers for Unusual Visitors

The tropics have many more weird flowers than are found in our northern climes. The reason is simple, some tropical flowers have really strange visitors. Let's consider a few examples. In Central America's rain forests you can sometimes see fairly large, greenish flowers hanging down from above, on a long stringlike vine. There may be a cluster of buds at the bottom of this hanging vine, but only one flower will be in bloom. Each flower will be two to three inches long, with an opening about an inch wide. Not only are they greenish—not easy to see—but they've got an unpleasant aroma. You'll sometimes see clusters of somewhat similar greenish or brownish flowers on the trunks of calabash and a few other trees. In all these cases the flowers are quite large (2–3 inches) and lacking vivid colors, though they may be white. In addition, these flowers and their parts are strong and thick; they're built tough. However, don't expect to see anybody visiting these flowers, unless you wait patiently with your flashlight through the long tropical night. These larger tough flowers are built that way for pollination by bats!

Small bats hang on to the flower while they're lapping nectar or chomping on pollen; this explains why these flowers are both larger and have such thick damage-resistant parts. No need for color here; in the darkness of night bats find the flowers mostly by aroma, and they seem to like those aromas. But why does the plant have only a single flower opening each night? It turns out that the bats usually fly the same route each evening, so each plant along the route will provide a newly opened nectar-filled flower every night. That's called "traplining." Some other pollinators, like large bees and hummingbirds, do much the same thing, repeating a favorite path through the forest each day or night.

Speaking of birds brings up another distinctive, largely tropical, pollination pattern. Bright red flowers with long, narrow corolla tubes are particularly noticeable in the American tropics. Despite the size and showiness of these flowers, they tend to have small corolla lobes, or lobes that are bent backward; there is no landing field. Also, the corolla tubes are usually held horizontally or pendant. More significantly, you won't find such flowers in the native vegetation of Africa, Europe, Asia, or Australia. The reason is simple: those parts of the world do not have *hummingbirds*. Narrow-tubed red flowers are a phenomenon restricted to the Americas and their hummingbirds. These flowers have no "landing field" because hummingbirds take nectar while *hovering in midair.*

Bird-pollinated flowers and inflorescences of Africa, Asia, and Australia are structured so their visitors have a place to perch while feeding. Thanks to their small size and rapid wing beats, hummingbirds don't need a perch. Indeed, one of the glories of the American tropics is how many different genera of flowering plants have responded to the presence of hummingbirds—by evolving flowers especially suited to these nimble fliers. Some of these flowers even have curved corolla tubes that conform to the curved beaks of the particular hummingbirds that visit them! Or is it the other way around? (A few butterflies can see red and are attracted to red flowers, but such flowers tend to be smaller, and they provide a perch for their visitor—as in the cardinal flower, *Lobelia cardinalis.*)

Hawk Moths, Agile Fliers of the Night

Hummingbirds aren't the only class of pollinators that can take nectar while hovering in midair almost motionless in front of the flower. Another group of agile fliers, with wing beats so fast that they can remain stationary or even fly backward, are the hawk moths. In fact, their wings beat so fast that, like hummingbirds, you can't see the wings when in flight. In this case, flowers tend to be white or pale colored, since these moths fly mostly at dusk, at dawn, or in the darkness of night. Like butterflies, these moths have long, slender tongues that roll up in a compact spiral when they're not feeding. And here, too, flowers have responded with narrow tubes. But in a few species these floral tubes have become more than four inches long! Why might floral tubes have become so long? This is a question that Charles Darwin responded to when he learned of the discovery of an orchid in Madagascar with a spur eight inches long. (In these orchids the base of the lower petal has a backward-pointing tube in which nectar is produced; this tube is called the *spur*. Not an orchid, but note the recurved spur in *Impatiens* in color insert D.) Darwin suggested that the orchid was pollinated by a hawk moth whose tongue was *equally long*. Some years later the long-tongued moth was collected and described. But why should either floral tube or insect tongue have become so long?

Darwin assumed that the orchid is always under selective pressure to position its pollen packets on the front of the moth, where they can be carried by the moth to another orchid for pollination. Without such pollination, and consequent seed production, the orchid species would go extinct. On the other side of the relationship there is a *contrary* selective pressure for the hawk moth: staying clear of the orchid so that its flight is not interrupted. Coming into contact with parts of the flower can damage the delicate wings of these moths or disrupt their flight. Thus, the eight-inch spur and eight-inch tongue would seem to be the result of the continuing interaction of two countervailing forces: contact for the orchid and avoiding contact for the moth. One can imagine how populations of both orchids and moths have been under continuous selective pressure, with both the nectar-containing spur and the hawk moth's tongue slowly becoming longer over thousands of generations. Logical enough; however, Darwin's interpretation has been challenged recently by a hypothesis based on observations in the field in Madagascar, where moths have been seen to use a "swing-hovering flight" when they feed at flowers with shorter nectar tubes. They fly up and down in front of the flower as they sip nectar. And why might they be flying in this way? Apparently, sometimes, there are big black spiders sitting on these orchids, waiting for the moths in the darkness of night. The swing-hovering flight seems to be a way of avoiding the grasp of these large nocturnal spiders.[7] As usual, nature may be more complicated than originally thought. Nevertheless, I suspect that Darwin's basic idea is essentially correct; tongue and spur have gradually grown longer over time.

Further support for Darwin's interpretation comes from a tropical American hawk moth (*Amphimoea walkeri*) with a tongue a full ten inches (25 cm) long. Though I'm not aware of anyone having seen these moths feeding, we do know of a few night-blooming flowers with slender floral tubes of equal length. They are *Posoqueria grandiflora*, a tree in the coffee family (Rubiaceae), and *Tanaecium jaroba*, a vine in the trumpet creeper or catalpa family (Bignoniaceae). Both have white flowers with real corolla tubes (not spurs as in the orchid example). Lacking direct observations, we can link this unusual moth and these long-tubed flowers by an interesting concordance. Both plant species and the moth share the same geographical range: from Nicaragua southward into the Amazon basin. More recent, well-documented studies seem to be telling us the same story regarding two groups of long-tongued flies in South Africa. Tongues four inches long are virtually unheard of among flies, but several South African fly species have such tongues, and the reason seems to be the same as in the case of the long-tongued moths.[8] Here, closely correlated coevoluton has resulted in several species of long-tubed flowers in four different families being pollinated by several species of flies with tongues of equal length. Clearly, what we have here is the product of an "evolutionary arms race" for longer tubes in the flowers and longer tongues in the pollinators.

Unfortunately, a few insects whose tongues are not long enough to reach the nectar within a tubular flower have chosen a perfidious solution to their problem: they chew their way through the tube to get at the nectar near the base. This is called "nectar robbing," and it does not result in pollination. You can sometimes see bumblebees engaged in this evil enterprise. Obviously the insect needs to be fairly strong to chew through a corolla, and, in most flowers, this might require almost as much energy as the nectar will provide. Thus, nectar robbing is not very common. But just as some insects have learned to "cheat" in dealing with flowers, a few flowers have learned how to "cheat" in dealing with insects.

Pollination by Deception

Both in our human world and in the world of nature, there's often an illegitimate way to come out ahead in a transaction. It turns out that some flowering plants "cheat" by using a clever form of deception. A few species of large rain-forest *Cydista* trees (Bignoniaceae family) produce a huge number of colorful flowers over a period of only a few days. These usually sparsely distributed trees are quite spectacular in the rain forest, as they suddenly become a blaze of color—amid a sea of dark greenery. Like our catalpa trees or the tropic's jacaranda and tabebuia trees (all are members of the Bignoniaceae family), the flowers are large and they cover the tree when in full flower. But there's a catch. These *Cydista* flowers produce no nectar; there is no reward for the busy pollinators. *Cydista* species rely on enough insects making a sufficient number of visits to various flowers to effect significant pollination before they "wise up" and go elsewhere for nectar. This is called *pollination by deception* and, like many other criminal activities, is rare. These nectarless species are uncommon; they are mimicking the many more tree species offering ample rewards to those who serve them. More than any other plant family, orchids demonstrate a great variety of deceptive patterns. The lovely lady slipper orchids (the genus *Cypripedium*) of our northern woodlands employ this strategy. They bloom early in spring but give no nectar. The brightly colored "slipper" entices the insect to enter, and though it is easy to get in, it isn't that easy to get out. To make its escape the insect must crawl past the pollinia and stigmatic surfaces. However, unless the insect goes through this routine at least twice, pollination cannot be achieved. These flowers appear to operate in the same way as do the *Cydista* trees, using the exploratory activities of insects to get themselves pollinated, as the insects search vainly for nectar. Such a strategy often requires newly hatched *naive* insects. These flowers, as Peter Bernhardt points out, ". . . must bloom at a time of the year when a sucker is born every minute."[9]

But why might orchids produce many more such deceptive nectarless species than other plant families? The answer seems to be that, unlike most flowers, orchids produce pollen not as loose powdery grains but in special packets (the pollinia) containing thousands of minute pollen grains. Thanks to this little detail, a single pollination event can produce thousands of seeds in orchids. True, these are minute seeds that need to germinate in just the right spot to create a new plantlet, but the seeds are about a hundred times more numerous than those of other plants. In fact, it's been estimated that in some orchid genera a single fruit (capsule) can produce up to three million seeds![10] What this means is that orchids don't have to win often, because when they do win (get pollinated), they win big, producing thousands of seeds. Not only do quite a number of orchids, such as our lovely springtime lady slippers, produce no nectar for their pollinators, a few species have gone much further and developed a truly bizarre form of deception.

There are a number of orchid genera in various corners of the world that have very strange-looking flowers. But though they may look strange to us, they apparently look extremely enticing to another group of creatures. These unusual little orchid flowers are so similar in appearance to the bodies of specific insect species that the love-hungry little males of these species actually attempt to mate with the flowers. Making this mistake more than once will result in having the myopic male carry pollinia from one orchid to another, with the potential of producing thousands of seeds. And it's not just the shape of the flower that inspires the passion of these males. A recent study of an insect-mimicking *Ophrys* orchid disclosed a mix of aromatic chemicals virtually identical to the pheromone mix of the female wasp. Here we have an example of both morphological mimicry (the flower's shape) and chemical mimicry (the aroma).[11]

Called *pseudocopulation*, this form of pollination was first described in an *Ophrys* species living in the Mediterranean region. Here, both orchid flowering and the emergence of male bees occur early in springtime, before the female wasps have emerged. These males are naive indeed. Until recently, this weird form of pollination has been found only in orchids, probably because it is a strategy that doesn't work very often. Again, orchids get away with it because they are the only plant family capable of producing a huge number of seeds with only a single pollination event. (See color insert O, depicting an Ecuadorian orchid believed to use pseudocopulation as a pollination strategy.)

Flowers vary in their accessibility. Some are easy to approach, alight on, and find nectar. Buttercups, daisies, and sunflowers are good examples of such flowers. Insects can fly in and land from any direction and, if their mouthparts are the right size, they can feed with no problem. Other kinds of flowers are more discriminating, especially those with a long, tubular corolla as mentioned earlier. Unless the insect or its mouthparts are the right size, it won't get at the goodies. Then there are the special flowers for special tastes. Bright red flowers, as we've noted, are often bird-pollinated. Dark reddish-purple blossoms that smell of rotting flesh are irresistible to flies that feed on the fluids of decaying bodies, so they are the principal pollinators of such flowers. But in a few lineages, "host specialization" has gone much further.

Pollination and the Evolution of Complete Interdependence

Let us return to orchids again. A very unusual pollination syndrome was first documented in the late 1950s, in the forests of Central and South America. Though their flowers are strongly aromatic, these orchid flowers produce zero nectar. More puzzling still, when the large- to medium-sized bees that pollinate them were examined, each and every one proved to be a male! Something strange was going on here; this didn't appear to be simple deception. The first thought was that these males had tanked up on nectar at other flowers, then visited these particular flowers to "get high" on the powerful fragrances. (They did seem to get a little dizzy when they engaged these flowers.) But Mother Nature doesn't approve of wasting energy; unless behavior patterns are useful, a drizzle of disruptive mutations should eliminate them. What was going on here? Further studies disclosed that these male bees were *gathering* the aromatic compounds and reprocessing them. Seems that meeting lady bees in the dark understory of the rain forest isn't easy. What some of these guys do, after gathering and processing the aromas, is to form a group at a specific site in the forest. With a small crowd of males buzzing about, and thanks to their newly acquired aromas, the female bees find it easier to find a potential mate. Better yet, the gals can check out a bunch of guys, and decide who's the most appealing.

In the case of these orchids, the bees may visit more than one orchid species, and the orchid species may be visited by a number of species of these specialized (euglossine) bees. By both mixing and processing the fragrances, each bee species can create its own specific aroma. It's like wearing their own designer cologne to attract the right females. Along other lines of specialization, some desert environments have bee species adapted to only a few flower species. Both flowers and bees are active for only a few weeks, due to the very short flowering periods in such stressful environments. And in a few instances Nature has gone even further, creating plant-insect relationships where one species of plant is pollinated by a single species of insect. In these extreme cases of coevolution, neither partner can survive without the other. Surely the most dramatic example of complete codependence between a group of plants and their pollinators is in the fig genus (*Ficus* of the mulberry family), and it is referred to as the "fig tree–fig wasp mutualism."

The fig tree's small flowers are unique in being completely

enclosed within the spherical fruitlike fig. The fig may look like a fruit but it is actually a hollow inflorescence. All the little flowers are contained inside the fig, with a cryptic entry at the top. A series of tightly overlapping bracts at the entry require that the flat-headed fig wasp squeeze her way through this impediment to enter the interior of the fig—losing her wings in the process. Once inside, she moves around among the female flowers of the fig with pollen she had collected before leaving the fig in which she had been born, thus effecting pollination. After laying an egg in each of many gall flowers where her young will develop, her life will have ended. She may be the only wasp to have entered this fig, or there may have been one or two others, thus the fig may house one or several sets of siblings.

After the wasp larvae have fed, matured, and pupated in the gall flowers, a new generation of blind and wingless males emerges. These raunchy fellows proceed to mate with the maturing but inactive females, still trapped within their gall flowers. The males then get together and chew a passageway through the thick wall of the fig allowing the females to escape with their wings undamaged. Having accomplished their life's work, the little males expire. Now the females become active just as the fig's male flowers are ready to release their pollen. Gathering pollen, the female wasps place it in special "pockets" alongside their abdomen. Now the pollen-packing mamas are ready to depart—through the passageway that the males had prepared for them. But these females face a formidable task. They must fly through the forest and find a fig tree that is both the correct species and whose figs are *ready for pollination*, in order to nurture a new brood of wasps.

To support all this activity, the fig tree produces three kinds of little flowers within the fig. The female flowers come in two forms: the gall flowers in which the wasp larvae develop, and the fertile female flowers that will become seeds. The third kind of flowers is male, producing pollen as the new females emerge and before they take flight. But what kind of signals might help orchestrate this flower-insect duet? It turns out that maturation of both the fig and its pollinators in tandem is controlled by a very simple factor: carbon dioxide. Tightly enclosed by wall and overlapping bracts at its only opening, the interior of the fig has little gas exchange with the outside world. Respiration of the wasp larvae as they feed and pupate makes the interior of the fig much richer in carbon dioxide. At the end of pupation, the little males are cued by the concentration of carbon dioxide to begin their work. But after they've chewed an escape channel and allowed outside air into the fig, carbon dioxide concentration within the fig plummets. Lowered carbon dioxide, in turn, activates the females and cues the male flowers to release pollen. In this way the pollen is ready for transport by the females. And there's more. Soon the wall of the fig becomes sweet and succulent. Nice timing: the female wasps depart, the seeds are mature and ready for dispersal, and the sweet juicy figs are ready to enhance a seed disperser's lunch.

Here we have a classic example of *mutualism*, a relationship in which both members of the association benefit. Fig wasps cannot grow or mate outside of the fig. And fig trees cannot produce seeds without the pollination services of the wasps. One might expect such a completely dependent mutualism to be a very risky business. After all, if the wasp goes extinct, its corresponding fig species is out of business. And if the tree species goes extinct, so goes the wasp. Reliance on a single species for survival is dangerous. Indeed, the fig tree–fig wasp mutualism is one of only a very few such tightly integrated relationships among tens of thousands of flowering plant genera. But with more than seven hundred species, the fig genus is surely one of the most successful genera of trees on the planet. The fig advantage, as Dan Janzen has pointed out, is simple: fig trees do not have to compete with other plants for pollinators because they've got their very own.[12]

But enough about flowers, let's look at the problem of fruit and seed dispersal. As we've just seen, the walls of figs become sweet and succulent, once the seeds are mature. Here's the basis for developing another kind of friendship.

Delivering the Goods: Fruit and Seed Dispersers

Flowers are about sex, and sex has profound consequences: new life. Petals may fall, stamens may shrivel, but a pollinated pistil will expand and develop. The ovary will grow into the fruit; the ovules will transform themselves into the seeds. Each seed holds the possibility of a new life, a new plant. To maximize the chance for successful growth, seeds need to be dispersed to locations where they can grow and flourish. The more widely the seeds can be dispersed, the better will be the chance for survival and success in an unpredictable world.

Again, as in pollination, many plants have elicited the services of animals to disperse their seeds. And, as in the flower, there is often a sweet enticement. As the seed reaches maturity it may be the wall of the ovary that becomes sweet and succulent; examples are the apple, avocado, cherry, and blueberry. In other fruits, it is the tissue around the seeds that we most enjoy, as in the case of the orange, watermelon, and tomato. And then there are plants in which the seeds themselves carry little yummy attachments that elicit animal activity. These are mostly very small seeds, gathered and distributed by ants. As with floral nectar, the payoff for the animal dispersers is *energy*, and the payoff for the plant is better seed dispersal.[13]

Bearing sweet fruits is not the only way in which flowering plants have gotten help in seed dispersal. There's another way: getting a free ride. With sharp little hooks, Velcro-like pubescence, or sticky surfaces, these fruits and seeds hitch a ride on mammal or bird. Our pants are often decorated with a variety of such fruits and seeds after we've been hiking in woods and fields. Both hair and feathers can carry these little "disseminules" long distances. In the case of small epiphytic peperomia plants, living on high branches, the seeds have a sticky surface. Though a bird may eat many such seeds, some will adhere to the side of the beak and need to be wiped off against a twig or branch; this is perfect to start a new life high in the treetops.

Finally, there are many plants that don't use animal friends at all; their disseminules take to the air. With thin, broad wings or long, slender hairs, they get carried away by the wind. Cotton, our most important textile, comes from the hairs on the cottonseed; these hairs help carry the seed by wind across the countryside. But wind dispersal can't really be called a friendship, so let's get back to animals.

Special Fruits for Special Friends

While most fruits appeal to a variety of birds and mammals, a few have adapted to a special clientele. An unusual example is the avocado. The avocado is the world's richest fruit as regards its overall food value. But why should it be so nutritious when many fruits offer little more than tasty sugar water? It turns out that the avocado tree and its relatives are dispersed almost entirely by a group of birds that eat little in the way of insects. These birds, the trogons, are large and colorful. Their most famous Mesoamerican species is the Quetzal, which was prized by pre-Columbian cultures for its beautiful tail feathers. Because these birds eat fruits almost exclusively, a close mutualism requires that the avocado and its relatives provide these birds with *all* their nutritional requirements. The birds, in turn, swallow the fruit, grind away the fruit's pulp in their crop, and then spit out the slightly scarified seed. This is what biologists call "high-quality seed dispersal." Of course, avocado trees are investing a lot more energy in producing such a rich bounty. And where there's a higher cost, there ought to be a substantial benefit; in this case, better dispersal. (Also, we should note that wild avocados aren't anywhere near as huge as our domestic favorites, which have been selectively enlarged over many generations.)

One of our favorite condiments, chocolate, comes from the

seeds of a small understory tree (*Theobroma cacao*) of American rain forests. The fruits of cacao are borne on the trunk and larger lower branches. They have the shape of an American football with about half the volume. The wall of the fruit has the texture of a small pumpkin and varies in color from pale green to yellow or purplish. When ripe, the soft rind can be broken into by monkeys, squirrels, or parrots. There they will find a sweet, juicy pulp surrounding the bitter almond-sized seeds (the "beans"). The yummy pulp makes breaking into the fruit worthwhile, but the bitter seeds will discourage chewing and the seeds (hopefully) will be dispersed. These bitter substances include caffeine and theobromine, the ingredients that make chocolate such a stimulating treat for us.

For larger animals, larger fruits give a more ample reward; again, with the "hope" that a small percentage of seeds will survive and be dispersed widely. In fact, some of the large-fruited trees that are having trouble replenishing their numbers nowadays may have lost their large dispersers thousands of years ago. Mastodons, mammoths, giant ground sloths, and indigenous horses went extinct in the Americas soon after human hunters arrived about twelve thousand years ago. These large wild animals are no longer here to devour and disperse such large fruits. Incidentally, a number of seeds with really hard seed coats require "scarification" before they can sprout. Lying around on the ground and waiting for mechanical damage or trampling can do the job, but being chewed on and swallowed by a large animal may be a much more reliable route.

To test the idea that larger mammals aided in the dispersal of some larger Central American fruits, Dan Janzen and his assistants did research using a donkey as the surrogate mammal. After feeding various large fruits and seeds to the front end of the donkey, the researchers kept careful vigil at the donkey's other end. There they waited, to retrieve seeds and find out if the passed seeds were still capable of germination. (So much for the idea that a lot of biological research is boring drudgery.) The osage orange tree (*Maclura pomifera*) of the central United States has large, greenish, grapefruit-sized fruit that has also been suggested as having evolved to be eaten and dispersed by mammoths and mastodons. As was the case in Central America, these animals went extinct shortly after stone spear points became part of North America's archaeological heritage. Similarly, a large-seeded tree of the Seychelles Islands in the Indian Ocean appears to be headed for extinction. Here again, it has been suggested that the extinct dodo (a large flightless bird) is no longer around to devour and disperse its seeds. These instances remind us that the loss of one species can have detrimental effects on many other species in the habitat.

As we noted earlier, some small seeds have yummy little fleshy appendages. It turns out that many of these are eagerly gathered by ants that bring them back to their nests. The tasty appendages are eaten by the ants and the seeds end up discarded in or near the ant nest. This is an especially vital relationship in dry shrubland where there may not be enough rainfall for smaller plants to produce larger energy-rich seeds. Ending up in or near an ant nest has special advantages for a small seed. Ants are not tidy creatures; their debris enriches the soil around their nests and makes a great place for a plant to start a new life. Seeds with such special appendages are especially common in the dry interior of Australia and in the American Southwest.

Fig trees, discussed earlier, are especially important food sources in many tropical forests. These trees, together with their wasps, maintain a pattern of fruiting that is continuous through the year in evergreen tropical forests. Because figs are available even in the dry season, they can be critical to the survival of many animals during this stressful time of year. This has been shown to be the case with smaller monkeys in evergreen Amazonian forests. Some of these fig species are thought to be "keystone" species; if they become extinct, a good many other species will also disappear. (Just as an arch may collapse if the keystone, at top center, is removed.)

Enemies of Enemies Are Friends

Ecologists have often wondered why so much of the world's terrestrial surface is green. Why is it that herbivores haven't eaten up all those tasty green plants? After all, the plants can't run. Plants do defend themselves in various ways. However, the creatures that eat plants, herbivores large and small, have their own predators and parasites to contend with. Think about the many herbivores of the African savanna. They range from the large, such as hippos, giraffes, rhinoceroses, elephants, elands, and buffalos, to little dik-diks and klipspringers, with zebras, gazelles, and warthogs in between. Surely these many grazers and browsers could do for the savanna what we do to our lawns. But think again. The African savanna also has lions, leopards, cheetahs, hunting dogs, hyenas, jackals, servals, and a variety of smaller cats—all of which feast on other animals, not plants. Add to these predators an army of insect parasites and microscopic pathogens and you can understand why herbivores haven't turned the African savannas, thornbush, or acacia woodlands into a wasteland. Many herbivores die because of things we rarely see: rinderpest, hoof-and-mouth disease, hog cholera, rabies, parasitic worms, and a variety of other ailments. Clearly, carnivores and pathogens put many herbivores out of business.

A dramatic example of predator removal and its effect on plant life has been documented in Yellowstone National Park. By 1926 wolves had been exterminated in the park, while herbivore populations, especially the elk, rose to new levels. Quickly, browse species, such as cottonwood, aspen, willows, and berry-producing shrubs, declined. And though coyotes, bears, and cougars were present in the park, their numbers were not sufficient to constrain the elk population. (Things would have gone from bad to worse if almost seventy-five thousand elk had not been removed between 1926 and 1968.) Then came the good news: reintroduction of thirty-one wolves during the winter of 1995/1996. Soon, cottonwoods and willows in the Lamar Valley were growing six to twelve feet high, instead of being continuously browsed down to three feet. Regrowth was especially strong near streams where denser vegetation made it easier for wolves to ambush their prey. Not only had the wolves brought significant predation back to the elk population, they had an additional effect: the "fear factor." Smart elk were now changing their browsing patterns, avoiding areas of denser growth. Soon the landscape began to resemble that of photographs taken in the early 1900s.[14] Clearly, top carnivores are key elements in helping maintain high levels of plant growth.

However, if you could add them all up and weigh them, the most significant herbivores would be the insects. Their vast numbers, both in species and in individuals, make up for their small size. Like the larger herbivores, insects are attacked by a variety of diseases and viruses. These pathogens have an easy time finding insects when insect numbers are high, but a difficult time when bugs are few and far between. Here's an automatic feedback mechanism that helps keep the ecosystem in balance. Insect-eating birds play a major role as well. The rain forest has many small birds, constantly flitting from plant to plant, searching for caterpillars and other insects. An exclosure experiment in a New England forest allowed insects ready access to shrubs and seedlings, while excluding birds. This experiment provided impressive data for the importance of birds in reducing plant damage. By excluding birds, the insect populations, recorded on foliage, increased 70 percent (over the controls where birds had access), and the leaf area missing at the end of the growing season increased from 22 to 55 percent in this study. Obviously, insectivorous birds are a major element in keeping herbivorous insects in check. No question about it: predators, parasites, and pathogens of plant-eating animals are among the most significant "friends" of flowering plants.

Right up there with insects, pathogenic fungi are the most serious enemies of plants. Surprisingly, recent research has shown that some plants "use fungi" to fight fungi. One of humankind's favorite plant products, noted earlier, comes from the bitter ground-up seeds (beans) of cacao (*Theobroma cacao*). Here is where chocolate originates. Unfortunately, cacao trees live in the dark, wet understory of rain forests where they can be damaged by *Phytophora* fungal disease. New studies have shown that the presence of other mutualistic fungi *within* cacao leaves diminishes the growth of the *Phytophora* pathogen! These other fungi help protect the cacao plant.[15] Needless to say, this is an additional cost to the plant, but such a cost can pay off in the long run—just as we bear the cost of medicines to counter diseases attacking us.

Clearly, flowering plants are one of the most successful lineages of living things because they've got so many "friends," whether fungi among their roots or within their leaves, animals visiting brightly colored flowers, various fruit and seed dispersers, or all those who help constrain the numbers of herbivores. These many relationships have helped angiosperms survive and prosper. But just as important, flowering plants have had to respond directly to their enemies through their own devices.

chapter 4

flowers and their enemies

During the 1970s and 1980s it became popular to deny the idea of "nature red in tooth and claw." Both popular writers and some ecologists imagined a more benign vision of nature; a world in which almost everybody and everything was living together in idyllic equilibrium. Unfortunately, common sense and more recent research suggest that the earlier ideas were correct: there's a very dangerous and competitive world out there. Neither prairie nor rain forest is any kind of Eden. If top predators aren't after you, then it's an army of microscopic enemies who will more likely end your days. Even the top predators, whether they are tigers or killer whales, have a host of parasites and pathogens to trouble them. And because plants translate sunshine into sustenance, they find themselves at the very bottom of the food chain.

Plants, like all other living things, are besieged by a wide variety of parasites and predators (we call plant predators *herbivores*). In fact, plants are the first course for much of life's hungry hordes. A large majority of animals, many fungi, and numerous bacteria use plants as their primary energy source. Photosynthetic life in the oceans (mostly microscopic) and plant life on land capture the energy of sunlight to sustain their own livelihoods. In so doing, they support a wealth of other creatures as well. Flowering plants are especially important in this regard, because there are so many of them, and they cover so much of the world's land surfaces. Stuck into one spot by their roots, plants are unable to run or hide from their enemies. This simple fact means that plants are particularly challenged to develop means for defending themselves.[1]

Herbivores

The chief enemies of plants are those that feed primarily on plants. Herbivores include grazers (that feed on grass), browsers (that feed on leaves), stem or root borers, leaf miners (that feed *within* a leaf), and a host of microscopic parasites and pathogens. Though they may not be as impressive as grazing bison or leaf-eating deer, it is surely among the insects that the largest numbers of plant eaters are found.

Aphids are a good example. These small insects suck fluids from the stems of plants. They're usually seen in small groups feeding together, and they are often "tended" by ants. Because they can be serious agricultural pests, aphids have been studied intensively. The problem for aphids is that the plant sap they are ingesting is rich in sugars but poor in some amino acids that are critical to the aphids' well-being. These little herbivores deal with this problem in two ways. The first is simple. They suck up an awful lot of plant sap and send sugar water out their rear ends. This is why you find all those little dried droplets on the surface of your car when you've parked under the wrong tree. Also, this sugary exudate is the reason why ants "care for" aphids; ants love sugar water. By pumping through a lot of sap, aphids can attempt to get the scarce amino acids they need. But this doesn't entirely solve the problem. Some essential amino acids are so low in concentration in plant sap that aphids have developed a second adaptation for dealing with this problem. Discovered only recently, aphids have special cells (bacteriocytes) in their hindgut containing millions of bacteria. These particular bacteria live nowhere else, and efforts to grow them in cultures outside the aphid have failed. Perhaps not surprisingly, the DNA of these bacteria have been found to contain many genes specifically devoted to the business of making just those amino acids which are in short supply in the aphid's food source. In effect, the aphids have their own little biosynthetic helpers to augment their nutrition. Comparative analysis of the DNA of aphids and the DNA of their bacterial symbionts indicates that this association first took place somewhere between 150 million and 250 million years ago, and the paired lineages of aphids and their bacteria have been running along together in close unity ever since their first mutual collaboration.

Another nice example of an herbivore is the guinea pig. During a trip to Peru, I wondered why a market in the Andean highlands had piles of fresh green barley plants; the plants had seed heads but the seeds were clearly not mature. Our guide informed us that the plants were here for purchasers who would take them home to feed the guinea pigs that run around in their houses. Sure enough, later on our tour, we visited a little stone house and there, on the dirt floor, were guinea pigs and fresh green barley. The next thing I saw quite amazed me. A guinea pig near my foot grabbed a barley stem, placed the bottom end in its mouth, and began to nibble. Slowly, the *entire barley plant* disappeared within the guinea pig. "Listen, you dumb rodent," I wanted to say. "Do you think you are a cow?" This animal ate the whole thing: stems, leaves, and seed head. Apparently, these creatures really can do what cows do. Like cows and other ruminants, these rodents undoubtedly have special bacteria in their intestinal tract to help them digest plant cellulose. Because guinea pigs can digest plant parts that we humans cannot, they are an important part of Andean village life, transforming stalks and leaves of grass into nutritious protein for the dinner table. That's right, these guinea pigs are an essential staple in the local cuisine.

Other Herbivore Strategies

Getting sufficient nutrition out of plant tissues to stay healthy and alive is a challenge many mammals face. There are several strategies. The first is to eat the most nutritious plant parts, like seeds, nectar-laden flowers, sweet juicy fruits, and nutritious tubers. This is what we humans have been doing for a very long time. But these plant parts may be seasonal, uncommon, and hard to find.

A more general strategy is to eat the plant parts that are out there in abundance all the time, like leaves and stems. Unfortunately, these plant parts are not as nutritious as seeds, fruits, and tubers. And again, as with the aphids, there are two general strategies. One is to eat a lot, digest it fairly rapidly, send it on out—and keep eating. Elephants, zebras, and horses do this. The other strategy is to take more time with one's food, and use many bacterial helpers in the digestive process. Both ruminant mammals with complex multiple stomachs (cows, goats, sheep, and antelopes) and some insects have symbiotic bacteria that live in their intestinal tracts and help them digest their food. These intestinal bacteria are able to do something animals *cannot* do: they can digest cellulose. Both starch and cellulose are made up of long strings of glucose molecules. But the glucose units are arranged differently in cellulose, and animals lack the enzymes that can pull those energy-rich glucose units apart. The primary reason that cattle, sheep, goats, and guinea pigs are so important to us is that they can digest widely available plant materials that we can't—and convert those plant materials into tasty meat and nutritious milk.

The great savanna faunas of tropical Africa give us fine examples of large herbivores sharing the same habitat, but using the two different digestive strategies. Zebras and rhinos eat more and more often, while trying to browse the more nutritious plants; they also send their food through their bodies fairly quickly. African buffalo and antelopes can eat poorer forage, but they keep it in their complex digestive tract a lot longer, allowing their bacterial symbionts more time to work it over. When cows are "chewing their cud," they are reprocessing some of the food they had eaten earlier as part of a more prolonged digestion. The very complex digestive system of ruminant animals, and their bacterial symbionts, is a very successful adaptation. They are the dominant herbivorous animals in today's world. These are the kinds of animals we humans use to get steaks, lamb chops, and milk products. Horses, on the other hand, who don't have such a complex digestive system and require higher quality pasture, have been used primarily in war, as draft animals, and at the racetrack.

Monkeys and their lifestyles are also closely dependent on their feeding habits. In tropical America, squirrel monkeys are always bouncing around the treetops. Burning so much energy requires higher-quality food: they eat bugs and fruits. Howler monkeys, on the other hand, move slowly and deliberately. They often sit around. Their leafy diet is lower in "energy" and they can't afford to waste it. Their slow and leisurely lifestyle is an accommodation to a diet that is not so rich in calories. Among the higher apes, it's clear that both chimps and gorillas have what look like "beer bellies." Healthy adult specimens of our own species (those that aren't overweight) *do not* have beer bellies. The bigger bellies of chimps and gorillas reflect the fact that they eat a much higher percentage of leaves than we humans do, and they need a longer intestinal tract to digest that "poorer quality" diet. We humans have a much shorter intestinal track and require richer foods. Clearly, one reason the world is so rich in animal species is that a variety of ways exist to make a living. And with so many different kinds of animals using plants to fill their dinner plates, it makes sense that plants have made major efforts to defend themselves.

Plant Defense, Part 1: Mechanical

All plants, and especially flowering plants, have developed a wide variety of defensive armaments. The most obvious are mechanical. Nasty spines or really hard tissues can keep hungry mouths and grinding teeth at bay. Hardwood, made of cellulose tightly glued together with lignin, is not digestible by most animals. Many grass tissues are defended by microscopic phytoliths (literally: plant stones). These little crystals are usually made of silica, the same stuff that composes sand and granite. Not to be outdone, many grazing animals have adapted to this inconvenience by developing much longer teeth. These deep-rooted teeth may wear down faster—thanks to the silica phytoliths—but they last longer because they are longer. In fact, paleontologists interpret the spread of deep-rooted (hypsodont) teeth in the fossil record as evidence for the spread of grasslands.

Another system of mechanical defense can be something as simple as hairiness. A dense covering of indigestible woolly hairs on leaves and stems may effectively deter a wide variety of small chewing insects. These hairs can be quite strong and stiff, presenting an effective barricade for small herbivores. Better than that, some of these small hairs have needlelike tips or sharp recurved hooks. Either can be deadly for little caterpillars, piercing or snagging and immobilizing them. Less than an eighth of an inch (3 mm) long, these little hairs are barely visible to us. However, such sharp little "grappling-hook" hairs can be useful in other settings as well. Wild chimpanzees carefully fold leaves with such hairs, and then swallow them with minimal chewing. These are not the leaves of plants they would normally eat. Rather than feeding, this behavior turns out to be a form of self-medication. They eat only a few of these leaves, and not very often. But the benefits are real. The minute hook-shaped hairs on these leaves apparently snag intestinal worms within the chimps' intestines, and help get rid of the parasites as the leaves pass through! A number of human groups in central Africa use these same leaves in much the same way, and for the same reason.

Protective hairs with little hooks at the ends are found in a variety of plants. Southwestern prickly pear cacti are doubly defended; they have two kinds of spines. The large, straight spines are easy to see and important to avoid. It is the tufts of smaller hairlike spines, which don't look very offensive, that can also give you real trouble. Brushing up against one of these cacti may provide you with a cluster of these thin, little devices firmly imbedded in your skin. Thanks to microscopic hooks at their tips, these hairs (called *glochids*) refuse to exit your epidermis. It takes careful effort and quite a bit of time to get rid of them. Such an experience is something you don't easily forget.[2]

Spines and thorns are larger defensive weapons; they come in a variety of different forms and sizes. Regions of drier vegetation and deserts, where vegetable dinners are sparse, are the home to a wide array of spiny and thorny plants. In Africa some species of picturesque flat-topped acacia trees have slender, whitish spines up to three inches long. That's impressive, but what caught me unawares was a very different kind of acacia. This shrubby beast doesn't get more than about twelve feet (4 m) high, and its dark thornlike spines seemed unimpressive. Only about a quarter of an inch high, these spines curved backward, pointing toward the base of the stem or to the interior of the bush. As a botanical collector, and encountering them in flower for the first time, I quickly learned a lot. The curved little spines didn't look particularly threatening, and they offered no resistance when I reached *into* the foliage to clip off a flowering stem. So far, no problem. Then came the moment of truth. As I began to *withdraw* my arm from within the foliage—those recurved spines had me! Both my skin and my shirt were now part of this damn bush. It seemed as if the tips of these spines must be the sharpest things in nature; no matter how carefully I was in extracting my arm, the slightest contact with those backward-curving tips had me hooked. English-speaking travelers in East Africa developed a special name for these miserable shrubs, calling them "abide-a-while." Once they take hold, it's going to take some time before you extract yourself with minimal damage.

There are other categories of nasty little spines. One group of plants that's especially nasty is that with very fine brittle-tipped hairs. Again, as in the acacia and cactus, they don't look too dangerous—until you've made contact. I discovered such hairs on the surface of palm fruits in Central America; the thin spine tips punctured my skin, and the tips then broke off under my skin! There they kept reminding me, painfully, of their presence until I dug them out. Again, excavating little spine tips from your epidermis produces long-lasting memories.

There's yet another kind of sharp, thin-pointed spine which can be equally memorable. Here the slender tips break as they enter your skin; only these little devils release a chemical that can sting for over an hour. The nettle family (Urticaceae) has a crowd of such species. My worst experience with such hairs occurred in northern Mexico, with an especially pernicious little vine. It seemed innocuous at first, but the very thin, straight, brownish hairs should have warned me; these hairs didn't look like ordinary hairs. And sure enough, when my bare legs brushed against them, *they stung*. But they didn't sting a lot; they stung just a little. Then came a slow revelation: the tingling sensation not only persisted, it transformed itself into a maddening itch—an itch that wouldn't stop! This was an itchiness of such subtle ferocity that I dashed into a nearby stream in desperate search of relief. This nasty little vine turned out to be a *Mucuna* species. *Mucuna* is a genus of the legume family, with a deservedly bad reputation throughout the tropics. In Ethiopia, I met a missionary whose children had collected the pretty lupinelike flowers of a local *Mucuna* species along a roadside while traveling. He related how he drove frantically to the nearest source of running water to alleviate his children's severe distress. These stinging hairs are examples of the plant kingdom's most important strategy for defense: chemistry.

Plant Defense, Part 2: Chemical

Flowering plants produce a huge variety of compounds and toxins to keep from being everybody's lunch. Many plants use tannins as a defense; acorns are full of them. Not only are tannins foul tasting, they bind to enzymes in energy-producing metabolism, which will put most animals out of business soon enough. If you are going to eat plant tissues with lots of tannins, you had better have a digestive system that can disassemble them. Among the legions of other defensive compounds produced by flowering plants are terpenes (including aromatic oils). Again, for most animals these are difficult to digest in higher concentrations. We use some of these fancy chemicals as our favorite spices and flavorings; but we use them only sparingly. An interesting exception is hot chili peppers; many people in the tropics eat these peppers despite the pain and discomfort. Why might they do that?

Let's take a closer look at these usually bright red chili peppers. Red is a color that birds like, and recent studies have shown that birds are effective dispersers of seeds. Experiments with wild chili peppers lacking the hot chemistry showed that mice love them, too, but the mice destroyed most of the seeds during digestion. In these experiments, the same wild-type peppers, both mild and hot, were eaten by the birds, and most of the seeds survived passage through the birds. In their natural tropical and subtropical habitats, hot chili peppers are avoided by wild mice, but birds feast on them—and serve as excellent dispersal agents. Wild chili peppers are mostly blueberry-sized, just right for small birds. Thousands of years of selection by farmers have given us much larger chilies, in a great variety of shapes and pungency.

Both in their American homeland and around the world, tropical people will often eat ferociously hot chilies as tears run down their faces. But why do so many tropical cultures celebrate the consumption of these fierce fruits? Human cultures have good reasons for eating superhot chili peppers. First, fresh or dried, chili peppers are a great source of vitamin C. For many tropical people eating boiled and fried foods, and with limited access to fresh vegetables, vitamin C is often a scarce but necessary item in their diet. Second, it seems that the "hot" compounds themselves can help reduce the numbers of intestinal parasites. So the chilies have a double benefit: vitamin C and a way of cleansing your innards!

Unfortunately, some plant defenses are a lot worse than powerful spices; many are lethal poisons. Foxglove, the lovely garden flower (*Digitalis purpurea*), harbors a very dangerous poison, one that can stop your heart from beating. That's the bad news. The good news is that this same poison from the foxglove, when carefully administered in low dosage, can be a highly beneficial medicine for people with certain heart irregularities. In fact, we've developed many medicines from the chemical arsenals of flowering plants, despite the fact that virtually all are poisonous at high concentrations. This is something to remember! It is a truism that applies to aspirin, herbal ingredients, and all our other medicines. Here's an instance where the idea that "if a little is good, more will be better" can actually kill you. Nearly all medicines are lethal at high dosage. But in addition to outright poisons, plants have also developed defensive compounds utilizing more subtle strategies for their defensive purposes.[3]

Some species of flowering plants produce compounds that *mimic* important compounds in our brains; these are the molecules to which we humans can become severely addicted. But why might a plant produce such a chemical? Why should a little coca bush produce cocaine, or why should the latex in the mature fruit wall of a poppy contain morphine and twenty other odd compounds? A half century ago it was suggested that these complex compounds might be some kind of chemical garbage, leftovers of plant metabolism. But this made little sense. These psychoactive substances are very complex molecules, composed of fancy five or six carbon ring structures, variously pasted together, and often with a nitrogen atom sitting in a corner. By no means are these "alkaloids" simple chemicals. Something as complex as cocaine or morphine requires special biochemical pathways for its synthesis, and its manufacture demands a lot of energy. Neither plant nor animal can afford to waste energy. So why are such compounds produced by these plants? The only reasonable assumption is that these complex molecules are part of the armament that plants use to defend themselves, in this particular case from mammals like us. For animals that have brain chemistry similar to our own, such chemicals need not poison the herbivore. All these defensive compounds need to do is to convince those who might devour them that taking more than a few nibbles is dangerous. Any plant molecule that mimics an important brain hormone will have clearly debilitating effects. Eating leaves containing such a molecule might slow down the animal's reflexes or compromise its perception. Losing alertness or slowing response time is not smart in a dangerous world. These are plants you want to keep off your menu. (The reason why some of these molecules have become so addictive to us is simple; we've learned how to extract and concentrate them, and we usually consume them in very private surroundings.) Further support for the defensive nature of such compounds is the interesting fact that *none of New Zealand's native plants or berries* is poisonous to us. Before people brought them, New Zealand had no mammals. Thus, the plants were under no selective pressure for developing nasty chemicals to deter mammalian herbivores.

Chemical defenses include other kinds of chemical mimics. Among the cleverer are plant compounds that mimic insect growth hormones. A larva eating such a compound derails its own development and cannot reach sexual maturity; it will never contribute to another generation. Such defensive compounds can deter a wide variety of insects; they are also found in ferns and conifers. Alternatively, compounds like nicotine or caffeine may have effects on insects or small mammals that differ from their effects on us. Interestingly, the pep-me-up quality of caffeine has been discovered and utilized by indigenous peoples all over the world, utilizing very different plants. The more famous examples are coffee from Ethiopia, tea from South Asia, cacao from Central America, and guarana from Brazil.

Plant Poisons: Defense and Counterdefense

But where there's defensive chemistry there can also be counteradaptations. For example, some lineages of insects have developed their own special counterdefense to plant poisons. Our common milkweed (*Asclepias syriaca*) has a milky sap laced with some very potent poisons. But the monarch butterfly caterpillar can feed on these plants, thanks to a digestive system that can *sequester* these poisons. By dealing with the poisons in this way, these insects become poisonous themselves.[4] Both the caterpillar, with black, white, and yellow stripes, and the bright orange butterfly are conspicuous and easy to see, apparently a warning to those who might eat them. When you examine other insects feeding on our common milkweed, you will notice two things. First and most obvious is that they are brightly colored; this includes a species of black-and-red bug and a red beetle. Second, you won't find any of these insects eating *other* plants. Apparently, their special adaptation now requires that they eat these plants, and only these plants. Obviously, to become poisonous themselves they must consume the plant's poisons. Thus, though the milkweed does prevent a wide range of insects and other animals from feeding on its tissues, there are a few insect lineages that have succeeded in dealing with this particular defense.

A similar case is that of *Heliconius* butterflies in the American tropics. Like the monarch, they are poisonous because they feed on a poisonous group of plants, the passionflowers (the *Passiflora* genus). *Heliconius* butterflies are also brightly marked with orange, black, and white, and, because they are poisonous, they have become the models for butterfly mimics that are not poisonous. Apparently, once a palatable butterfly species begins to look like a poisonous species, birds avoid it just as they do the poisonous ones. Such mimicry studies have played an important role in discussions regarding the efficacy of natural selection.

Plant poisons and the insects that tolerate these poisons can also have deadly *indirect* effects. Wild cherry trees are defended by having amygdalin (a cyanogenic glycoside) in their stems and leaves, giving them a sharp aroma and bitter taste. When digested, this compound gives rise to hydrocyanic acid, a nasty poison. But here again, there are a few insect lineages that have learned how to ingest these poisons and become poisonous themselves. Among these are the tent caterpillars of the eastern United States. They aggregate in considerable numbers and make their tents among cherry branches in early spring. In certain years, when the caterpillars reach high numbers, cherry trees can suffer massive defoliation. Fortunately, the tent caterpillar populations are usually held in check by their own pathogens, so heavy infestations are not frequent. However, an infestation of tent caterpillars in the early spring of 2001 had profound effects on one of our most elegant domestic animals.

Breeders of racehorses in Kentucky suddenly realized that something was going terribly wrong in the spring of 2001. Mares were aborting a much higher than normal number of pregnancies, and many foals were born dead or were dying shortly after birth. At first, it wasn't clear what was causing the loss of so many young among these highly prized animals. The mares looked to be normal and perfectly healthy. After considerable effort, the cause was finally discovered: tent caterpillars! It seems that a warm spring, followed by cold weather and drought, had caused the cherry trees to produce especially high concentrations of the cyanide-containing poison. The many caterpillars, having fed voraciously, and after reaching full maturity as larvae, came down from the trees and began to wander. They do this before they enter the ground to pupate—a prelude to reaching adulthood as a moth. It was during this wandering phase that grazing mares inadvertently ate the caterpillars. The cyanide contents of the caterpillars was not enough to make the large mares ill, but it was devastating for their little unborn and newborn offspring. The loss of so many carefully bred foals was a costly setback to the thoroughbred industry. And it is safe to surmise that horse breeders now keep a keen eye out for the numbers of tent caterpillars around their farms.

Plant Enemies That Are Hard to See

It's not only insects and herbivores that plants have to defend against; they have many other smaller enemies. Nematodes are another major plant pest; these thin, little worms attack the underground root system. They are a very serious problem in world agriculture, and a major focus of agricultural science. Unseen and microscopic, there are thousands of other enemies among the army of plant killers; among them are the bacteria, plant viruses, and fungi. All together, these pathogens are so threatening to our own economies that agricultural science has created a major subdiscipline—plant pathology—to deal with them. An old admonition circulating among students of agriculture was that if you were looking for a really secure job you ought to be a wheat breeder. This notion was based on the fact that the wheat rust disease mutates incessantly. And since these diseases are constantly changing, the argument went, you'll always have a job developing new varieties of resistant wheat.

As in the case of wheat rust, a wide range of fungi is responsible for a majority of plant diseases, and this is the biggest single cause of plant death. An interesting example of the battle between plant and fungal pathogens comes from an unusual tree of tropical American rain forests. Forest ecologist Robin Foster noticed something very strange about a particular tree species (*Tachigali versicolor*) on Barro Colorado Island in central Panama. These trees were one hundred feet tall, and years had gone by without any of them flowering, even though all appeared to be fully mature and in good health. Then, quite suddenly, a small number of the trees burst into bloom at exactly the same time, in the same year, covering their crowns with flowers. The pink and yellow flowers were spectacular, the fruiting prodigious. But at the very same time, these trees were becoming sickly—and then they died! After some months, their trunks would weaken, and the dead trees came crashing to the ground. Calling this species a "suicide tree," Foster suggested that the mother tree, by dying and then crashing to the ground, was providing a sunny opening in the canopy under which her seedlings could grow. Many small plants and some bamboos have such a "flower-fruit-and-die" lifestyle, but it is very unusual for trees. What was especially noteworthy in this drama was how the bark of the mother tree began to deteriorate, even as her flowers bloomed. Here was a smooth, brown-barked trunk, which had stood tall in the forest for several decades, that was suddenly attacked by fungi. In this case the tree was apparently gathering all its energy resources for a grand finale of flowering and fruiting. With its defensive armament withdrawn to support a reproductive extravaganza, the bark was now being attacked by fungi. Apparently, these same trees had been maintaining powerful antifungal compounds in their bark for self-protection over many years. Once those defenses were withdrawn, the forces of decay closed in.

Supporting this interpretation is the fact that Amerindian forest peoples use the bark of healthy *Tachigali* trees as remedies for fungal skin infections. Also, leaf-cutter ants that grow fungi in their underground abodes *do not collect* the leaves of these trees. Clearly, the moist rain forest is no idyllic paradise; it's populated with many nasty pathogens. Surely, fungi must be a threat for all the other trees of the rain forest as well, and they must also maintain defensive systems to carry on their life activities.

Deception and Reward in Plant Defense

A simple kind of "deception" that plants use to defend themselves is either to have leaves of varying toxicity or to appear identical to toxic individuals. Leaves of the same plant can vary greatly in their palatability but look the same. Likewise, individual plants of the same species can differ in the levels of their chemical armament. This presents insect pests with a problem:

how to find the most palatable leaf or individual plant. The only solution to this quandary is to search. Not only does searching waste valuable time and energy, it subjects the searching insect to the keen eyesight of birds ready to make a meal of them. Thus there is ample reason for the leaves on a tree, or the individuals in a population, to vary in their toxicity but look identical. This is a kind of mimicry, and though it's a statistical gamble, making fewer defensive compounds does save energy.

Earlier, we noted how a few plants have used deceit as a means of getting themselves pollinated. Chance mutations and natural selection have, in a few cases, achieved the same goal in the business of defense. In this instance the "deceit device" is, again, a form of mimicry. A few species of herbs and vines have leaves bearing little, rounded, white projections on their surfaces that look like insect eggs. And it seems that's exactly what they're supposed to look like. The "reason" these plants have such "eggs" is tied to the behavior of the butterflies whose leaf-devouring caterpillars feed on them. (The plants, of course, haven't figured this out; it is probably a small morphological feature that came along accidentally and is now maintained by natural selection.) A crucial point in this story is the nasty feeding habits of the caterpillars themselves. In these species, the bigger caterpillars will eat more than just the leaves; given the opportunity, they'll eat smaller members of their own species. This cannibalistic trait has, in turn, resulted in an unusual behavior pattern of the adults. Before laying her eggs, the butterfly will first check the plant for eggs already deposited by a previous butterfly. Who could have imagined that butterflies might be this smart? Again, they don't think about it; this is a genetically based inborn behavior pattern. An ecologist researcher tested the "egg-mimicry" hypothesis by carefully removing the fake eggs from marked plants. Sure enough, there was a statistically significant increase in the number of real eggs laid by the butterflies on the "treated" plants (with the fake eggs removed). In contrast, the ungroomed control plants, which still possessed all their dummy eggs, had fewer eggs deposited on their leaves.[5] While an unusual defensive strategy, it is the product of rather special parameters. The plants are small and have few leaves, and the caterpillars eat members of their own species.

Supporting a Protective "Police Force"

The little white "dummy eggs" bring to mind another kind of structure seen in a wide variety of plants. These are often in the form of little knobs or cups, only about a tenth of an inch (2.5 mm) wide. They often seem shiny, as if they are exuding a liquid, and, indeed, that's what they do. Producing sweet secretions, these devices are called "extrafloral nectaries." They produce sweet secretions and tend to be found on distal stems and leaves, not on or in flowers (hence the term *extrafloral*). Such nectaries are usually associated with ants that feed off the secretion and can often be seen actively running up and down the stems and over the leaves. Here again, we have a major energy expense on the part of the plant, paying for what appears to be a "police force" of ants. The best example of this kind of defense is one of the most highly evolved cases: the ant acacias of Mexico and Central America.

In Central America, you can often spot these small trees (*Acacia cornigera* and allies) from a distance in the woods; they've got only sparse vegetation beneath them, and there are no vines growing over them. If you grab hold of a twig, you'll quickly find out why, as fierce little ants rush out to sting you. These same ants girdle stems of vines and nearby plants, allowing the acacia more access to light. The "reason" the ants behave in this way may actually be for the ants' own welfare, making it difficult for predators to reach them by coming across on the vines. (Again, it isn't likely that the ants have figured this out themselves. These are behavior patterns that probably arose by accidental mutations, and have been maintained over time by natural selection.)[6]

But too much protection could be a bad thing. How can pollinators visit flowering stems that are protected by fierce ants? Nature has apparently worked things out. The young acacia flowers give off a volatile chemical, signaling the ants to keep their distance from the flowers during the time the flowers are awaiting pollinators! However, all this police work doesn't come free; these acacia trees produce *two* kinds of rewards. One kind is the protein bodies (Beltian bodies) on the leaves. The second reward is produced by special nectaries along the stems, supplying the ants with energy-rich sucrose. And that's not all, the spines of these acacia trees are greatly inflated near their base and hollow; these provide housing for the ants. All this sounds like a great example of mutual benefits: the ants get food and housing while the trees have a mob of vigilant defenders. Again, the bad news is that maintaining a small ant army takes a great deal of energy. (This may explain why only a few acacias out of more than a thousand acacia species have got themselves tied up in this expensive syndrome.) But there are some energy savings. A careful comparison of the leaf chemistry of an ant-acacia and its antless relatives in Mexico has shown that most of the ant-supporting species do not produce "expensive" cyanogenic compounds. The related acacia species synthesize these chemicals to make their leaves less palatable to herbivores. Thanks to their ants, the ant-acacias do not need as much defensive chemistry. Surely, the costly energetic expense of supporting ants or building chemical deterrents is clear evidence that there's a real "war" between plants and herbivores going on out there.

Since we're talking about a Central American species and its fierce ants, I'm reminded of another uncomfortable experience while doing fieldwork in Costa Rica. *Cecropia* is a very distinctive genus of tropical American trees. Its species have large and umbrella-like leaves (one to three feet wide) with deep lobes all radiating out from a central point. Because the leaves are so large, the trees have only a few major branches. These are usually fast-growing trees with whitish trunks in open, sunny sites, which makes them even more conspicuous. But it's not only these distinctions that have given them a special reputation. Like the acacia we just mentioned, most *Cecropia* species house fearsome ants that come out to bite you when you mess with the tree. And here again, the plant provides both hollow stems in which the ants can live, and food bodies at the base of the leaves to nourish them. I knew all this before I began hacking away with my machete to bring down a small *Cecropia* tree from which to make specimens. The ants running around the trunk were quite small and probably wouldn't hurt too much, I figured. They'd have to run to the ground and then to my boots to get at my legs. My plan was to cut as much as I could, until the ants started biting my legs; then I'd move away, coming back later to finish the job and gather specimens after the ants had run out of energy. What I hadn't figured on was how aggressive these little beasts might be. Ignoring my strategy, these little devils ran across my boots and up my legs with such speed and enthusiasm that, by the time they slowed down to start biting, they had reached a very sensitive region. I have not made that mistake again.

Directly or indirectly, many other animals help defend plants from their enemies. As we noted earlier, insectivorous birds are among the most important defenders of plants, constantly flitting around foliage looking for tasty insects to devour. And here, too, mimicry is part of the playbill. Many plant-devouring insects mimic leaves, stems, bark, lichens, and even bird droppings, to avoid being found by keen-eyed birds. Here's an unusual example from our southwestern deserts. *Neomaria arizonaria* is a geometrid moth whose larvae feed on oak leaves. The first brood of caterpillars in the year is rough in texture and yellowish brown; they resemble the oak's flowering catkins in springtime. Hatching later, when the catkins are gone and the leaves have expanded, the second brood of caterpillars comes onto the scene. These creatures are smoother and grayish in color, resembling short stems. Careful rearing experiments have shown that it was not temperature or day length that caused the caterpillars to produce the two different forms; the differences depended on whether the young caterpillars were fed early spring leaves or early summer foliage. Clearly, the caterpillars were responding to subtle chemical changes within the leaves to keep their mimicry in tune with their surroundings and to escape hungry birds.

Predatory and parasitic insects, as well as birds and spiders, are also essential in keeping an army of insect herbivores in check. However, there are other plant defenders that we rarely see because they are so very small. Among the trees of evergreen tropical forests one often finds leaves with rather unusual little "structures" on their undersides. These devices are found in the angle where the secondary veins meet the primary veins (the vein axils) near the base of the blade. They are always on the underside of the leaf. Most are made up of nothing more than tufted little hairs, but a few have tissues that produce what looks like a little enclosure. These arrangements were inferred to be little "homes" (called *domatia*, after the Latin word for "house"). Usually a bit smaller than a lowercase *v* in this text, they are much too small to house ants. Some skeptics doubted that these structures had any usefulness at all. But if they had no use, why were they there? Nature appears to have rules. One of them is: use it or lose it! If these structures had no use, deleterious mutations should have gotten rid of the genes that determined their construction many generations ago. But that clearly hasn't happened. What were these structures for?

Recently, careful research has disclosed that these domatia are indeed domiciles—for mites. (Mites are minute relatives of spiders and ticks, generally less than a tenth of an inch long.) Apparently, in the wet forest environment, there's a big advantage for leaves having teeny-weeny mites on board. These mites eat fungal spores, insect eggs, and other *herbivorous* mites. Regular cleaning by the little mites should do much to maintain the health of leaves in a wet tropical environment, where leaves may remain functional for several years.

Signaling for Help!

Though domatia have been known for a long time, recent research has divulged other examples of plant defense we hadn't even imagined. Because maize (what we Americans call corn) is such an important commercial crop, this species has been studied intensively over many years. Recent research on maize has uncovered something quite surprising. When voracious corn-borer caterpillars start chomping on their favorite meal, the corn stem, the plant responds by producing a "cocktail" of aromatic chemicals. These chemicals diffuse into the air and are spread by the wind. They appear to have a special purpose: they attract small wasps. Not ordinary wasps, but species of wasps that parasitize corn-borer caterpillars! The female wasps insert their eggs into the bodies of the caterpillars, and that's where new wasp larvae will develop—slowly devouring the caterpillar from within. These wasp larvae are clever; they're programmed to eat the least important parts first, saving the vital organs for dessert. Pretty awful, but hey, that's the real world.

A parasitized caterpillar may not do the corn plant a lot of good in the short run, since the caterpillars will continue feeding. But over the long haul this is a major defensive strategy. Parasitized caterpillars cannot survive to become adult moths with the potential of producing hundreds of new caterpillars. Over time, and with the help of these parasitic wasps, the corn plant's aromatic distress call helps keep the corn-borer population under control. What's especially clever about this system is that when scientists cut the corn leaf with a knife, the plant *does not* produce the volatile aromatics. Likewise, mechanical damage by a windstorm does not invoke the production of a distress signal. It takes the small damage over time by the caterpillar to set off this defensive system. Like the aromatic scents and colors of flowers, these plant volatiles are signals for the wasps; in this case helping wasps locate their caterpillar hosts, who are not easy to find, in a rich and complex green world.

The story of flowering plants sending out signals when being chewed on can be even more complex. Experiments have shown that lima bean plants also emit volatile substances when herbivorous mites are attacking their leaves. But in this case the experiments showed that these volatiles also induced distant and *uninfected* lima bean plants to start producing volatile compounds that are attractive to predatory mites. As mentioned above, predatory mites like to eat herbivorous mites, so they're good to have around when there's an infestation of these mites. In this case, leaves on the uninfected lima bean plant activate five different genes—after getting the signal from an infected plant. These genes are presumed to initiate the production of the chemicals that will attract the predatory mites. Again, as in the corn plants, mechanical damage to the plant does not result in the production of the signaling molecules. Instead, what we have here appears to be evidence of plant-to-plant communication. Pretty cool, though still a bit controversial; do the laboratory findings really resemble what's going on in nature?

In southern Africa, kudu antelopes (*Tragelaphus strepsiceros*) are known to browse *Acacia caffra* trees for a few minutes and then move upwind to browse on another *Acacia caffra* for a few minutes, and then continue this pattern of feeding. The implication here is that, as the kudu begins to feed, the tree cranks up its chemical defenses, explaining why the kudu feeds for only a few minutes. But there may be more to it than that. It has been hypothesized that the reason the kudu moves upwind to feed on another tree is that these chemical defenses become volatilized and signal other *Acacia caffra* trees to also build up their chemical defenses. Since the volatile vapors move downwind, the kudu moves upwind to avoid trees that may have been "warned" and are building up their defenses. Unfortunately, this interpretation may be difficult to substantiate. Can acacia trees really mobilize defensive chemicals that quickly? And moving upwind is how the kudu avoids keen-scented predators. Perhaps we're dealing with a "just-so story" here, a scenario that sounds good but may not be actually happening in nature.[7]

Along similar lines, it has been suggested that some species of northern trees signal each other in a similar manner. In this case, as in the corn plant, caterpillars get things started. Among these trees, caterpillar damage induces the trees to produce an increased concentration of chemical repellents as a means of self-defense. These same chemicals become volatilized from the surfaces of the chewed-up leaves and, it is claimed, can be sensed by other trees in the community—as the wind wafts these chemicals around. As in the kudu story, other plants of the same community respond—it is claimed—to distress signals from their besieged brethren. Perhaps they do this in a fashion similar to what the lima bean experiments indicated. Unfortunately, many of these ideas are still quite speculative—and difficult to verify. While there's no doubt that plants can emit a variety of chemical signals, the question remains: are other plants really listening? One problem is that the atmosphere of a laboratory may have concentrations of chemicals far higher than those to be found in a windy woodland. Stay tuned, there's much we need to learn before this issue is settled.

Lots of Genes for Lots of Problems

More important, these varied studies indicate that plants are not merely sitting around, full of chemicals, in a static state, waiting to be chewed on. It is now clear that plants can respond and mobilize defensive compounds when they are challenged. Though plants have neither a nervous system nor a brain, they do have connections between their cells that can transfer chemical signals and produce a coordinated response. Though nowhere near as fast as electrical charges running along nerve fibers, the various parts of a plant can respond to trauma and communicate with other parts of the plant. This fits in with a huge surprise of only a few years ago.

Detailed DNA studies have shown that our favorite little experimental plant, *Arabidopsis thaliana*, has far more genes in its genome than does that other laboratory favorite, the fruit fly *Drosophila melanogaster*! (The genome is the full genetic complement of the organism. Both species have been the organisms of choice for intensive genetic and other kinds of research.) The little crucifer plant, which rarely exceeds two feet in height, is thought to have about twenty-two thousand genes, while the fruit fly numbers about fifteen thousand. This revelation took everyone by surprise. Surely the little insect is a far more complex being. Not only can it fly around and see the world, the male even has mating dances to impress the females. Why should a simple little stationary plant, with no behavior to speak of, have so many genes?

One reason might be that plants are sitting ducks for voracious herbivores and nasty diseases, and these challenges require a sophisticated arsenal of both defensive compounds *and dynamic strategies*. We are now learning that plants can sense and respond to the chemical signals of animals. To do all these things requires sophisticated information systems, and the large genome may be doing just that. One report suggests that a single plant's response to a single insect can involve between eight hundred and fifteen hundred different genes within the plant.

And it's not only animals that can act as "enemies." Other plants can be antagonists as well. Being shaded by a neighbor can seriously reduce the life processes of a plant. In many instances there is nothing the plant can do. Seeds of a sun-loving species sprouting in a shaded woodland will perish. But once some stature is achieved, a plant can begin to exude chemicals to deter other species from germinating nearby. Some eucalyptus trees are notorious for having few other species grow beneath them. Whether exudates from their roots, or products of their decomposing leaves, these trees create a "chemical neighborhood" that helps keep others out. Actually, this is a kind of chemical pollution that prevents the seeds of other species from crowding the species that's already established—in a process called *allelopathy*. This may be another aspect of a plant's defensive strategies, and part of a large genome.

Another fundamental reason why plants have a large genome is the complex chloroplast and its intricate biochemistry. The chloroplast is where photosynthesis takes place, and keeping that system running requires a big bunch of genes. In addition, plant genes may not be as versatile as animal genes, with the result that animal gene products can be modified for a greater variety of functions than can plant gene products. By having greater flexibility in RNA transcription, animal gene products can be modified to perform in a greater variety of functions—with the result that animals simply do not need as many genes as do plants.

No one ever thought that corn plants could send out messages for help, or that trees might be "communicating" distress signals with each other. Nor did anyone imagine that plants might be mounting quick biosynthetic responses to being attacked. Continuing revelations such as these help make research in the plant sciences so rewarding. Better yet, there's the possibility that a few of these discoveries will provide us new ways of helping sustain and protect those plants that provide us with food, fibers, flavors, and colorful flowers.

A *(left)*. Bent over, the Michigan lily, with its curled-back tepals, displays six brown anthers and a reddish style.

B *(below)*. Beetle visiting a wild geranium flower. Note the five separate petals, ten stamens, and central stigmas.

C. Hoverfly on a meadow rose. Note the five separate petals, many stamens, and central mound of stigmas.

D. The tubular corolla forms a curved spur in the *Impatiens* flower, holding nectar for long-tongued insects.

E. Rapid "buzzing" allows bees to extract pollen from the distal pores of the yellow stamens of *Solanum dulcamara.*

F. Spikelet of a grass (*Lolium* sp.). Only an anther and two filamentous stigmas reveal the florets within.

G. Only the upper third of a *Hydnora* flower emerges above the soil. Dark objects within the flower are beetles.

H. Two pseudoflowers of skunk cabbage; each spathe encloses a spherical spadix bearing many congested flowers.

I. Pseudoflowers of the flowering spurge. Near the ovary base, a tiny fly sips nectar from a greenish nectary.

J. The head or pseudoflower of a New England aster. The bright yellow disk flowers form a highly visible center.

K. Inflorescence of Queen Anne's lace. Here, a large umbel of small umbels helps little flowers make a big show.

L. The many flowers of *Banksia* are congested on a thick spike that becomes woody, protecting the seeds from fire.

M. Frontal view of *Tabebuia rosea* flowers; note the broad landing field, with yellow "honey guides."

N. *Cymbidium* orchid flower; note the upper column, and the lower lip with landing field and honey guides.

O *(above).* This *Trichoceros* orchid mimics the hairy body of a tachnid fly; an example of pollination by pseudocopulation.

P *(left).* Over time, flowers and their pollinators have created biotas far richer than any that had come before.

chapter 5

how are the flowering plants distinguished?

What is it about flowering plants that clearly distinguishes them from all other land plants? Trying to answer this question is a problem of classification, or systematics. For some readers this may prove to be the most tedious chapter in the book. However, this topic is basic to our understanding of the flowering plants, and a necessary part of the larger story. We'll start by reviewing more primitive photosynthetic lineages, move on to land plants and seed plants, and then tackle the angiosperms themselves.

With the exception of very dry and very cold areas, our planet's land surfaces are embroidered with a great variety of greenery. This diverse vegetation includes plants that range from the very small and very simple to trees that reach three hundred feet into the sky. The simplest plantlike organisms that inhabit the land are green algae and blue-green bacteria. Very simple in structure, they are mostly seen as thin filaments, aggregated in larger hairlike arrangements, or as thicker "gunk." They are usually found in moist depressions, and along the edges of ponds and streams, where they lead a semiaquatic life. Many of these simple beings can also live on moist surfaces where they are protected from the desiccating sun or blustery wind. According to a few biologists, however, neither the blue-green bacteria nor the algae are "plants." For these experts, the designation *plant* and the plant kingdom should be restricted to more complex multicellular organisms.

According to these biologists, we should restrict "plants" to those photosynthesizing organisms in which the fertilized egg cell forms an embryo that begins the development of the young plant. From that point of view, all "plants" are land plants—and things like green algae are simply dumped into a large grab bag of simpler organisms (called *Protists* or *Protoctista*). I disagree with that viewpoint, but it doesn't really matter since we're only talking about land plants in this book. And since all land plants develop from well-defined embryos, they are *plants* by any definition.

The successful invasion of dry land by larger, more complex, green plants was one of the greatest triumphs in the history of life. While puddles of blue-green slop may have supported a few creepy-crawlies in earlier times, the development of a real land flora changed the world in a way similar to that of the "Cambrian explosion," when animal life in the sea erupted into a wealth of new and diverse creatures. But whereas the Cambrian marine diversification of 540 million years ago included the sudden appearance of a diverse crowd of different animal lineages, the invasion of land by more complex green plants was restricted to a solitary group of pioneers over a substantial period of time. Eventually, a number of different animal lineages ventured out onto the land, but in the case of plants, this bold new frontier appears to have been crossed only once, and by a single lineage.

In today's world, the simplest land plants are the liverworts. They have little leaves only one cell thick, or flattened bodies only a few cells thick. They usually don't grow more than a few inches long, and are only found in wet sites, or places that are likely to remain moist for at least a few months each year. These plants are, as far as we can tell, the living descendants of the very first land plants; plants that left their mark in the fossil record only as spores, over 450 million years ago. Many millions of years would pass before actual tissues of these plants would become part of the fossil record. However, to become larger and more diverse, land plants needed a more complex body than the liverworts possessed.[1]

The Development of a Land Flora

In the case of animals, as just mentioned, a variety of different critters managed to crawl out of the sea and become adapted to life on land. The ancestors of spiders, insects, millipedes, worms, and the land vertebrates (tetrapods) all crawled out of their watery worlds *independently*. Each of these lineages came from a well-differentiated aquatic life form, not closely related to the others that had made the same transition to the land. In strong contrast, all land plants appear to have originated from a single lineage of complex green algae. Lacking the movement and versatility of animals, plants had a far more difficult task in adapting to the land environment, and it appears that only one such attempt was successful.[2]

Today's most primitive land plants, the liverworts and hornworts, have patterns in their DNA that indicate a connection to a group of complex green algae, the Charophytes, which are found *only* in freshwater. This suggests that land plants arose in freshwater lakes and streams or from river deltas—not from the ocean shore. But what might have caused a plant, any plant, to adapt to the harshness of a land environment? Surely, staying immersed in a liquid environment seems much the safer thing to do. Keep in mind, natural selection can operate only in the here and now; plants must remain adapted to current conditions. Natural selection doesn't prepare a lineage for the future—except by random accident. So, why live on land?

The most reasonable scenario suggests that early land plants lived along streams or in ponds *that dried up* during part of the year. Their first adaptation, for living on land, was a little disseminule or propagule that could survive this dry period. That little disseminule was the spore. Spores can survive severe drying, and travel through the air as well. Carried by the wind, the microscopic spore could be scattered across wide areas, germinating—if lucky—in a distant pool of quiet water or moist mud. Such spores are, as we've noted, the first evidence of plants living on land.

Earlier, blue-green bacteria probably lived on damp soil and in shallow pools over hundreds of millions of years. They were probably joined by green algae and fungi to form puddles of green glop in moist depressions and along the edges of ponds and streams. These lineages may have populated land surfaces for many millions of years *before* more complex land plants arose. And they, too, had to develop sporelike stages to survive desiccation and effect dispersal. Unfortunately, because they did not possess resistant tissues, these early forms of a land flora left no trace in the fossil record. The first "true" land plants were, as noted, probably flat liverwort-like plants with little leaves only one cell layer thick. Again, these "first land plants" are remembered in the rocks only by their spores. Being able to withstand the physical stresses of drying and rehydration, the tough spore wall is also resistant to decomposition. Thus, spores became an important element in the fossil record. (Algal and fungal spores are quite different.) These early spores are first recorded in the rocks dated at around 470 million years ago, and they are usually found as groups of four (called tetrads). By about 430 million years ago, spore tetrads declined in abundance and separate individual spores with a Y-shaped (trilete) configuration on one side became common.

About 410 million years ago, with plant spores diversifying, the rocks preserved the first fragmentary evidence of larger plants. It was not until more complex land plants had evolved desiccation-resistant surfaces and a strong erect architecture that a really three-dimensional terrestrial vegetation came into being. Waxy surfaces were a major innovation in the advent of larger terrestrial plants. In addition to reducing water loss through outer surfaces, a waxy layer protected against ultraviolet radiation, microbial attack, and corrosive chemicals. Unfortunately, such protection also presents a problem: how can a plant absorb carbon dioxide from the air if it's covered with wax? Moist interior cell surfaces must be exposed to air in order to absorb carbon dioxide for photosynthesis. Inevitably, exposing moist cells to air will also result in water loss by evaporation from within the plant. And because carbon dioxide concentrations are low in air, a great deal of water will be lost in the process of absorbing carbon dioxide. (This is why plants require a lot more water to survive than do animals with a similar volume of active tissue.) The solution to this plant problem came in the form of special little pores (called *stomates*) on the surfaces of the plant: pores that could be opened for gas exchange, and closed when conditions became too dry.[3]

The other major adaptation for life on land was a "plumbing system" to conduct water from the roots to the parts of the plant active in photosynthesis. The vascular/plumbing system includes thick-walled cells that quickly become dead, empty, and open at the ends. Laid out end-to-end these empty cells resemble a system of pipes, facilitating the flow of water through the plant.[4] These same dead thick-walled cells would also serve as architectural bracing. With a vascular system and stomates in place, the earliest erect plants were able to draw water up to the aerial parts of the plant, and provide strength in the face of wind and weather. Nevertheless, these early vascular plants were only a few inches tall. Then came another major innovation: "secondary growth." With a tubular layer of continually dividing meristem cells (the cambium), which added new cells both on the inside and the outside, stems with secondary growth could get thicker each year. That increasing thickness provided additional vascular tissue and increased physical support for an ever-larger plant. That's how a great majority of trees are built. By about 375 million years ago, the first forests appeared on the surface of the earth, with trees soon reaching a hundred feet in height.

Living descendants of the early vascular plants include the club mosses (Lycophyta), as well as the horsetails, ferns, and their allies (Pterophyta). Today the lycophytes and horsetails rarely get more than a few feet tall; but, during earlier times, some members of these lineages grew to become tall trees. In contrast, ferns vary from a few inches in height to tree ferns over fifty feet tall. Slender tree ferns with broad and graceful leafy fronds are found only in wet tropical forests today, but they are quite similar to plants that lived over 300 million years ago. Together with giant lycopods, horsetails, and smaller ferns, tree ferns helped create ancient swamp forests. Dead and buried along rivers and deltas, these plants became the great coal deposits of the Carboniferous period, around 280 million years ago. During those early times, a few land plants developed yet another new and important innovation.

The Seed: A Major Evolutionary Advance

The next major step in plant evolution was the development of the seed. In a way, the seed represents for plants what the tough desiccation-resistant amniote egg represented for land vertebrates: *reproduction without water*. Even today, liverworts, mosses, ferns, and frogs all require external moisture for fertilization. Pollen gave the seed plants an alternative to sperm having to swim their way to the egg cell.

The more "primitive" nonseed plants may be simple in overall structure but they are by no means simple in their normal life trajectory. In these lineages, spores produce a new generation of plantlets built from cells having only one set of chromosomes. And that makes sense because the spores had only one set of chromosomes to begin with. We call these little plantlets *haploid* because they have half the number of chromosomes (two sets) that normal diploid plants or animals possess. It is these little haploid plants that produce the sex cells or gametes, and that's why these small plants are called *gametophytes*. Each gamete, whether sperm or egg cell, has only one set of chromosomes, as did the plant that produced it. After the sperm and egg cell come together in fertilization, they produce a diploid zygote with two sets of chromosomes. That zygote then begins dividing to form an embryo, and eventually a new plant. Sexual union creates the new diploid organism with two sets of chromosomes; one set from the sperm cell and the other from the egg cell. This diploid plant is called the *sporophyte* because it will produce the haploid spores after the process of chromosome reduction (called *meiosis*). In turn, the haploid spores will give rise to a new haploid plant as the cycling continues. This continuing cycle from haploid plant to diploid plant—from gametophyte to sporophyte and back again—is called the *alternation of generations*. Today, ferns, lycopods, liverworts, and mosses still have alternation of generations as a significant aspect of their life cycles. Like their aquatic ancestors, these plants require thin films of liquid water for their sperm to wriggle their way to the egg cells. Just as in the case of animals, these sperm cells have long wiggly tails that propel them toward the chemical signal of the female egg cell. Because sperm cells are so small, the sperm need only a thin film of water to swim to the egg cell where fertilization can take place. But now back to pollen and seed plants.

Carried by the wind through the air or transported by animal agents, pollen grains emancipated seed plants from the necessity for water as a medium for fertilization. Germinating near the ovule, most pollen grains split open and develop into a microscopic tube that grows toward the egg cell; once there, the male nucleus is able to join a female nucleus in the act of fertilization. This creates the diploid zygote, which begins to divide and form the embryo. Because seed plants no longer required a watery environment for sexual reproduction, they paralleled the origin of amniote vertebrates that didn't have to find a nice pond for fertilization—as most frogs and toads do. Seed plants now became major players in a wider array of terrestrial ecosystems where water was scarce. The seed plants include the extinct seed ferns, cycads, the conifers, ginkgo trees, the strange gnetophytes (*Ephedra*, *Gnetum*, and *Welwitschia*), and the flowering plants (Anthophyta or Magnoliopsida).

The triumph of the seed was twofold. First, pollen grains did away with the necessity for liquid water in fertilization. And second, the seed got rid of the two-stage life cycle by extreme reduction of the gametophytic plantlet. Technically speaking, pollen is a greatly transformed male gametophyte. Instead of forming a haploid little plant, this stage of the life cycle now produced a tough disseminule, the pollen grain. (In cycads and ginkgo, the pollen tube actually releases sperm that swim a short distance to the egg cell.) Likewise, the female gametophyte was reduced to a very small size, snuggling within the ovule of the diploid mother plant; there the developing embryo would be nurtured, as the ovule matured into the seed.

Seed plants were a major evolutionary advance in the history of life on land. And then came another. The last major innovation in land plant evolution becomes manifest in the fossil record in rocks between 130 and 120 million years old. This is the time period in which we find the first unequivocal fossil evidence for the presence of flowering plants. Expanding rapidly during the latter part of the age of dinosaurs (the Late Cretaceous), the flowering plants have been diversifying busily ever since. Today, flowering plants dominate most of the world's tropical and temperate land surfaces. But exactly how do we recognize them?

Distinguishing the Flowering Plants

Our system of organizing the living world is still largely based on the work of Carolus Linnaeus (1707–1776) of Sweden. He is most famous for instituting the "binomial system" of scientific names for plants and animals. Thus, our eastern white oak is given the scientific name *Quercus alba*, italicized because it is a Latin name. The first name of the binomial is that of the oak genus, from the Latin for "oak," *Quercus*; the second is the species name, *alba* for "white." *Geranium maculatum* and *Rosa carolina* are the scientific names for the first two wildflowers we discussed. This simple two-word species epithet replaced phrase names that could run to a paragraph in length, and greatly simplified scientific discourse regarding plants and animals. This Linnaean system of classification has been accepted and used by biologists around the globe. We even continue to describe new species with a short Latin "diagnosis," regardless of the language of the author or journal. Having an international system of plant and animal nomenclature is essential to our studies of the biosphere, and our ability to communicate precisely with colleagues in faraway lands.

More importantly, Linnaean classification places plants and animals within a hierarchical framework of ever-larger categories, in effect a system of "nested sets." For example, the genus is a group of species more closely related to each other than to any species outside of the genus. Likewise, related genera (plural of "genus") are part of the same family and are more closely related to each other than to genera outside that family. In turn, related families are placed in the same order, and related orders are placed in the same class, with related classes in the same division. Finally, divisions are placed within one of the major kingdoms: animals, plants, or fungi. (Zoologists use phylum plural: "phyla," for the rank that botanists call *division*.) Remember our wild rose; here's how it was conventionally placed in Linnaeus's hierarchy. Rose species are placed in the genus *Rosa*, and that genus with related genera are placed in the family Rosaceae. The Rosaceae and a number of other families are placed in the order Rosales, in the class Dicotyledones (about which shortly) in the division containing all flowering plants.

Linnaeus's system proved grandly pragmatic: simple binomial species in a clear hierarchy of larger and larger groupings. It is an effective way of organizing plants and animals on the basis of their overall similarities. Linnaeus saw himself as cataloging God's natural creations, whereas modern evolutionary biologists use this same system and view the hierarchy as the product of a long history. From this perspective, species very similar to each other are closely related because their ancestors diverged only a short time ago. The ancestors of very different lineages diverged much further back in time. Clearly, pigs look more like antelopes than they do monkeys; for one thing, they walk on cloven hooves. And we humans look a lot more like monkeys than do either pigs or antelopes. An evolutionary framework suggests that these degrees of similarity are a reflection of history. Seems simple enough, however, turning the diversity of life and its various degrees of relatedness into a classification can be quite arbitrary. How different do genera really have to be to be placed in a different family? How different ought families be to be excluded from the same order? Experts can, and do, disagree. Nevertheless, each and every category requires anatomical, structural, and chemical features to describe, delimit, and distinguish them. How then do we go about distinguishing the flowering plants from all other seed plants?

We have already seen that flowers themselves can range from the conspicuous meadow rose to minute florets protected within the bracts of a grass spikelet. Overall, flower sizes range over a wide spectrum. The largest flower belongs to *Rafflesia*, a root parasite growing in the rain forests of Borneo. More than two feet across, these flowers emerge from their subterranean home to flower on the forest floor, much like *Hydnora* in Africa. The smallest flowers probably belong to the sea grasses and duckweed (*Lemna* and its allies); they are about a millimeter (a twenty-fourth of an inch) long. Some of these itty-bitty "flowers" are hardly more than a few stamens or a solitary pistil, and they seem to range well beyond our definition of what it means to be a flower. But if *flowers* are so variable, how can we circumscribe the plants that bear them? How do we recognize the lineage of plants that we call the flowering plants?

Willi Hennig, a German entomologist, proposed that we use "uniquely derived" characteristics to define "monophyletic lineages" of plants and animals. Monophyly implies that all the members of the lineage came from the same single common ancestor, however long ago. Unique and complex derived characteristics, restricted to special groupings of plants or animals, are not likely to have been developed more than once. Hence, such features should be very useful in defining natural lineages. Mammals, for example, are distinguished by nursing their young, having three little bones inside their ears, growing hair, and lots more. Such a concordance of unique features suggests that all living mammals came from a single original lineage; that is, mammals are monophyletic.

Because plant parts, especially flowers, quickly decompose,

the fossil record of plants is particularly poor. With that in mind, we are forced into studying primarily plants that are alive today. By using their most distinctive characteristics, we can endeavor to recognize "natural" or monophyletic lineages among living plants. If we can characterize a group or lineage with *several* concordant, unique features, we should be quite confident that the group or lineage is, like the mammals, descended from a single ancestral population. The probability that two *different* lineages developed the same suite of unique characteristics independently is highly unlikely, and gives us confidence that we really are working with a unitary lineage. But what are the defining traits that might distinguish the flowering plants from all other plants?

The most fundamental trait of flowering plants is embodied in one of their traditional scientific names: the angiosperms. *Sperm* means seed, and *angio* implies that they are enclosed or hidden. As in other seed plants, the ovules of flowering plants are ellipsoid structures within which the egg cell and accessory cells reside. As we noted, there are many other seed plants in today's world, such as the cycads, ginkgo trees, and conifers. The last group includes pines, firs, spruces, giant redwoods, and many others. All these are called *gymnosperms* because of their "naked" seeds. What makes flowering plants different is that their ovules originate and develop within a chamber: the locule of the ovary (see fig. 2B, p. 25). The ovules are enclosed or "hidden" within the ovary, and differ in this way from the "naked" ovules of gymnosperms, which have no such enclosure. Because no other seed plants have an ovary-like enclosure for their developing seeds, we can use this unique characteristic to identify exactly which plants belong to the category of flowering plants.[5]

Flowering plants have another significant feature, and it is also associated with reproduction. When the pollen grain germinates on the stigmatic surface and begins to grow, it forms a tube. The pollen tube then grows down the style toward the ovule. As it grows toward the ovule, the pollen tube usually carries two or three pollen nuclei. Growth of the pollen tube will bring these nuclei into contact with the ovule. One of these pollen nuclei then enters the ovule to unite with the egg cell, in the manner of a sperm cell—resulting in fertilization. The fertilized egg will grow into the embryo that develops further to become the seedling, the early stages of a new plant.

So far so good, and not much different from most other plants and animals. But here's where things get interesting. In flowering plants, *a second pollen nucleus also enters* the ovule and usually joins with two other female nuclei within the ovule in what is called *triple fusion*. In a few flowering plants only two nuclei come together, but, either way, this *second union* begins the development of nourishing tissue for the developing seed. That nutritive tissue is called *endosperm* ("within the seed"). So here we've got another "unique and derived" feature; we call it *double fertilization and triple fusion*. Together with the carpel with its locule and enclosed ovules, these traits help define the flowering plants.[6]

The Angiosperm Carpel and the Genesis of Fruiting Complexity

Though there is considerable diversity among the different floral organs, neither the sepals, petals, or stamens exhibit the range of morphological diversity found among the female parts of the flower. You'll recall that the female parts of the flower, whether solitary or multiple, are collectively called the *gynoecium* (plural: gynoecia). As we mentioned, angiosperm gynoecia are thought to have been built up over evolutionary time by leaflike carpels that may have been solitary or several, and variously joined together. We've just noted that the "hidden seeds" within locules of the ovary are a unique feature of angiosperms. It is within these locules that the seeds of flowering plants are hidden, protected, nurtured, and, finally, matured. To appreciate the evolution of the ovary and its locules we need to understand the theory of the carpel, because it is the carpel that forms the locules.

Consider an ordinary sweet pea pod. At maturity it is considerably longer than wide, thin-walled, and almost round in cross section, with all its peas neatly arranged in a row down one side within the pod. To get at the peas you squeeze the pod, splitting it open into two longitudinal halves; spreading the halves apart makes it easy to strip out the peas. Once you've popped the pea pod open, it doesn't take much imagination to see that the opened pod resembles a leaf; a leaf whose curved lateral sides—folded around and pasted together—easily becomes the "carpel." In effect, the carpel is a hypothetical leaf that has come to enclose the ovules. Union of the sides of the leaf created a single enclosure (the locule) with the ovules neatly enclosed and protected. All that was required—over evolutionary time—was to fold the leaf around the ovules, paste together the openings, and, *voilà*, you've got a carpel and its locule. Because the pea flower has a single solitary pistil with a single locule, single style, and single terminal stigma, its simple pistil is a fine example of a carpel.

Pea pods and other legumes with simple pistils are easy to understand, but how did Nature create something like a tomato or an orange? That's where botanists get more imaginative. Here we think of three, four, or five carpels (in-rolled leaves) closely appressed and becoming united—with the three, four, or five locules all that are now apparent of the original leaves. In some families the carpels seem to have fused not along their sides but *along their edges* to produce one larger locule with three, four, or five longitudinal rows of ovules down the sides. Pumpkins, cucumbers, and passion fruits are products of this kind of ovary. And that's not all; add evolutionary loss, duplication, or fusion with other floral parts (as in an inferior ovary), and you've got a huge variety of gynoecia.

The wide diversity of fruits among flowering plants is a direct consequence of the many varieties of gynoecia. Whether from a simple pistil like a pea pod or string bean, to more complex pistils like those of the watermelon, grapefruit, and apple, the fruits of angiosperms are grandly diverse. Better yet, the outer walls of angiosperm fruits have been gloriously and variously modified. Some fruit walls are hard and protective, as in hazel nuts, pistachios, and coconuts. Here the stuff we want to eat is the seed itself, protected within the tough fruit. Some fruit walls are tough but easy enough for an animal to open when ripe, as in the watermelon, pumpkin, grapefruit, and cacao pod. In these it is the tissue around the seeds that is so eminently edible. In other lineages most all of the fruit wall becomes succulent and delicious, as in the grape, blueberry, and avocado. In others the fruit wall is a hard but temporary protection, often covered with spines, and splitting open only when the seeds are ready for dispersal.

And there's more to this cornucopia of fruit diversity. In some lineages other parts of the flower or floral axis get into the act. Mulberries are a clustering of little flowers in which the minute perianth parts also become succulent. In some flowers the floral axis (receptacle) on which the carpels are borne becomes an important part of the succulent fruit, as in the strawberry, soursop, and custard apple. Inferior ovaries may have parts of the outer wall contribute to the fruit, as in the apple, peach, and coffee fruit. (Our "coffee beans" are actually the dried seeds.) Tightly clustered fruits can become united with the floral axis to become a larger "collective fruit" as in the breadfruit and pineapple. Similarly, the inflorescence-axis of Australia's *Banksia* (color insert L) becomes a thick cylindrical woody cone that seems to have "lips" on its surface. After a fire, these lips open, allowing the seeds to disperse. And finally, the fruit wall can remain thin, unite with the wall of the seed, and become almost indistinguishable from the seed coat, as in a grain of wheat or kernel of corn. All told, the huge variety in fruit and seed morphology has played a key role in making the angiosperms so successful.

Other important "unique and derived" characteristics that help us distinguish angiosperms are found in various microscopic structures. Most angiosperm pollen grains have a complex wall where the outer surface is borne on special projections (columellae), quite unlike the pollen of other seed plants. Likewise, the vascular system has highly specialized cells (vessels in the xylem and sieve tubes in the phloem) unlike the cells in most other seed plants. The wood of angiosperms is characterized by these and other complex cells. In fact, flowering plants display a greater variety of wood anatomy than all other seed plants put together. But none of these characteristics are as visible as flowers; why aren't we using flowers to *define* the flowering plants?

What's the Problem with Flowers?

One would think that an obvious candidate as a unique feature of the angiosperms might be the flower itself. Unfortunately, flowers present us with two problems. For one thing, a few gymnosperms, and some nonangiospermous fossils, have flowerlike configurations for their sexual parts, and it becomes difficult to create a definition that can exclude them from the flowering plants. Second, there's all that variation in floral configuration among flowering plants as mentioned earlier and no clear indication of what an "original" flower might have looked like. In fact, there are a few flowering plants whose floral arrangements are so simple that they may represent ancient "prefloral" arrangements lacking any perianth. (This may be the case in *Chloranthus* of the Chloranthaceae family and *Peperomia* of the Piperaceae family.)

A long-standing scenario of floral evolution assumed that the "primitive angiosperm flower" had lots of parts in a spiral arrangement, rather like a magnolia flower. Resembling a stem with an elongate spiral of leaves, this "primitive flower" was thought to have its parts in an ascending spiral of perianth, stamens, and pistils. That's quite different from the majority we see in the world today, such as our geraniums, roses, and lilies. The problem with a magnolia-like "evolutionary ancestor" is that it is not clear how most flowers were derived from such an ancestor. How might floral parts have been "reduced down" from flowers with many parts in a spiral, to form flowers with parts neatly arranged in whorls of threes or fives? The magnolia-as-ancestor scenario would require "losing" lots of parts—and then having the few remaining parts arrange themselves in whorls of the same number. Another problem is that the magnolia scenario implied that *all* small and simple angiosperm flowers had become "reduced" over evolutionary time through loss of parts. The fossil record doesn't really jibe with this idea. Indeed, there are large ancient fossil flowers resembling *Magnolia* flowers, but there are even earlier smaller and simpler flowers as well.

A terrific recent find from China, in fact, presents us with something that doesn't look flowerlike at all. Apparently growing along the edge of a lake, this plant was two to three feet tall and had highly dissected, probably submerged, leaves. At the apex of the plant was a slender vertical axis with a spiral of separate simple pistils at the top; farther down the same axis were small groups of paired stamens. Lacking bracts, sepals, or petals, this may have been an arrangement of carpels and stamens that preceded what we nowadays think of as flowers. (It makes sense that such open and exposed stamens and pistils would benefit from becoming enclosed in protective bracts or perianth over evolutionary time. Perhaps they survived in those ancient times because there weren't as many herbivores around.) What's especially significant about this fossil, called *Archaefructus*, is that it is dated around 120 million years old; the oldest well-preserved fossil angiosperm ever found! We're pretty sure it is an angiosperm because of its pealike carpels, but, clearly, this plant doesn't really have flowers.[7]

Another way to look at evolutionary origins is to study early development. The human embryo develops and then loses both grooves that resemble gill slits and a curved tail. These features are "relics" of earlier stages in our evolution; they come early in development—and are lost in later stages. (This concept was enshrined in the phrase "ontogeny recapitulates phylogeny.") Surely, similar patterns can be found in flowers. And indeed they are. Many developing flowers begin with three or five little bumps in successive series. The first series produces three or five sepals; the second three or five petals. The next series may produce three or five stamens or divide to produce multiples of three or five. Surely, this kind of development in many living flowers fails to give any evidence for an early spiral configuration, or for the loss of parts. More significantly, the variety we find in floral development suggests that flowers were formed in different ways in different lineages. That is to say, flowers seem to be *polyphyletic*; they did not originate from a single ancestral type. What happened, it seems, is that different lineages within the angiosperms found a variety of ways to construct what we now call flowers. This needn't challenge the monophyly of angiosperms however; the other characteristics we mentioned make it clear that angiosperms are a single coherent lineage. Rather, it appears that different lineages of flowering plants put their flowers together in a variety of ways. But what was the ancestry of angiosperms—where did they come from?

Where Did Flowering Plants Come From?

In a letter to a botanical colleague, Charles Darwin described the origin of flowering plants as "an abominable mystery." Today, 150 years later, the origin of angiosperms is still a profound enigma. Plant fossils, especially of delicate herbs and flowers, are so rare and of such poor quality that we still have not identified an ancestral lineage for the flowering plants. Much recent DNA work has been problematic; different analyses give different patterns for early divisions, and every few years a new "most basal living angiosperm" is announced. Interestingly, zoologists studying the origin of birds or mammals don't waste their time on such an enterprise. They know that there are no "basal" birds or mammals alive today. But zoologists do have a huge advantage: vertebrates have bones—and bones have left a rich fossil legacy. Even though early birds and mammals are rare fossil discoveries, their ancient bones are hugely informative. From those bones we know that birds were derived from small bipedal meat-eating dinosaurs. In fact they share a bone unique in vertebrates, one we are all familiar with;

we call it the wishbone (from a fusion of clavicles). In the case of mammals, the fossils indicate a derivation from mammal-like reptiles 200 million years ago.

Because plant fossils are so rare and so uninformative, much emphasis in the last two decades has focused on finding DNA relationships among living plants. These analyses have been especially helpful in more recent relationships—over the last 50 million years. But unraveling the early differentiation of flowering plant lineages on the basis of today's DNA has proven difficult. One hundred million years is plenty of time for random mutation to mix up the genetic code. Historic relationships can be obscured by duplication, loss, and repeated changes in the strands of DNA. Because the earliest angiosperms were, in all likelihood, delicate herbs or small shrubs, their probability of becoming fossils was very small. (*Archaefructus*, mentioned above, appears to be a small semi-aquatic herb with dissected leaves, probably preserved because of its muddy lakeside environment.) The latest DNA studies, nonetheless, suggest that the flowering plants are sister to *all other seed plants*; that is to say, we cannot identify a sister lineage for flowering plants among the seed plants that are alive today. In fact, the problem may be more profound: seed plants themselves may not be monophyletic. Instead, the seed may have developed independently in several lineages, and the origin of angiosperms may reach back into those early times. Because it is so contrary to current thinking, this idea has not been given serious attention for almost a century. Thus, with no clear "sister lineage" we'll just have to wait for new fossils or some exceptionally informative DNA to clear up this matter.

Dividing Up the Angiosperms

It is important to point out that the angiosperms have traditionally been divided into two great divisions on the basis of the first leaf or leaves (cotyledons or seed leaves) formed by the embryo. The two groups are: dicotyledons ("dicots" for short) and the monocotyledons (or "monocots"). First discovered over three hundred years ago, this division has no significant exceptions; it really worked well in dividing the angiosperms.[8]

Monocotyledons have only a single terminal leaf on the embryo, with the shoot apex along the side. These plants include the lilies, sedges, grasses, orchids, aroids, bananas, palms, and all their relations. Many monocots have narrow leaves with parallel venation; relatively few have broad leaves with stalks (petioles) or netted venation. Virtually all monocots have their floral parts arranged in whorls of three, as we saw in our wild lilies. Trees are very rare among monocots, and they tend to have strange wood structure in their trunks, as we see in palms and pandans. Only a very few have a tubular vasculature that gets thicker every year, as is so common among dicots. The monocots are very limited in their branching patterns; palm stems rarely branch. And at their base, monocots produce a bunch of fibrous roots, instead of one deep taproot. Though they have two huge families—the orchids and the grasses—monocots account for only about 20 percent of the species of flowering plants. Despite this lower number of species, monocots are quite conspicuous in botanical conservatories and display houses. The big-leaved bananas and aroids give an in-your-face impression of what tropical plants can look like. Tall and slender palms add grace and stylish leaves. Lots of lilies and their relatives fill in at ground level, while a remarkable diversity of orchid flowers hang in baskets or sit on tree trunks. These plants make a visit to a botanical conservatory a special experience; very different from the plants we are used to in our temperate zone surroundings.

Dicotyledons have two paired opposite cotyledons terminating their little embryos, with the shoot apex between the cotyledons. Dicots comprise the great majority of angiosperm species. They include all the flowering trees and shrubs of our northern forests, and most of those in the tropics, as well as many herbs. Some of our largest plant families, such as the composites, the legumes, and the coffee family, are dicots. Most dicot families have trees whose trunks get thicker each year as their cambium expands. (Recall that the cambium is a tubular meristem that divides to produce additional new tissue around the stem.) Nearly all dicots have broad leaves borne on slender petioles, and leaf blades with complex reticulate venation. Dicotyledonous plants are often highly branched, whether trees, shrubs, or tumbleweeds. Their flowers come in a huge variety, but a majority has its floral parts in fives or multiples of five—as we saw in the wild geranium and pasture rose.

The two classes, Monocotyledones and Dicotyledones, were then placed in the division Anthophyta (also often cited as Angiospermae or Magnoliopsida) and, together with other divisions, comprise the kingdom Plantae. However, scientists are always busy discovering new things and the traditional classification of flowering plants is undergoing some dramatic changes. These changes are by no means final, but they indicate how the larger classification of flowering plants is currently viewed.

Problems with the accepted classification began when botanists realized that, though most dicots have tricolpate pollen (with three grooves), a small number of dicotyledonous families have monosulcate pollen (with a single furrow) and also pollen derived from the monosulcate type. Most botanists now believe that this is a difference that far outweighs the differences between cotyledon numbers in the embryo. For one thing, tricolpate pollen shows up much later in the fossil record than does monosulcate pollen. In addition, those families with the more ancient form of pollen were also found to have primitive wood anatomy, and a few even had the "ancient" magnolia-like flowers with many parts in spirals. Surely, these lineages seemed to be the most ancient of living angiosperms. The fact that a large majority of these plants were treelike fit in with the fact that nearly all other living seed plants were trees. For these reasons, the "woody Magnoliales" and their associates were considered to be the living descendants of the earliest flowering plants.

Thus, though angiosperms had been divided into two major groupings—dicots and monocots—for over two hundred years, they are no longer divided in so simple a fashion. DNA similarities have supported this new division of the angiosperms, in line with the pollen differences. Angiosperms are now arranged into a basal group of magnolia relatives, together with other early derivatives—the monocots among them. These plants all have pollen grains with a single furrow, or pollen derived from such early forms. None has the more modern three-grooved pollen. In addition, most have three-parted flowers or flowers with variably numbered whorls. In contrast, the "advanced dicots" (Eudicots) are characterized by having the more modern three-furrowed (tricolpate) pollen grains or derivative kinds. A great majority of these have floral whorls with five elements or multiples of five. The nice thing about this arrangement is that the fossil record clearly tells us when tricolpate pollen first appeared, about 100 million years ago, thus serving as a clear benchmark in early angiosperm evolution. Thus, the new arrangement of flowering plants begins with what are believed to be the most ancient lineages, none with modern tricolpate pollen, followed by the second more modern crowd with more modern pollen types.

Within these broad outlines, DNA base-pair sequences have provided "gene trees" that are especially helpful in relating families to each other and clarifying the position of genera that had been difficult to classify. These DNA studies have also given us a consensus regarding the "basal" living lineages. *Amborella trichopoda*, a small shrub surviving only on the island of New Caledonia in the southwest Pacific, is currently thought to be the sole living representative of the earliest angiosperm lineage. However, the water lilies (Nymphaeales), black peppers and their relatives (the Piperales), the laurel alliance (Laurales), the magnolia alliance (Magnoliales), the monocotyledons, and a few other groups are also part of early angiosperm diversification. Trying to untangle this early history, working entirely from DNA sequences within the genes of *plants living today*, continues to be a very difficult task. I believe that our current understanding of early angiosperm diversification is very limited, and will require more and better fossil material.[9]

Clearly we haven't solved Darwin's abominable mystery: we cannot identify the ancestors—or even the sister group—of angiosperms! That separation probably took place over 150 million years ago, and the close relatives of those early times are long gone. Though not a positive note on which to end the chapter, we can rest assured that flowering plants are a natural, clearly circumscribed group. But now, after maneuvering through the intricacies of classification, it's time to ask: why did they become so successful?

chapter 6

what makes the flowering plants so special?

When we survey vegetated land surfaces on our planet, we find that flowering plants are often the dominant players. However, if the fossil record is to be taken seriously, then flowers are a relatively recent innovation on our planet. Nothing like them has been found in the extensive swamp forests that formed so many coal deposits around 300 million years ago. Likewise, around 200 million years ago, when cycads and conifers were becoming a major part of the landscape, there is no evidence of flowering plants. The earliest substantial fossil evidence for plants that are clearly angiosperms is dated at around 120 million years ago. Apparently, angiosperms began to make themselves known on the landscape at this time, during the age of dinosaurs. A great extinction wiped out the dinosaurs and a lot of other lineages 65 million years ago, but it had little effect on the flowering plants. Since then they have continued to proliferate.

Some years ago, paleobotanist Norman Hughes published a series of estimates regarding the increasing numbers of species of vascular plants over the last 300 million years.[1] Vascular plants, you'll recall, include the flowering plants, conifers, cycads, ferns, and fern allies. Larger vascular plants give terrestrial vegetation its complex three-dimensional structure, and they are a land habitat's primary energy source. They are the major players in creating the world's terrestrial vegetation. Hughes suggested that there were only about 500 species of vascular plants living at any one time in the Carboniferous period throughout the world. That was 300 million years ago; by 150 million years ago, he estimated a world flora of 3,000 species. At the end of the Cretaceous, 65 million years ago, Hughes's estimate rises to 25,000 species. Today's total of vascular plants numbers about 275,000 species, of which around 260,000 (or 94.5 percent) are flowering plants. This is a pretty amazing increase.

If Hughes's estimates are anywhere near correct, they suggest that there has been a remarkable increase in land plant diversity over time, and especially over the last 150 million years. Though the ferns also proliferated over this time period, the recent increase is almost entirely due to flowering plants. By adding moss and liverwort species (the nonvascular land plants) to today's number of vascular plants, we get a total estimate of all land plants just shy of 300,000 species. In this case the percentage of flowering plants is 87 percent—still pretty impressive. So what is it about flowering plants that has made them so successful? As usual, biological questions such as these have no quick and simple answers. There are undoubtedly a number of features that account for angiosperm success. The most obvious is an extraordinary variety of growth forms.

Why Are Flowering Plants So Varied in Size and Lifestyle?

For starters, let's compare plants and animals. If you have to move around looking for something to eat, it's great to have a front end and a rear end—as most animals do. The need to search for and find food has resulted in a horizontal forward-pointing body style for most animals. This body form puts most of the sensory organs up front, just where you want them when searching for something to eat. With immobile plants, rooted to the ground and reaching for light, the structural body orientation is completely different. Here, the axis of the plant is up and down, not front to back as in most animals. Plant architecture requires a big top where the leaves can intercept light, and a big bottom to provide a firm base that will allow the roots to absorb water and nutrients. That's why most plants are bipolar, with a widely branching aerial top and an elaborately divided root system.

Plants, in effect, live in two different worlds. Their tops exist in an open, sunny environment of desiccating breezes, while their bottoms live in a dark and crowded world of soil and rocks. They may look static but plants actually live a very vibrant life. The leaves, flexing in the breeze, are absorbing carbon dioxide, while inevitably losing water vapor and oxygen. At the same time, root hairs are constantly growing and probing the soil to absorb mineral nutrients and to replace water being lost from the leaves.

These features characterize virtually all the larger land plants, from small ferns to giant sequoias. But to get really big, to achieve real height, plants had to invent the tree. A number of different lineages of land plants have "learned" how to build tall trees over the last 300 million years. These include fossil lycopods and horsetails, tree ferns, most all the gymnosperms, and many lineages of flowering plants. But flowering plants have produced huge trees *as well as* diminutive floating aquatics and short-lived desert ephemerals.

Why is it that flowering plants exhibit a much wider range of sizes and lifestyles than do any other plant lineages? By "lifestyles" I mean how plants live their lives, whether short-lived desert herbs, northern shrubs at the edge of the tundra, seasonal water lilies living in lakes, or long-lived rain-forest trees. No other group of plants includes moss-sized epiphytes, barrel-shaped cacti, small herbs, and giant forest trees.[2] How did flowering plants manage such a feat over evolutionary time?

To answer this question we need to return to our discussion of double fertilization. This is one of the characteristics unique to the flowering plants, and it has a very special significance. The ovule, remember, is where fertilization takes place in seed plants, and here is where the embryo begins its growth.

As it develops, the ovule will become the seed, increasing in size with stored food or a much-enlarged embryo. But forming a larger seed requires a lot of energy, and here is where the angiosperms are different. Double fertilization *initiates* the production of food reserves (called *endosperm*) within the seed *only after* fertilization has taken place.

That's not how things work in most gymnosperms (which lack double fertilization). In these plants seeds are outfitted with food reserves *before fertilization has taken place.* Not smart! If fertilization fails to take place—and in an unpredictable environment this can happen often—these plants produce energy-filled seeds that cannot sprout. This is a huge waste of resources. And here we have the fundamental reason why, among all of the living seed plants, only the flowering plants include little herbs and small shrubs. For a large tree, the production of infertile seeds may be a tolerable energy drain; but for a small plant it is an impossible burden.

Thanks to double fertilization, flowering plants have been able to produce a greater variety of growth forms than any other seed-plant lineage that ever lived. By linking endosperm production directly with fertilization, flowering plants build energy-rich seeds *only for seeds that have been fertilized.* Unlike gymnosperms, healthy flowering plants do not produce seeds incapable of sprouting. Whether in a small herb growing on the dark floor of the rain forest, or in a desert shrub flowering and fruiting over a short rainy season, energy is something that must not be squandered. Double fertilization made flowering plants more efficient in the primary job of life: reproduction.[3]

Why Are Flowering Plants So Diverse in Structure?

Flowering plants have lots of other characteristics that have contributed to their structural diversity. For one thing they've got more genes that control growth; they have more *developmental* genes than do mosses, ferns, or even gymnosperms. These additional genes carry the extra instructions needed to put together something as complex as a flower. And they help in fashioning everything from Spanish moss to baobab trees and crab grass. Flowering plants also have a greater variety of cell types than do other plants. For example, quite a number of flowering plants are densely hairy. Hairiness is a feature that's uncommon in other plant lineages. As we've already seen, it can be very useful in deterring insects from chewing on plants. But hairiness has other uses. On high tropical mountains, hairiness helps insulate the stems and leaves during the long frigid night. And in deserts, hairiness may be a way of capturing early morning dew.

Compared to animals, however, even flowering plants have very few cell types or tissue systems. And, speaking of animals, we should consider another fundamental plant-animal distinction. Plant cells have strong stiff cellulose walls, very different from the thin membranous walls of animal cells. Not only can animal cells slither around and move to other parts of the animal during early development, they can easily disassemble themselves during the processes of growth and development. Our fetal fingers had webbing between them before we were born; that tissue was resorbed as our fingers developed further.

In contrast, plant cells resemble little bricks that get glued in place to create the larger plant body. Once in place they stay in place. That's not as flexible or versatile as animal construction; but it is strong and serviceable. This explains why plant forms are far more limited than are the forms and parts of animals, both externally and internally. (Because plant cells are "stuck in place," they cannot cause cancerous growths to spread around the body as in animals and us.)

Not only do flowering plants have more cell types than other plants, they also build a much greater variety of structural forms than do the ferns or gymnosperms. Angiosperms do include fernlike herbs and gymnosperm-like trees, but they also encompass cacti in the form of little barrels, widely branched trees, tall palms with only a single stem, tumbleweeds with a multitude of stems, and cabbagelike forms (some on little trunks). The growth and anatomy of a palm trunk is completely different from that of an oak or a mahogany trunk. Likewise, the light wood of the balsa tree is very different from the dense wood of a rosewood or walnut. Most angiosperm trees, from oaks to acacias, build their trunks the same way that conifers build theirs: with a growing region called the *cambium*. As we've noted before, by adding new cells on both the inside and the outside, the tubular cambium makes the trunk a little thicker each year. This accomplishes two things. First, it produces a larger trunk to support greater weight and stress. Second, it adds vasculature (xylem on the inside, phloem on the outside) to sustain an increasing burden of leaves and stems. (Because the trunk grows in girth, a nail in the trunk soon gets buried in new growth, and the nail does not change in height, even as the top of the tree continues growing upward.)

Flowering plants can do things with their growing tips that few other plants can duplicate. Recall that plants grow from primordia, whether from the tip of the embryo as it begins its life trajectory, or from the minute tips at the ends of growing points. These growing tips, in turn, can develop smaller lobes that result in three-parted leaves or five-parted flowers. These primordia lobes can remain separate and give us the five petals of the rose flower or the six tepals of the lilies. In other flowers the primordia can fuse into a ring to grow a long tube and *then* split to create five spreading petals at the end of the tube, as in a phlox or frangipani. As mentioned, the ovary wall has become fantastically diverse in different lineages of flowering plants to produce a wide array of fruiting types. No other plant lineage possesses so rich a variety of growth patterns.

Leaves also exhibit a huge variety of forms within the flowering plants. Grass leaves are narrow with parallel veins, growing mostly from the base—which is why we can keep on mowing them and cattle can keep on eating them. The leaves of most trees have spreading veins in broad blades, to support an expanse of thin green tissue. Some aquatics have circular leaves that float on water, while their underwater leaves are slender and finely dissected. Leaves on trees within the rain forest are often narrowly elliptic with prolonged "drip tips." Such a huge variety of forms demands versatility in growth patterns, and here angiosperms excel as well.

In addition, flowering plants can grow relatively rapidly.

This is important in desert regions where rainy seasons are short, and in wet rain forests where, if a plant doesn't grow fast, another will grow over it. Fast growth and high turnover rates are characteristic of many tropical evergreen forests. Here again, angiosperm species hugely outnumber all other seed-plant lineages. Ferns may also be well represented in such moist forests, but remember, they are not seed plants. In the tropics, it is only under very special conditions, such as on the higher mountains, that the gymnosperms play a dominating role. Gymnosperms also thrive in some temperate forest situations (such as the Pacific Northwest, our western highlands, and our southeastern swamps). The conifers are especially suited for colder short-season boreal forests that range across huge areas of northern Canada, Russia, and Alaska. Nevertheless, the understory of these forests and the tundra flora in these cold climes are largely made up of flowering plants.

What's So Special about the Deciduous Leaf?

One of the most familiar characteristics of flowering trees is surely one of their most clever adaptations: leaves that are shed each year. For people who have to rake them up every fall, this may not seem like a great idea. However, for plants dealing with either a long and brutal cold season, or a long, hot dry season, the easily disposable (deciduous) leaf became a major evolutionary innovation. Not a new development by any means—ginkgos, larches, and others developed similar devices—the readily disposable leaf played a major role in allowing the flowering plants to survive in strongly seasonal environments.

Ideally, a disposable leaf ought to be a "cheap" leaf. After all, it's going to be thrown away each and every year. Such a leaf should have two main elements. First, it needs a broad surface to maximize the absorption of sunlight (we call this the *blade* or *lamina*). Since sunlight is the basic energy source green plants depend on, a broad flat surface will absorb a lot of light. Second, it should be thin, making it cheaper and quicker to construct. Despite being thin, it should remain strong in the face of wind. A system of interconnected veins provides both strong mechanical bracing and plumbing to move water and nutrients throughout the blade. And the broad blade should also have a long flexible stalk (the *petiole*).

A slender but strong petiole serves several important functions. Its strength and flexibility will allow the blade to twist and turn in high winds—lessening the chance of having the blade torn apart. Tests of leaves in wind tunnels, anchored by their petioles, have shown how the petiole functions in exactly this way. In many plants the petiole also functions to *orient* the leaf blade for maximum exposure to the sun. In some prairie plants that means holding the leaf vertical to catch the light of early morning, and in trees it may mean holding the leaves horizontal. Of course, the petiole must also act as a conduit for water and soil nutrients entering the leaf, as well as photosynthetic products *leaving* the blade. In fact, before the leaf is shed, the plant must salvage carbohydrates, proteins, lipids, and as many essential elements as possible *from the leaf*; this is why so many leaves change color as they prepare to fall. Finally, the petiole must develop a basal abscission layer along which the leaf will easily split off. By dying, becoming fragile, and breaking, these abscission-layer cells will prevent ripping tissue from the stem as the leaf is shed. (Some palm trees haven't figured this out and their trunks are covered with the thick, broken bases of leaves long gone.) All told, the easily disposable, petiolate, thin, and broad-bladed leaf was an important innovation for angiosperms—first appearing in the fossil record about 100 million years ago.

However, as mentioned, there are a few habitats in which flowering trees are in the minority; these are the cold boreal forests of the far north and those of higher elevations in mountainous regions. Here is where many conifers are both successful and dominant. Frigid temperatures over more than half the year are part of the problem, but ice and snow are an even more serious danger. Conifers have adapted with slender needlelike leaves and a *narrowly conical* growth form. With this growth form (think of Christmas trees) the lower branches can help support the upper as they become laden with heavy snow or ice. Flowering trees simply aren't built this way. Searching for more sunshine, flowering trees tend to spread their upper branches widely. They grow tall and straight only within the forest where light comes from above. Having originated in a tropical homeland, angiosperms can't grow as narrow and as tall in northern forests as do the pines, spruces, firs, and hemlocks. But, in less frigid temperate regions and in warm tropical climates, "cheaper" deciduous leaves work well. Dropping them reduces the loss of water through transpiration, an advantage in both hot and cool climates. In the cold of winter, roots can't pull up enough water to balance evaporation from broad leaves, so it's best to get rid of them. Likewise, in hot climates, dropping the leaves at the beginning of the dry season reduces water loss for the plant. In warm climates with short rainy seasons, leaf blades are usually less than four inches (10 cm) long, and less than two inches wide. In fact, some tropical leaf blades are less than a half inch long, as in the acacias. Here the tiny leaflets are borne on slender axes that are actually parts of a single larger *dissected* leaf (with its abscission layer at the stem). Such little leaflets borne on broader "compound leaves" may provide maximum light interception with minimal investment in tissue. Better yet, the multitude of little leaf edges provides additional cooling in a hot tropical sun. (Edges dissipate heat better than flat surfaces.) All of that, together with a flat broadly spreading tree crown, gives us the characteristic trees of the African savanna. Clearly, both in the seasonally dry tropics and seasonally frigid temperate zone, deciduous leaves have contributed much to the success of flowering plants.

What Might Be So Special about Grass?

Like deciduous leaves, grasses may not seem impressive but they were a hugely important innovation on the face of our planet. Characteristic of dry and sunny open landscapes with short growing seasons, grasses grow and reproduce quickly, before the rains cease or the cold season arrives. Their quick growth gives them a head start after the rainy season returns to fire-scorched landscapes. Grasses do this by storing energy *underground*, both in more extensive root systems and rhizomes that travel laterally just below the surface. These underground parts are safe both from fires and grazing herbivores. A long, cold season and scant rain gave us the steppes of central Asia and the short-grass prairies of North America. Warmer climates with short rainy seasons produced the grasslands of Africa, Australia, and parts of South America. On the highest mountains, limited rainfall and strong temperature fluctuations produced the grassy alpine meadows of the Alps, the Himalayas, and the Rockies, as well as the grassy alpine *paramo* and *puna* formations of the high Andes. In all these many environments, the grasses are dominant, ready to sprout as soon as conditions become favorable.

Early grasses may have reduced their investment in defensive chemicals in order to spend more energy on growing, and on growing fast. With such an emphasis on growth, grasses provided a bonanza for another group of living beings: large herbivores. Around 30 million years ago, temperatures grew cooler and climates became drier. It was around this time that the ruminants (such as cattle, bison, antelopes, gazelles, and their relatives) began diversifying. By browsing shrubs and small trees, some of these animals may have aided the early spread of grasslands over drier landscapes. But soon they turned to eating the grasses themselves. Now, grazing animals themselves became a severe selective pressure on grasses. Forced to protect themselves, the grasses elaborated a microscopic defense: phytoliths within their tissues. These are minute crystals of silica; the same stuff you find in sand and granite—and just as hard. As we mentioned earlier, these hard little crystals caused the grazing mammals to produce deep-rooted teeth. Such teeth would wear down just like other teeth, but because they were so long, they endowed their owners with a longer life.

In the continuing struggle for survival, grasses developed other evolutionary tricks. They now exhibit more chromosome doubling (called *polyploidy*) than any other flowering plant family. Though extremely rare among animals, chromosome doubling has been a significant evolutionary device among the flowering plants, and especially the grasses. When different species hybridize, the resultant animal or plant may be hale and hearty, as exemplified by the mule, which is a strong and useful hybrid of the horse and donkey. However, because their chromosomes are different (one set from a horse, the other from a donkey), these chromosomes can't line up properly to produce functional sex cells—and this is why mules are sterile. But if you do have a hybrid between two rather different plant species, and the chromosomes (somehow) get doubled, then the chromosomes can align properly—and produce effective sex cells. Not only did such a process take place in the history of wheat, it happened twice! Wheat is both a grass and our most important food grain. The once-doubled wheats include those we use to make pasta; the second-doubling gave us the bread wheats. These accidental changes in chromosome numbers, picked up by early farmers in the Middle East, provided a major staple for people ranging from the eastern Mediterranean to northern China.

Grasses are the plants that have produced savannas,

steppes, and grasslands covering so many of Earth's land surfaces. These are the plants that gave us humans the cereal grains—food with which to build civilizations. No other group of plants is so essential to our domestic animals and us. One of the reasons we don't appreciate them more may be that grasses have such inconspicuous flowers. So let's get back to pretty flowers, and see how flowers themselves helped make their plants so special.

The *Real* Advantage of Colorful Flowers!

As already discussed, colorful flowers put on a show to attract animal pollinators. But all that fancy color, structure, aroma, and nectar does add up to a big expense. And those flowering plants that use wind for pollination—like grasses and oak trees—must produce huge amounts of pollen in order to have a chance for pollination to be effective. Either way, there's a big energy expense. In fact, total energy requirements may not be that different between plants that produce lots of pollen from little flowers and those that have fewer large, aromatic, nectar-filled flowers. If that's the case, why might colorful, animal-pollinated flowers have made a significant difference?

Wind pollination has been a very successful survival strategy for a great many plants over many millions of years. After all, there's usually plenty of wind around to do the work. That's the good news. The bad news is that the wind doesn't know where it's going; wind pollination is hugely inefficient. This form of pollination requires not only a great amount of pollen, but it also demands something else: high local population densities. The wind simply cannot pollinate isolated or widely scattered individual plants effectively. Think of broad grasslands with only a few dozen grass species; here high population densities make wind pollination effective. And this gives us a clue to why colorful flowers were such a big advantage.

Colorful flowers put on an extravagant show in order to attract their animal pollinators. Thanks to these sentient "friends," the *real* advantage of animal pollination is that it allows for gene transfer between isolated plants of the same species that may be far apart; in such situations the statistical probability of effective pollination by the wind is utterly unlikely. Only animal pollination can be effective between individuals that may be half a mile apart—as within a tropical rain forest or a sparsely vegetated desert. Animal pollinators can maintain gene flow for species with scattered individuals, and this helps small isolated populations from going extinct (as we saw in the wolf story). But there were other more general results as well.

Having scattered plants over wide areas has another vital advantage: their enemies have trouble finding them. Parasites and predators need to locate their prey. Plants that are concentrated in dense groupings of a few species are easy targets for pests and parasites. But not when they are few and far between. Clearly, the central advantage of animal pollination is maintaining gene flow between populations of widely scattered individuals or clumps of individuals. Wind-pollinated gymnosperms with concentrated stands of fewer species must invest a lot of energy in defensive chemistry. Angiosperms that are more widely scattered may be able to invest less energy in defense and more on growth. Clearly, colorful flowers have played a major role in angiosperm success.

And Lastly, the Most Fundamental Reason for Being So Special

The most fundamental reason for flowering plants being special is simple: *all green plants are special*. They are special because green plants capture the energy of sunlight through the process of photosynthesis. *They use the Sun's energy to construct and reproduce themselves, and thereby provide the sustenance for nearly all the other living things on Earth!* In fact, it is the photosynthesis of green plants, algae, and bacteria, both on the land and in the sea, that sustains 99 percent of the living things that grace our planet. To help us understand this aspect of green plants, we'll have to do a little chemistry.[4]

First, let's get to the fundamentals: what really keeps us alive, awake, and running around at a regular clip? The answer to this question is simple: *it is energy*. The oxygen we inhale and the food we eat is "burned" within microscopic organelles within our cells called *mitochondria*. (Organelles are complex little structures that carry on vital functions within the complex nucleated cell.) It is the mitochondria (singular: mitochondrion) that act as little "fuel cells" within each cell. Here is where energy is extracted from the breakdown of our food; this is the energy that keeps us going. Most of our foods are carbohydrates, such as sugars and starch (which animals can transform into fat). Carbohydrates are complex molecules built of carbon, oxygen, and hydrogen. To gain energy our mitochondria tear these compounds apart, giving off carbon dioxide as a byproduct. But, after you've pulled carbon dioxide out of a carbohydrate, you've got something important left over: hydrogen. This is important because this hydrogen happens to pack a lot of energy. To capture that energy, hydrogen atoms are transferred through a series of enzymes within the mitochondrion to be united with the oxygen we breathe, in the final reactions of respiration.

What's especially interesting about this very basic aspect of our lives is that most people don't realize that the oxygen we breathe ends up as *water*—not carbon dioxide! As we've just pointed out, the first part of respiration rips carbon dioxide out of the energy-rich carbohydrate molecules; this provides us with a small amount of energy. It is the second step—combining the oxygen we breathe with hydrogen—that provides us with the majority of the energy that sustains us. A cascade of hydrogen atoms tumbling through the electron-transfer enzymes concludes in a union with oxygen atoms to produce water and vital energy for the work of the cell. Thus, nearly all the oxygen we breathe ends up as water. This process is called *respiration*, powering all living things built of larger cells with nuclei (eukaryotic cells); such cells are the building blocks of plants, animals, fungi, and protozoa-like creatures. But what is the ultimate source of all this food energy?

what makes the flowering plants so special?

Photosynthesis

Here's why green plants are so significant. Green chlorophyll molecules and their associated reaction centers use the energy of sunlight to *break apart* the water molecule (H_2O). The oxygen gets tossed away and ends up in our atmosphere. The hydrogen, together with carbon dioxide, is then built into energy-rich carbohydrates. These compounds can power the further creation of oils, fats, proteins, and other complex substances that make up the living organism. Splitting the water molecule is a central aspect of photosynthesis. If we humans could do this cheaply our energy problems would be over. We could "burn" hydrogen in fuel cells to generate electricity, and the only pollutant produced would be water.

Hydrogen and oxygen really like each other; that's why there's such a hot time when they get together (as in the *Hindenburg* dirigible disaster). The trouble is, once together, they don't want to come apart. Just as they give up a lot of energy when they get together, pulling the hydrogen and oxygen atoms of water apart requires a great deal of energy. (Keep in mind, boiling water vigorously *does not* separate the hydrogen from the oxygen atoms; all it does is transform liquid water into water vapor.)

Basically, photosynthesis rips apart the water molecule, while respiration puts it back together again. Here is a perfectly balanced set of reactions beginning and ending with water and carbon dioxide. The first part of this system, photosynthesis, requires a lot of energy to pull apart the water molecule. To drive photosynthesis, chlorophyll and its tightly structured molecular associates capture the energy of sunlight. This complex system, involving several pigments, absorbs the yellow, red, and blue portions of the Sun's spectrum to do its work. And because green light is *not absorbed*, the world of photosynthesizing plants looks green. Like little solar panels, green leaves capture the energy of sunlight as they inhale carbon dioxide and exhale oxygen, with water being sucked up from the roots—helping to build a richer world.

To put it more simply: green plants pull the water molecule apart through photosynthesis, and then unite hydrogen with carbon dioxide from air to build energy-rich compounds. We animals eat and digest those compounds and send their remains to the mitochondria within our cells, where energy-producing respiration takes place, which gives off carbon dioxide and water. Perfectly symmetrical, hydrogen and oxygen are separated by plants in photosynthesis and put together again in the process of respiration by both plants and animals. Because plants are nowhere near as metabolically active as animals, they use relatively little oxygen in respiration, and so there's a lot left over for the rest of the world. Best of all, this is a sustainable system of reactions, powered by the light of our favorite star.

The simple fact is that without photosynthesis, 99 percent of the life on Earth simply wouldn't be here. Whether trees and grasses on land, or microscopic life floating near the surface of the oceans, green plants give life the power to proliferate. And this is one of the reasons why flowering plants are so special; they are the power sources of most terrestrial ecosystems. Part of their importance is that there are so many flowering plants.

As noted earlier, of an estimated 300,000 species of land plants alive in the world today, about 260,000 are flowering plants. These are the primary species on our farms, in our pastures, and gracing our gardens. And they are the majority of species in much of the natural vegetation that surrounds us. However, that vegetation differs greatly in different parts of the world.

Flowering Plants around the Globe

Though flowering plants may be a dominant element in most of the world's vegetation, there are clearly many differing kinds of vegetation across the globe. Constrained by temperatures and rainfall, different regions of the Earth present us with a variety of distinct floras. (A flora is a listing of a region's plant species.) The most obvious examples are where conditions are most severe. In deserts, at the poles, and on the highest mountains, plant life becomes both diminished and diminutive. In fact, under really extreme conditions there simply is nothing that can be called vegetation. Arabia's Empty Quarter, much of the Sahara, tropical mountains above 16,000 feet elevation, and polar mountains at lower elevations comprise such areas. On Antarctica, the most significant "plant life" is made up of lichens, or lichenized fungi as they are now called. Often looking like nothing more than a dab of paint on a rock, these tough little life forms harbor photosynthesizing algae within a fungal matrix. Compared to about four hundred lichen species, and perhaps twenty mosses, only two species of flowering plants manage to survive in Antarctica. However, when we move to more equable regions, it is the angiosperms that predominate.

From a global perspective, the most obvious distinction in the world's vegetation zones is that between the tropics and the seasonally colder regions. One flowering plant family makes this graphically clear; it is the palms (Arecaeae or Palmae). No palm can survive a really severe cold spell, and only a few can survive a mild winter. In the southeastern United States we have palmetto palms growing as far north as coastal North Carolina. And in our southwest, *Washingtonia* palms live in protected valleys. But when you look out a window and see palm-lined avenues, you *know* you're in the tropics. A number of other striking monocotyledonous families belonging to the banana and ginger alliance (Zingiberales) show a similar absence in cooler climates. And this can be seen under certain circumstances in the tropics as well. Ascending through evergreen vegetation on a tropical mountain, you will notice that banana-like species and their relations become very few in number around 5,000 feet above sea level, and by 6,500 feet (2,000 meters), they are gone. An interesting exception to this pattern is enset (the false banana, *Ensete ventricosum*) of Ethiopia. These large banana-like plants are grown at 8,000 feet elevation for the production of starch in their corms. But that's exceptional. In fact, it is the banana-like and gingerlike plants, together with the palms, and their foliage that give the understory of the rain forest much of its distinctive appearance.

Speaking of bananas and their relatives, we should note how they've assorted themselves around the tropics. (Bananas and their relatives have had their family relations revised in recent years but all are in the ginger order: Zingiberales.) The genus *Heliconia* (the lobster claw plant) is part of this alliance, but is confined to the Americas and a few Pacific islands. The bird-of-paradise flower and its congeners (*Strelitzia*) are confined to Africa, but the closely related but highly distinctive traveler's palm (*Ravenala*, which is not a palm at all) is a native of Madagascar. The true bananas (the genus *Musa*) were probably restricted to southeastern Asia before humans began planting them around the tropics. These are very distinctive genera. Each was originally nicely confined to its own geographic region. Few plant alliances have their genera so neatly arranged around the world.

All the plants we've just mentioned—the palms, the bananas, and the gingers—are monocotyledons. A number of dicot families are also tightly restricted to the tropics. Some of the most distinctive trees in tropical forests belong to the Brazil nut family (Lecythidacae), the cacao family (Sterculiaceae), the dipterocarp family (Dipterocarpaceae), and the baobab family (Bombacaeae). None of these families has species living in colder climates. However, most larger families of flowering plants, well represented in the tropics, do have genera that live in northern climates. In the northeastern United States the buttonball bush (*Cephalanthus occidentalis*) is our only woody member of the coffee family (Rubiacae)—a family with literally thousands of trees and shrubs living throughout the tropics. And there are, of course, counterexamples; a few families have the majority of their species in temperate and subtropical climates. The oak or beech family (Fagaceae) has most of its species either in the temperate zone or on higher mountains in a few tropical areas. The willow and cottonwood family (Salicaceae) is another that has most of its species in northerly regions.

Just as there are some families with clear predilections for hot or cool climates, there are families that prefer particular kinds of habitats. (By habitat I mean the plant's usual location and surroundings, such as swamp, forest understory, open grassland, or windy mountaintop.) Though ranging over a great variety of habitats, the blueberry and rhododendron family (Ericaceae) has an affinity for acidic soils. This family ranges from the tundra and boreal forests of the far north to mountainous areas in both temperate and tropical regions, wherever the soils are acidic. Cacti (Cactaceae), with their barrel-like forms and succulent stems, can't compete for sunlight in anything but dry open landscapes. The only cacti you'll find in the rain forest are slender-stemmed epiphytes high in the canopy. Cacti, incidentally, are an essentially American family; only one species is native elsewhere, in western Africa. Thus, around the world, we have a combination of factors, both environmental and historical, that determine which plants thrive where.

After temperature, rainfall is the most critical determinant of vegetation cover. In the tropics, an evergreen broadleaf forest usually requires more than five feet (1.5 meters) of rain evenly spaced through the year. With a pronounced dry season, the tropical forest can range from being partly deciduous to completely deciduous or forming open woodland. Less rain and we venture into the world of open savanna, grassland, or subdesert thornbush. We have similar moisture ranges in cooler areas;

from temperate rain forests (as on the Olympic peninsula) to broad-leaved deciduous forest, and from open conifer forests to grasslands. Closer to the poles, or higher on the mountains, where temperatures get really cold, the conifers often become the dominant tree cover. Nevertheless, flowering plants make up virtually all of the herbs and shrubs in these cold habitats as well. But then, it isn't just climate and plant preferences that determine the composition of a flora. There's another very important factor, and that is history.

A Biogeographic Disjunction

When Asa Gray, professor of botany at Harvard University in the mid-1800s, began studying recently shipped plant specimens from Japan, he noticed something very unusual. He was familiar with many of the Japanese specimens because they were genera that also had species in the eastern United States. After further study something seemed quite astonishing: our eastern plants seemed more similar to the plants of Japan than to those of the western United States! From Professor Gray's studies, the plants of Oregon differed more from the plants of our eastern forests than did those of Japan, even though the plants from Oregon shared the same continent as our eastern flora. This pattern received further confirmation over succeeding decades as new plant collections were also made in China. Eastern Asia and eastern North America share a number of unusual genera found nowhere else on Earth. Our tall and stately tulip tree, *Liriodendron tulipifera*, has only a single relative, *Liriodendron chinense* of central China. Our sassafras, *Sassafras albidum*, with its distinctive foliage—some leaves resemble mittens in outline—has two closely related species; one grows in China, the other in Japan. Our earlier acquaintance, the skunk cabbage, is part of this pattern, belonging to a genus whose only other species grows in eastern Asia. The ginseng genus (*Panax*) has a group of species in eastern North America and a larger group of species in China. The sweetgum genus (*Liquidambar*) and the witch hazel (*Hamamelis*) also fall into this pattern, and there are more. Again, what's especially noteworthy is that these particular genera are found nowhere else on Earth. And that's not all.

It turns out that alligators are found in only two areas on our planet, the southeastern United States and in southeastern Asia. The strange paddlefish of the Mississippi River system belongs to an old lineage with only one surviving relative; that relative swims in China's Yangtze River. One of the very strange animals of the northern Ozark region and Ohio River drainage is the hellbender (*Cryptobranchus alleganiensis*). An unusual amphibian, one to two feet long, this animal looks as if it got run over by a truck. And again, this animal has only one close relative living in the world today, in southeastern Asia. So what's been going on here?

What we've just been discussing is called the "eastern Asia–eastern North American biogeographic disjunction." These closely related plants and animals are separated not just by the Pacific Ocean but also by western North America and much of central North America. This is a huge separation (disjunction) in range. Fortunately, fossil evidence provides telling data regarding this enigma. Fossils suggest that most of these plants and animals were widespread around the Northern Hemisphere tens of millions of years ago. What appears to have happened is that the ice ages of the last two million years (the Pleistocene epoch) exterminated these creatures over much of their former areas. With drastic cooling during glacial episodes, animals and plants in Europe suffered major losses. Blocked by the Pyrenees, the Alps, and the Balkans, European species could not migrate southward. (Recall that plants can migrate by seed dispersal over time.) In contrast, plant and animal species could move southward in North America, at least as far as the Gulf of Mexico. (With the earlier rise of the Rocky Mountains, the interior of central North America became dry and inhospitable to these moisture-loving species.) In eastern Asia the story was very different. First of all, the ice sheets in northern Asia were much smaller in extent than those in Europe and North America. More important, plants and animals had no impediment there to southward migration. Thus the greatest number of northern species to survive the ice ages is found in eastern Asia, with a lesser number surviving in eastern North America, and far fewer in Europe. Differential extinction during the ice ages explains the unusual eastern Asia–eastern North American disjunction. Here's a clear example of how history has determined the mix of plants and animals found in a particular region of the world today.

Interestingly, there is a major division in the animal faunas of southeastern Asia and those of New Guinea and Australia. That narrow spatial division lies east of Borneo and west of New Guinea, with monkeys and leopards on the western side and kangaroos and other marsupials on the eastern side. In this case the historical explanation involves plate tectonics. (A major revolution in earth science, plate tectonics sees the surface of our planet divided into plate systems that can move and abut each other.) As the Australian plate moved from a south-temperate location northward, it contacted the Asian plate, bringing two very different faunas into close proximity. Plants, however, do not exhibit this marked contrast as clearly as do the animals. Thanks to seeds and spores, plants travel more easily over open water than do most animals.

Australia's northward movement, however, had other devastating effects. As that island continent moved from a zone of cool temperate moisture into a much drier subtropical seasonality, many moisture-loving species became extinct. Something similar appears to have happened to Africa, with a flora much poorer in species than comparable tropical areas. For example, there are about three times as many species of palms living on the island of Madagascar as there are found in all of Africa, despite the fact that Africa is fifty times larger in area. Tectonic movements are not involved here. Instead, it seems that Africa has suffered from severe dry periods over the last few million years and, consequently, lost many species. This leads us to the question: where do we find the regions of the highest numbers of plant and animal species, or "biodiversity" as we like to call it?

One thing is clear, for high species numbers you must go to the tropics. Habitats in the cooler temperate zone simply cannot support as many species. The second factor for promoting species numbers is having habitat diversity, and that means mountains. And when we add abundant rainfall to the mix, we've got a surefire way of finding a biodiversity "hotspot." Little Costa Rica in Central America, with about half the area of Ohio—on a flat map—has twice as many species of ferns as all of North America north of Mexico, and twice as many species of mammals as well. Divided by a number of mountain ranges, and with rainfall ranging from three feet to fifteen feet of rain per year, Costa Rica has about 9,000 species of flowering plants. All of North America north of Mexico is estimated to have about 17,000 species, scarcely twice as many as little Costa Rica. And when one considers the largest mountain range in the tropics, the South American Andes, numbers escalate even further. There are far more species on the rugged Amazonian slopes of the Andes mountains than there are across the Amazonian lowlands. The reason is simple; complex mountain topography with many isolated valleys and peaks afford plenty of opportunities for habitat specialists (plants and animals that live only in very specific environmental circumstances).

Species numbers are not the only way of measuring floristic richness. There is another way of looking at species' hot spots, and that is by counting the "endemics"—the plants or animals that live nowhere else on Earth. Using this criterion, tropical islands rank high. Among these are the Hawaiian Islands, the most isolated island group in the world. Our concept of islands with many indigenous species includes some of the largest islands. Madagascar, New Zealand, New Guinea, and Borneo all have high percentages of plant species uniquely their own. Beyond islands, unusual climatic conditions can produce floras with high numbers of endemic species. Though seasonally dry, the Mediterranean floras we mentioned earlier are also rich in endemic species. All told, flowering plants have given us a world richly diverse in floristic regions as well as in numbers of species.

Now that we've seen how special flowering plants really are, and how they've distributed themselves around the world, we can begin to discuss how they actually changed the world. Let's start with our favorite subject: our own species and those animals most closely related to us.

chapter 7

primates, people, and the flowering plants

The advent of the angiosperms introduced vital new resources into the world's biosphere. Here were plants that produced fragrant flowers and much nectar, together with a huge variety of nutritious fruits and seeds. Flowering plants probably started as smaller herbs and little shrubs, growing in moist tropical habitats. Once they learned how to build trees, they became an important part of forest and woodland, at first in the tropics and later in cooler climes. With flowering plants expanding over the globe, several animal lineages also proliferated—especially the insects. Thanks to flowering trees, many more insects became active and numerous in the treetops.

Nectar and pollen, foliage, fruits, and seeds all provided fodder for an increasing number of *bugs*. Here then, with the expansion of angiosperms, was a greatly enlarged resource for those who enjoyed eating bugs. As a consequence, treetops now became home to a wider variety of hungry insectivorous animals, including birds, reptiles, mammals, and even amphibians. Among these many different creatures there emerged a very distinctive lineage of smaller mammals: the primates.

Flowering Plants and Primates

Fossil evidence of teeth and jaw structure indicates that the earliest primates were insect eaters. Today the primates include the lemurs, bush babies, monkeys, apes, and ourselves—an unusually smart lineage of mostly climbing mammals. It seems likely that a treetop proliferation of insects inspired the earliest primates to pursue their prey high above the ground.

Since foliage, flowers, and fruits were largely confined to the ends of branchlets, that's also where most of the bugs would be. And to reach those bugs, primates evolved longer limbs with grasping digits to hang on tight. Luckily, the earliest primates had retained all five digits on both hands and feet—very useful for clambering around in the treetops. With time, claws became transformed into more useful nails, and the grasping surfaces of hands and feet developed a ridged "friction skin" that could hold tight to smooth stems. (These little ridges give us our fingerprints.)[1]

Over many millions of years the flowering trees responded

to the presence of their new arboreal residents by providing monkeys with something to complement, or even replace, their buggy diet: larger nutritious fruits. The monkeys, in turn, developed flexible wrists and nimble fingers for the careful examination of such fruits; all primates use hand-to-mouth feeding after careful examination. They also developed better color vision to find and distinguish colorful ripe fruit within the leafy greens of the canopy. Advanced primates have *three* sets of color-sensitive cells in their eyes, allowing for the easy discrimination of many subtle color differences. We humans share these three classes of cells, each with a slightly different spectral sensitivity, allowing us to distinguish an enormous range of hues.

In addition to color vision, climbing around in the treetops and jumping from branch to branch demanded better three-dimensional vision. This gave rise, over time, to a flatter face, with eyes pointing in the same direction. Now the primate brain had to deal with two overlapping fields of vision and translate that information into an accurate representation of depth and distance. Unfortunately, with both eyes pointed forward, peripheral vision was reduced. Monkeys could no longer see as widely, which made them more vulnerable to predators, such as eagles, arboreal cats, and snakes. Primates responded to this problem by living in small troops where extra pairs of eyes can survey all directions. In turn, these larger groupings resulted in more complex social interactions. The demands of climbing around in a complex three-dimensional treetop environment, together with a more complex social life, required a bigger brain. And that's why monkeys got so smart.

But building a bigger brain is expensive. The cost of having a bigger brain is reflected in many aspects of an animal's life and lifestyle. By comparing small mammals that share the same body weight but have different brain weights, scientists have discovered significant generalities. A rabbit may weigh the same as a small monkey but its brain is much smaller and its life expectancy is much shorter. Because the small-brained rabbit is "cheaper-to-build" and requires relatively brief parental care, female rabbits produce more babies with each pregnancy and have litters more frequently than is the case with monkeys of equivalent weight. Not-so-bright rabbits keep ahead of extinction with a high reproductive rate, balanced against a short life span and a high predation rate. Despite their lower reproductive rate, clever monkeys fend off extinction with longer individual life spans and a reduced predation rate, thanks to a bigger brain and, for most, a treetop habitat. Monkeys use about 9 percent of their resting energy to support their brain activities, compared to about 5 percent for most other mammals. Clearly, carrying around a bigger computer requires more energy. All told, the primates have become the "brainiest" lineage of smaller mammals—all thanks to the flowering plants.

Flowering Trees and Our "Swinging" Ancestors

Moving into treetops was the first step in a continuing trajectory of getting smarter. For primates, the second step came about 20 million years ago with the appearance of the apes, a new lineage of primates with even larger brains. The living apes include the gibbons, orangutan, chimpanzees, and gorillas. These animals are larger and heavier than most monkeys; they have broader shoulders, a flatter chest, long arms, flexible elbows, and strong wrists. They also have a stronger back and legs to support a more upright posture. And, most notably, they've lost their tails! Some of these structural body features appear to be correlated with the business of *swinging* through the treetops (called *brachiating*). If you haven't seen gibbons swinging through the branches, whether in a zoo or in nature, you've missed something very special. Their rapid travel through the branches is a swift ballet in three dimensions. Our own broad shoulders, highly rotatable arms, and long fingers are the legacy of a swinging ancestry. Without such flexible arms and wrists, we humans could not carry, throw, manipulate, or gesture in the many ways we do. Surely, without that "swinging" ancestry we simply wouldn't have such versatile forward appendages, and, here again, I would argue, that flowering plants have played an absolutely critical role.[2]

Most angiosperm trees have broadly spreading branches in their upper halves; not down below like so many conifers. Not only do the conifers lack flowers and such a variety of tasty fruits, few have widely spreading branches in their upper crowns. Did you ever try to climb a pine tree? Not a lot of fun. To swing from branch to branch and tree to tree you need strong lateral branches that create an open but closely branched canopy. The flowering plants include many such trees, creating a broadly three-dimensional canopy of many branches. This is especially the case in evergreen tropical forests where primates live. Think of the spreading crowns of African acacias in open savannas; similar (but much taller) trees grow in tropical evergreen forests. Without such trees, our close relatives, the larger more upright swinging apes, would simply not have evolved. Not only did angiosperms foster the origin of primates, they helped create the important elements of our upper body that have made us so successful. Few other animals on Earth have arms as flexible as ours; arms that can be rotated at the shoulder rather like a windmill. Think about that. Neither your dog nor your cat, nor many monkeys, can rotate their forward appendages the way we can. It was our swinging ancestors who developed these useful abilities. And no animal has such strong and versatile hands as we do. These characteristics are another legacy of our swinging arboreal ancestry.

Then, once we came down out of the trees and began to walk on two feet, our wonderfully flexible arms were free! These arms, flexible wrists, and dexterous digits helped make us the world's dominant land animal. Now, using a few simple tools, we could throw, dig, pound, chip, and cut. Reminiscent of the bipedal carnivorous dinosaurs that ruled the world over 100 million years, bipedal humans now became the world's most dangerous predators. But why did we become two-legged? What ecological factors might have contributed to such a dramatic change? Here, I believe, was another stage in the human epic where flowering plants played a central role.

Flowering Plants and a Walking Primate

Flowering plants provided the habitats that helped us learn to walk. As climates cooled and dried over the last ten million years, grasslands expanded. Fossil evidence for grasslands first becomes apparent in South America about thirty million years ago. Grasslands developed a bit later in Africa and other parts of the world, and by ten million years ago were widespread in seasonally dry environments in both the tropics and temperate zone. In Africa, grasslands sustained the diversification of an impressive number of large mammals. These included a number of species of the horse lineage, many species of antelope and pigs, as well as buffalo, giraffe, rhinoceros, hippopotamus, and elephant. Those were the larger vegetarians. Lions, leopards, hyenas, wild dogs, jackal, and smaller cats made up a rich array of carnivores. Despite long and severe dry seasons, grasslands feed many more species of large mammals than do equivalent areas of evergreen forest or deciduous woodland. Though rain forests may support a huge number of different animal species, most of these are insects and small vertebrates. Despite lower rainfall and more limited species diversity, it is in the grasslands that you find lots of large herbivores. If our ancient ancestors were wondering "Where's the beef?" they found it in the savannas and grasslands of Africa. And that "beef" was of paramount importance to our ancestors. Animal protein and fat are hugely nutritious, just what you need to sustain you over long treks or for nursing your baby. Rich animal food helped humans build a larger brain, a brain that increased threefold in volume between 3.5

million years ago and 50,000 years ago. Lacking a rich diet of animal fats and protein, it seems unlikely that we would have become as brainy a species as we have.

Our human lineage first lived in moist evergreen forests, just as our closest living relatives do today. Both chimpanzees and gorillas are large animals in evergreen forests, and they spend considerable time on the ground using their arms to "walk on their knuckles." Subtle aspects of our own wrist and arm bones suggest that our ancestors may have walked in a similar way—before we became upright and two-legged. But what might have made our ancestors become two-legged? The most reasonable scenario posits that the upright, bipedal posture was an effective adaptation for making a living in open woodlands, thornbush, and grass savannas during times when forests were shrinking and drier vegetation was expanding. While great areas of tropical rain forest and evergreen forest remained in central Africa, the eastern flank of the continent was rifting and rising. These geological changes, together with changing rainfall patterns, expanded drier vegetation over wide areas of eastern Africa. With evergreen forests becoming more restricted, our lineage got up off its knuckles to explore drier, more open vegetation. Walking on two legs allowed us to range over greater distances and gave us a better view of our surroundings; it is also reasonably energy efficient. An upright posture also reduced our exposure to the hot tropical sun overhead, and exposed us to cooler breezes. Though food may have been scarce and more widely scattered in the grass savannas, those many mammalian herbivores were surely a terrific source of potential nourishment. Leaving dense moist evergreen forest had an additional advantage: we would be plagued by fewer diseases and parasites in a drier environment.[3]

Because of its varied topography, considerable elevation, and north-to-south axis, the geological rift system of eastern Africa allowed vegetation to move up and down altitudinal slopes as climates got hotter or cooler, moister or drier. Here was a corner of the world where the dramatic climatic cycles of the last few million years—the ice ages—were least severe. Here was the part of the world where our ancestors became erect, wandered out into drier vegetation, and soon began building a bigger brain. As we just mentioned, freed arms and hands—that weren't busy holding us up—produced a host of new possibilities. Making tools, throwing rocks, carrying food, and building shelter was a lot easier with free arms and hands. But the most important new activity may have been subtler: *gesturing* to our comrades with hands and fingers. Gesturing silently to comrades while hunting may have helped early humans bring home precious protein. Later, gestures, augmented by vocal utterances, provided the start-up system for learning how to talk. And walking upright helped develop that talent. Now, breathing was freed from the rhythmic gate of one's forward legs. That greater freedom in breathing meant that we could use our vocal cords for making more noise, more often. This may have been fundamental to acquiring our premier talent: language.

For small groups of two-legged primates feeding on both plants and animals, the diverse landscapes of seasonally dry Africa were the ideal environment to explore a new lifestyle. Hills, plains, mountains, and river channels along the east African Rift system afforded a wide variety of habitats. Even in dry areas, rivers were edged with evergreen forest, while valleys and floodplains were supporting open woodlands. Rocky slopes were dominated by small trees and thorn scrub; broad flat or gently undulating landscapes supported fire-prone grassland and savanna. In such dry landscapes, you will rarely encounter ferns, conifers, or cycads; virtually the entire terrestrial biomass of these dry habitats is made up of flowering plants. Overall, Africa has about 40,000 species of flowering plants, and it has been estimated that, of these, around 5 percent offer some nutrition for humans. Needless to say, on a continent with only 10,000 species, 5 percent might not have been sufficient to sustain early humans. And even with a rich flora of flowering plants, local environments often had a limited variety of tubers, fruits, and seeds. Also, such resources were not in abundance throughout the year. However, as we've already noted, these same open grassland habitats did have a form of nutrition of the highest quality: larger mammals, and lots of them. It was in such a world, supported almost entirely by the flowering plants, that a lineage of omnivorous two-legged primates became human.

Nevertheless, and despite having larger brains and expanding our range across Eurasia, the human lifestyle continued to be a very risky proposition. Though smart, modern humans (*Homo sapiens sapiens*) remained hunter-gatherers living in bands of fewer than a hundred mature individuals, just as their ancestors had over several million years. Then, about 30,000 years ago in Europe (and perhaps as much as 80,000 years ago in Africa), there is evidence that our ancestors began making carefully crafted new tools. Fishhooks, needles, and stone spear points were being fashioned for the first time in the history of our planet. But still we were small wandering bands of few people. Then, quite suddenly, between about 10,000 and 7,000 years ago, and in several different corners of the globe, our species made a major cultural advance.

Flowering Plants and People: The Beginnings of Agriculture

Beginning about 100,000 years ago in Africa, and considerably later elsewhere, humans were making their presence felt in the landscape. Larger mammals and flightless birds were going extinct in close synchrony as our ancestors' innovative hunting technologies swept across the world. Humans reached Australia by about 50,000 years ago, the Americas by 13,000 years ago, Madagascar and the distant islands of the Pacific around 1,500 years ago. During each of these *invasions*, we managed to exterminate some very impressive animals. Beginning with a few larger mammals in Africa, following along with mammoths and mastodons in the Northern Hemisphere, we moved on to rid Australia, New Zealand, and Madagascar of their giant flightless birds. Carefully crafted *stone* spear points, and the spear-thrower, made this expansive onslaught possible.[4] But despite new hunting techniques, most habitats could support human bands of only 30 to 150 people—not much different from a band of baboons.

The problem is, human beings require lots of high-quality food—and we need it *all the time*. Despite our extraordinary minds, having to procure sustenance in seasonal and unpredictable environments meant that starvation was a constant threat. Our big brains burn up 20 percent of the energy needed to sustain us—even while we sleep. That requires a great deal of fuel! Interestingly, fossil evidence suggests that our big brains haven't been getting any bigger over the last 50,000 years. Rather than enlarging, it seems that our big brains were now developing a new evolutionary dynamic: progressive cultural change! We were now elaborating new tools in ways never seen before, communicating more effectively, and expressing ourselves in artful ways. Finely fluted spear points, carefully carved fishing hooks, and needles were early expressions of this new creativity. Also, we were learning more about our environments and sharing that knowledge through sophisticated language abilities. Then, we took that knowledge of our local environments and its many plants and animals and did something really special.

Quite suddenly, and in several different areas of the world, we humans selected a few species of plants and animals to become our close partners in the business of staying alive. We call these new cultural innovations *agriculture* and *animal husbandry*. By developing a special relationship with a number of plants and a few animal species, we now produced something extremely important: more to eat. And not just more food, but food that could be held over for bad times. Wandering over a landscape, trying to find something to eat *all the time*, was a never-ending challenge for our early ancestors. Humans can't eat grass or leaves and hope to survive. We need "high-quality" foods with many easily digestible calories. In most environments such food is scarce and scattered, and, even when abundant, it is usually seasonal. Berries, seeds, nuts, green-and-tender shoots, all have the habit of disappearing quickly or becoming indigestible as they age. Worse yet, in bad years these menu items may not show up at all. Fortunately, most landscapes include another form of food of the very highest quality. Unfortunately, this source of nutrition is alert, elusive, and not easy to catch. These especially delicious calories (mostly fish, birds, and mammals) also become scarce with drought or the onset of winter. Thus, for wandering hunters and gatherers, starvation was a constant threat. By discovering ways to nurture a few special species of plants and animals, we humans suddenly expanded our ability to sustain ourselves. But interestingly, we humans were not our planet's first farmers.

The First (Six-legged) Farmers

Humans were not the first agriculturalists. That distinction goes to the world's most diverse lineage: little six-legged creatures we call insects. And not just one but three different groups of insects developed "farming" practices. Among them is a lineage belonging to the world's most species-rich order, the beetles. Related to the weevils, a few genera of ambrosia beetles excavate tunnels into the wood of trees where they tend special strains of fungi on which they and their larvae feed. A second group of "farmers" are a specialized lineage of termites that eat wood and plant debris, and then use their droppings to cultivate a special kind of fungus on which to feed.

The third and last group of insect agriculturalists produces the most spectacular display. These are ants that bring plant materials into their nests, and cultivate fungal symbionts on the decaying plant material. Most of these are little ants with small inconspicuous underground nests, but a few are really impressive. Some of these "leaf-cutters" build underground nests more than fifteen feet in diameter. They march considerable distances across the forest floor to reach a variety of trees, and then climb high into the canopy to cut sections from the leaves. Long columns of these ants can be seen marching their leaf sections, looking like little green banners, down from the canopy. Witnessing the wiggly procession of hundreds of leaf fragments hurrying inexorably toward the nest gives one a sense of determined purpose like few other scenes in the world of nature. I had no idea how significant these little animals were to the dynamics of the rain forest, or how long such colonies might be active, until I noticed that these often-repeated "long marches" had actually created clear, debris-free little avenues, two to three inches wide, across the forest floor! Who would have thought that the many little footsteps of ants might create visible corridors through the debris at the bottom of a rain forest?[5]

Each of these three forms of insect agriculture is based on the cultivation of special strains of three unrelated fungi. And because each of these three farming lineages is related to allied insect species that *do not* farm fungi, we can infer that all three cases arose independently. The termites appear to have begun their fungi-tending activities in Africa, and moved outward to Europe and Asia with their symbionts. The leaf-cutters are restricted to the American tropics, while the tunneling beetles are more widespread. But what's also especially noteworthy is that all these "agricultural systems" farm a single species; they are *monocultures*. You may have heard criticisms of our own use of monocultures as being unnatural and vulnerable to pathogen or parasite. But here are similar systems run by "Mother Nature" herself. How have these insects dealt with the problems of pathogens attacking their fungi? Surely, if pathogens or parasites can attack our monocultures, they'll figure out a way of attacking other systems as well.

We haven't examined these other systems carefully enough to know what's going on in each of them. But in the case of the more studied—leaf-cutters—there is indeed a nasty little fungus that attacks the cultivated fungus of the leaf-cutters. When this parasite gets too numerous, the ant colony is doomed. How might the ants combat this pest? It turns out the ants also support the growth of a *symbiotic bacterium*. This bacterial species doesn't bother the cultivated fungus, but does attack the parasite of the cultigen! Here is a great solution to the problem of relying completely on one species for your sustenance. The bacterial species, because it is a living lineage, can change its genetic characteristics and adapt to the changes that might arise in the pathogenic fungal pest. We humans are doing something similar, as we maintain genetic diversity for our crops with the aim of providing plant breeders the wherewithal to combat new or modified plant pathogens. But enough about insects, fungi, and pathogens. Let's take a closer look at how flowering plants and people came together to create agriculture.

The Origins of Human Farming

The exact way in which humans first domesticated plants and animals has been the subject of much conjecture and many theories. An increasing number of careful archaeological studies now suggest that settled farming villages developed in differing ways under a variety of environmental conditions, and over long periods of time. People in at least five parts of the world appear to have developed agriculture early and independently—in the Middle East, China-Southeast Asia, New Guinea, Mesoamerica, and South America. They did this in areas with rather different environments, over a period of perhaps less than 4,000 years. Two general factors have been invoked for explaining these distant but concurrent trends; these were: (1) climate-forced environmental changes at the end of the last glacial cycle; and (2) ongoing cultural advancement, as small seasonally settled communities came under pressure to acquire more reliable food resources.[6]

The earliest generally accepted dates for plant domestication have recently been documented for rice in the central Yangtze valley of China. Here, archaeological evidence shows a transition from a limited diet of wild rice about 12,000 years ago, to a diet rich in cultivated rice by 10,000 years ago.[7] In contrast, the earliest substantial evidence for *settled agriculture* in the "Fertile Crescent" of the Middle East is about 9,000 years ago. This crescent sweeps northward from Israel and western Jordan through Syria into and eastward across Turkey, then southeastward into the drainage basins of the Tigris and Euphrates rivers in Iraq. Here, einkorn wheat, emmer wheat, barley, peas, chickpeas, and lentils were among the earliest of our domesticated plants. However, a very long period of earlier plant gathering predated plant domestication in this region.

An exciting new discovery from deposits in the Sea of Galilee has shed new light on these earlier human activities. An analysis of starch grains adhering to a grinding stone identifies ground and baked barley. But what makes this find special is that the grinding stone was resting in a depositional stratum dated at around 23,000 years of age! Clearly, these people were gathering wild grains and using them in sophisticated ways 14,000 years before there is evidence for settled villages. Another recent archaeological study from this region indicated that early hunter-gatherers had been using a wide variety of small wild grass seeds as a regular part of their diet. Once the major grains had been domesticated, these other species were no longer used. In the case of the grains, domestication may have been a slow process of selecting those plants whose grain heads did not shatter (as wild species do in order to disperse their seeds). Thus, intact heads could now be harvested with minimal loss of seeds. Further selection resulted in larger plants producing plumper, more nutritious grains—plants that could survive only by human propagation. But why did plant domestication and the establishment of larger human communities occur at this particular time?

Other recent studies indicate that squash domestication

may have begun in Mexico as early as 9,000 years ago (4,000 years before maize became an essential staple). By 7,000 years ago, bananas and taro (a source of starch) were being cultivated in the mountains of New Guinea. All these studies provide evidence for settled agriculture—a big step beyond hunter-gatherers grinding and baking wild grain—and all within a period of about 4,000 years. We still have no satisfactory explanation of why people in far-distant corners of the world embarked on the business of agriculture within only a few thousand years of each other. Perhaps agriculture required a state of cultural organization that was first achieved only around this time.

In the Middle East, animals may have been domesticated before the first local grains became regularly tended field crops. Animal domestication might have been an easy cultural innovation for hunter-gatherers, traveling with captured young goats and sheep. Because they lived naturally in close social hierarchies dominated by a single animal, goats, sheep, pigs, and cattle captured as young animals required little modification to become part of either a nomadic or settled human community. By around 8,000 years ago, sheep, goats, cattle, and pigs had become regular members of the human community. It now seems likely that these four species were first domesticated in different areas, but, once they became our partners, were quickly adopted by people living elsewhere.

Geographical "Centers" of Agricultural Innovation

Some specific areas of the world are distinguished by having a suite of important indigenous crops, and they are thought to have been critical areas of agricultural innovation. The Americas have four such areas. Mexico and Central America gave the world maize, tomatoes, grain amaranths, cacao, vanilla, avocado, papaya, and sunflowers (for their oil-rich seeds), as well as specific varieties of chili peppers, squashes, kidney beans, and cotton. The high Andes are the original home of potatoes, oca, ulluca (all are root crops), quinoa (a grainlike seed), and lima beans. From lower elevations, South America gave us cassava (manioc), sweet potatoes, peanuts, pineapples, cashews, tobacco, and coca leaves (the source of cocaine). And paralleling the innovations in the north, South America developed local varieties of cotton, chili peppers, maize, and beans. Finally, and at a considerably later time, the Mississippi and Ohio valleys of the United States witnessed the domestication of the marsh elder and native goosefoot, together with new varieties of maize, pumpkins, and sunflowers.[8]

Similarly, there were a number of significant areas of local domestication in the Old World. As noted, the earliest areas were the Fertile Crescent of the Middle East (southwestern Asia) and China. Tropical southern and southeastern Asia were also important as the original locales for the domestication of certain varieties of rice, plantains, mangos, many citrus fruits, cucumbers, black pepper, Old World cotton, eggplant, yams, and many kinds of beans. From a little farther away, New

Guinea and the western Pacific gave us coconuts, taro, bread-fruit, wing beans, sago palm, sugar cane, and bananas. Africa, south of the Sahara, gave us pearl millet, African rice, sorghum, cowpeas, indigenous yams, okra, the oil palm, and a variety of melons. The high mountains of Ethiopia are the exclusive home of teff (the world's smallest cereal grain), ensete (a banana relative with starchy base), noog (an oilseed), coffee, and distinctive varieties of sorghum and barley.

Areas of the Old World also gave us our earliest domestic animals, including goats, sheep, pigs, chickens, cattle, donkeys, and water buffalo. Horses and camels came a bit later. As was the case for certain plants in the Americas, particular wild animal species may have been domesticated independently in different regions of the Old World. Recent DNA evidence suggests that cattle may have been domesticated independently in three regions: India, the eastern Mediterranean, and the Sahara —before it became the inhospitable desert it is today. Likewise, pigs were probably domesticated both in western Asia and in southeastern Asia, just as rice may have been developed both in South Asia and in China independently. Even before the advent of agriculture, perhaps 15,000 years ago, dogs had become our close comrades in several areas of the Old World. Dogs were not only our partners in hunting, they also became sanitary engineers and security guards once larger villages were established (and in many cultures, part of the dinner menu). And after Middle Eastern people began storing large amounts of grain, rodents became a problem and cats became our friends.

What might have been the factors that fostered such "regional centers of agricultural innovation"? Climate surely played a role; most all of the areas mentioned above are seasonally dry or seasonally cold. A nongrowing season "rests the soil," reduces erosion, limits pathogens, makes harvesting and replanting easier, and gives the populace time to build and sustain a more complex society. Such societies, stable over a longer period of time, could bring together useful species domesticated over a much wider region. (In fact, these may have been centers of agricultural concentration, and not centers of actual agricultural innovation.) Another important factor regarding these major centers was their specific *geography*.

The people of the eastern Mediterranean and Fertile Crescent, for example, were surrounded by major mountain chains with a hugely rich flora. The many montane habitats provided a wide array of plants with useful qualities, and some tasty animals as well. Also, because they were part of the same Eurasian supercontinent, and had similar day lengths, they were able to acquire plant and animal domesticates from other cultures across this wide region. It was in this region that the early farmers of the Fertile Crescent first grew wheat, barley, lentils, peas, chickpeas, onions, and flax (for oil and fibers). A little later they were growing olives, date palms, figs, and grapes (for both raisins and wine). Within a few thousand years the peoples of the Mediterranean and western Asia also had asparagus, beets, cabbage, celery, cucumbers, lettuce, radishes, spinach, and turnips among their foodstuffs. Their fruits included peaches, cherries, apples, plums, apricots, and pomegranates. And for flavoring, there was anise, basil, bay leaves, capers, coriander, cumin, fennel, garlic, leek, mint, mustard seed, rosemary, sage, thyme, and sesame seed. These plants, coming from a variety of habitats and regions, greatly enriched the lives of Near Eastern civilizations.

Other centers of plant domestication around the world had similarly rich stocks of vegetables and flavorings. But here's the main point: *every* plant we've just listed is a flowering plant! Whether in the Middle East, India, China, Mexico, or the high Andes, the story was the same: flowering plants gave us the wherewithal to build grand civilizations.

More importantly, these new human crops and domestic animals brought about a major change in the human epic, regardless of their geographic origins. Hunter-gatherer encampments had averaged about half an acre in area before the agricultural revolution. With the advent of agriculture, human settlements now averaged two acres in area or larger. Many parts of the world could now support ten times as many agricultural people as there had been hunter-gatherers before. A bit later, with carefully planned hydraulic engineering along great rivers, agriculture could support cities numbering in the tens of thousands.

Regrettably, agriculture also had serious negative consequences. Many people were now *shorter* than their hunter-gatherer antecedents, perhaps because their diets were too heavily dependent on the cereal grains. Both because of the crowding of people in villages, and close association with domestic animals, diseases ran rampant in a way that was impossible among sparsely scattered wanderers. Nevertheless, humans increased their numbers and developed more complex societies. Together with our new "symbiotic" partners, the human species now became a force that would change the face of the planet.[9]

The Basics: Grains, Beans, and Tubers

Energy keeps us alive and active. As mentioned earlier, it is within the little mitochondria inside our cells where respiration gives us the energy we cannot live without. With the invention of agriculture most of the energy for our mitochondria now came from the cereal grains. They give us—and our domestic animals—the majority of the energy that keeps us running. The most significant grains (all from the grass family) are: wheat, rice, and maize—in that order—as well as sorghum, barley, oats, rye, and the millets. (Actually, maize is number one in world production, but a good portion of that harvest goes to animal feed and other derivative products.) However, we need more than just energy to stay healthy; we also need critical elements and vitamins.

One of our most important nutritional necessities is *nitrogen*. We need this element to build proteins, nucleic acids, and other body components. This is the single most essential element in modern fertilizers; no other nutrient is so often deficient in soils that have been farmed for any length of time. But if about 79 percent of the air we breathe is made up of nitrogen, why is getting nitrogen such a problem?

Nitrogen in air is in the form of *molecular* nitrogen (N_2), not elemental nitrogen. Molecular nitrogen has its two atoms locked in a tight embrace. The problem is that when two nitrogen atoms get together, they unite with three strong covalent chemical bonds—and they like to stay that way. That's why aerial nitrogen is so unreactive, and useless to most life processes. To become available as nourishment for plants and animals, the two nitrogen atoms must first be ripped apart and then built into ammonia or nitrates. Pulling apart the N_2 molecule takes a great deal of energy. Fortunately, a few bacteria can do this; they "fix" atmospheric nitrogen by pulling it apart and building the solitary nitrogen atom into compounds available to other organisms. Today, we modern humans employ the "Haber-Bosch" process to do the same on an industrial scale, using catalysts together with high temperatures and high pressures. This process is the source of our commercial nitrogen fertilizers, helping feed many of the world's six billion people. In fact, through this energy-demanding technology, we humans are currently fixing about as much nitrogen each year as all the world's natural processes. Natural sources include lightning strikes, nitrogen-fixing microbes, and a very special group of flowering plants: the legumes.

Our premier food source, the grains, contains only modest amounts of protein, and they are deficient in lysine, one of the essential amino acids that we cannot make ourselves. Here is why the pulses—members of the legume family—have proven to be so vital. The pulses or legumes include some of our most valuable foods, such as soybeans, peas, kidney beans, lentils, string beans, lima beans, cowpeas, and peanuts. As in the grass family, it is the hard, dry seeds that are our most important foods. Legume seeds tend to be considerably larger than the grass grains and much richer in proteins and lipids. They are richer in proteins because legumes participate in a unique symbiosis with nitrogen-fixing bacteria.[10]

The roots of legumes have "nodules" containing special little organelles (called *symbiosomes*) and that's where nitrogen-fixing bacteria live. Here the bacteria are provided an oxygen-free environment where they can work their biochemical magic, transforming inert nitrogen molecules into available nitrogen. In addition, the legume plant provides special proteins (called *nodulins*) that help feed the bacteria. Unfortunately, all of this effort in the support of their symbiotic partners requires a lot of energy. It is for this reason that a field of legumes simply cannot produce the same yield of seeds as does a field of corn, wheat, or rice. By using the grains for "energy" and the legumes for proteins, cultures around the world have been able to obtain a more balanced and nutritious diet.

The significance of the legumes in human history rests on another factor as well. Not only are their seeds (beans) richer in proteins, legume plants themselves are richer in nitrogen than are grasses and most other plants. Thus, when they expire and decompose, legumes add nitrogen to the soil in which they grow. This fact was essential to maintaining soil fertility before the invention of industrial fertilizers. By rotating crops—one or a few seasons of grain and one season of legumes—early farmers were able to continue working the same land over thousands of years.

Like the cereal grains, our favorite "beans" are small, hard, and low in moisture. These qualities are crucial for long-term storage. How long can you store a tomato, an avocado, or an apple? Not very long. The low moisture content of grains and beans is a critical aspect of their usefulness. Stored in a dry place, away from vermin, they can be kept indefinitely. Together, the cereal grains and the legumes have been the basis of all the great civilizations. In addition, these two families—

the grasses and the legumes—are key elements in pasture management and feedstock production, providing us meat, milk, hides, and wool.

Root crops have also been essential staples in many cultures, especially in tropical lowlands where most grains do not grow as well. Although not all are roots in a technical sense, these crops include potatoes, cassava (manioc), sweet potato, taro, yams, turnips, and a variety of others. Because root crops contain more moisture than the grains, they are less concentrated in energy value and more susceptible to spoilage. They are also lower in protein content; but many have the advantage of being stored by simply leaving them in the ground, to be dug up when needed. The potato, developed in the cool highlands of the Andes, is now the world's fourth most important food crop. While people often speak of the great Irish famine, caused by the potato blight disease, few are aware that the human population of Ireland had *more than tripled* as a consequence of introducing potato cultivation to the Emerald Isle. Potatoes provided much more nourishment than did oats and rye in these cool northern climates. Until that devastating blight, potatoes and buttermilk provided the Irish with a very nutritious diet.

Clearly, agriculture and animal husbandry provided humans with more reliable sustenance in a world of unpredictable weather and rainfall.[11] Larger settled societies created an environment in which food storage and ceramic firing was followed by metalworking; and where religion was a precursor to literature, philosophy, and science. All these complex societies were based on the cultivation of not just plants, but *flowering plants*. Except for a few mushrooms and algae, fish and shellfish, *all the energy that sustains us comes from sunshine first absorbed by flowering plants.*

Agriculture: The Supreme Symbiosis

Agriculture gave humans the power to proliferate. Because agriculture is so commonplace, we no longer appreciate the ways in which it transformed humankind. Never has there been a major civilization without a grain, a bean, or a root crop to sustain its citizenry. This profound cultural innovation grew out of knowledge people had accumulated over an immense period of time. As noted earlier, starch grains adhering to an ancient grinding stone indicate that people in the Middle East had been grinding and baking wild barley 23,000 years ago. That's 14,000 years before the advent of larger settled farming in this same region. Clearly, centuries of accumulated knowledge preceded the development of settled agriculture. Nowadays we have universities and county agents to investigate and solve agricultural problems. But in times gone by, it was only the trial and error of local farmers that developed the know-how of what to do and when to do it. That knowledge continues to sustain large numbers of people in many corners of the world. Such knowledge had to be especially precise in regard to the best time to begin planting for the growing season.

In a strongly seasonal environment farmers are faced with a narrow "window of opportunity" in which they must plant their crops to be successful. Plant too early, and the northern farmer flirts with a late spring frost that destroys his seedlings.

Plant too late, and an early autumn cold snap may destroy the crop before it fully matures. Likewise, in the seasonally dry tropics, the farmer's crop may fail if planted too late and the rains cease before the seeds are fully mature. For these reasons, all the great agricultural civilizations made the annual calendar a central aspect of their cultures. Watching the sun and stars carefully, they tracked the seasons. Just how clever some native agriculturalists have been in the business of knowing when to plant is epitomized by a recent study from Peru and Bolivia.

For perhaps thousands of years, potato farmers, high in the Andes Mountains, stayed up all night during the evening of their shortest day, June 21. They waited through the frigid mountaintop evening to view the Pleiades as they rise in the east, just before sunrise. The Pleiades are a group of stars of varying brightness. These stars are not only brighter on a really clear evening, but it seems as if there are more of them. But why spend a frigid night waiting for the Pleiades to rise? The reason, the farmers gave, is to determine *when to plant* their potatoes. In normal years, they plant their potatoes early; and their knowledge of the seasons tells them when to do that. But in bad years, in El Niño years, the rains are unreliable and sparser. For such a rainy season the farmers should plant their potatoes *later* than normal. Watching the rising stars will tell them if such a year is imminent.

Attempting to understand these ancient practices, anthropologists, working with climatologists and with satellite data, recently tried to find a scientific explanation for what these farmers had been doing over centuries. Sure enough, during an El Niño year, very thin high-altitude clouds form in this part of the world, clouds that are so tenuous that they are not visible to the naked eye. But by looking through the atmosphere at the horizon from their high-elevation observation points, the farmers can see the effects of these clouds in the diminished brilliance of the Pleiades. During normal years the Pleiades rise clear and bright in the east; but in El Niño years the Pleiades are not so clear, and it appears that there are fewer of them. This is a signal to plant potatoes later in the year. These farmers have been predicting a diminished El Niño rainy season *five months in advance*! Only acute observation, good memories, and a persistent oral tradition are likely to account for such an extraordinary feat of environmental awareness.[12]

Modern scientific agriculture is continuing in the footsteps of these traditional farmers. We mentioned how bacterial symbionts help the legumes create nitrogen-rich seeds and tissues. But there are many bacteria that cause significant plant diseases and loss to crops. As noted in chapter 4, the study of plant pathology is devoted to countering plant diseases. The crown gall story proves to be an illuminating example of the importance of *basic research*. I had thought that plant scientists were wasting their time when they were trying to uncover the mechanism for crown gall disease of carrots. It seemed to me that they could just breed more resistant carrots to get around the problem. Funding for this research in the late 1950s was supported in part by the idea that we would learn something about cancer by studying these cancerlike growths in carrots. This seemed naive to me at the time; plant tissues and animal tissues are profoundly different—how could these studies help us in the battle against cancer? It had long been known that crown gall disease was caused by a bacterium (*Agrobacterium tumefaciens*). But the plant scientists got their funding and by the early 1970s discovered that what the bacterium did was to send a packet of genes (the *plasmid*) into the nucleus of a carrot cell, which would then grow and divide to form the crown gall. Not only did the gall provide a nice place for the bacteria to live, but the "transformed" gall cells produced compounds not formed by carrot cells—compounds that served as "food" for the proliferating bacteria. These bacterial genes, inserted into the nucleus of a carrot cell, caused the cell to create both the gall and to make compounds not otherwise produced by carrot tissues. This was a huge discovery. Plant geneticists have since learned how to send genes that we select into crop plants that we are trying to improve—via bacterial plasmids! Clearly, I was right about the cancer aspects of this research, but completely off base regarding the significance of the research. This is, surely, an outstanding example of how "basic research"—trying to understand crown gall disease—has indirectly led to an important and extremely useful discovery.

But there were significant exceptions. Plasmids work well for dicotyledonous flowering plants (like the carrot) but not so well in monocotyledonous plants, which include our most important grains. So, would you believe, plant geneticists have since developed a means for "shooting genes" into embryonic cells of grasses to try and improve our grain harvests? That's right; they placed genes on particles of gold and blasted them into plant cells. This may be a very inefficient way of doing things, but, when you can screen tens of thousands of seedlings, you may get the gene transfer you are looking for. Nevertheless, plasmids are still our most effective way of introducing foreign genes into the plants we want to improve or protect. Clearly, it was only after we understood how *Agrobacterium tumifaciens* transformed cells to produce crown gall that modern genetic engineering of crop plants really began. With these techniques we are able to introduce valuable new traits into virtually any crop. Some geneticists even dream of having a plant produce those specific molecules that initiate an immune response in humans. Such a bioengineered plant product might allow us to inoculate people in impoverished tropical countries without having to inject them with refrigerated serum. Clearly, the future of agriculture, despite critics and activists, appears to be one of extraordinary new possibilities.[13]

Flowering plants have been central to the evolution of primates and swinging apes, to the origin of bipedal humans, and to the development of agriculture. Today, flowering plants provide a bit more than 90 percent of our caloric intake; and they are the primary food for our domesticated animals as well. In 1990 it was estimated that we humans were raising 1,294 million head of cattle, 856 million pigs, and 10,770 million chickens—mostly fed with flowering plants.[14] Clearly, flowering plants are the foundation for larger human communities over the entire planet. Putting all these observations together, one can easily claim that *without flowering plants* we humans and our grand civilizations simply wouldn't be here.

But is it possible that flowering plants transformed terrestrial ecosystems much earlier, and maybe even changed the weather? Finally, let's find out how flowering plants have, quite literally, changed the world.

chapter 8

how flowers changed the world

With the advent of the flower, whole new levels of complexity came into the world; more interdependence, more information, more communication, more experimentation.

—Michael Pollon[1]

An interesting anomaly in contemporary biological science is the avoidance of any mention of "progress." Things develop, floras change, morphological complexity may increase, but nothing seems to *progress* in the literature of modern evolutionary science. During the middle of the nineteenth century there had been a lot of talk about progress. After all, these were the times in which the Industrial Revolution came into being; dramatic new inventions were creating profound changes in a dynamic progression that continues into the present. People saw progress everywhere. In addition, many people were eager to ascribe such changes to a divine plan or purposeful universe. Unfortunately, because the notion of progress had become entwined with predestination and divine purpose, scientists of the twentieth century avoided both the term and the concept. In addition, the notion of progress often implies better conditions or better adaptations, and such value-laden judgments are something most biologists wanted to avoid. The result has been a major gap in the research literature regarding life on our planet. Prominent paleontologists like Stephen Jay Gould and Niles Eldredge both authored popular books proclaiming that there were no real signs of "progress" in the fossil record.[2] I still haven't figured out why two such capable scientists insisted on contradicting a phenomenon that seems to be utterly self-evident, both in the world around us and in the history of life.

Certainly, our current land flora—festooned with rain forests, acacia savannas, and flower-filled prairies—is evidence of dramatic *progress* beyond the puddles of blue-green slop that represented the only life on land 500 million years ago. Life has become more complex and species have increased greatly in numbers over time. Surely, this is progress in any sense of the term. Gould and Eldredge, as most other biologists, insisted that natural selection *lacks* any inherent progressive drive. True enough; but that doesn't mean that competitive environments and special circumstances cannot create conditions that *foster the emergence* of progressive trends. Clearly, being pursued by faster predators or threatened by nastier diseases is going to result in *progressive* counteradaptations. Though not a direct product of mutation or survivorship, over time, progress appears to be a clearly emergent phenomenon among complex living things in richly diverse and intensely competitive environments.

One of the most obvious progressive tendencies to be found in the history of life is the increase in species numbers over time. We noted earlier Hughes's estimate of a nearly thousandfold increase in species numbers among vascular plants over the last 300 million years. The numbers among insects must be just as compelling. Greater species numbers foster more interactions, which, in turn, result in more complex ecosystems. More significantly, the escalation of species numbers on our planet's land surfaces over the last 60 million years seems to have been greatly accelerated by the flowering plants themselves. Here was a progressive lineage like no other. Certainly, bird and mammal species may display many "progressive advances" beyond their reptilian antecedents, but they do not comprise a quarter of a million species. In addition, neither millions of insect species nor thousands of land vertebrates can *sustain* an ecosystem. Green plants do, and the flowering plants have done it in spades! Let's take a closer look at what they've done and how they did it.

How Flowers Made More Flowering Plants

As noted earlier, the *real* advantage of animal pollination is that it allows for gene transfer between isolated plants of the same species that may be far apart—where the statistical probability of effective pollination by the wind would be unlikely. Also, we argued that animal pollination could counteract gene loss in small isolated populations. But we did not focus on how animal pollination itself might have increased diversity within ecosystems.

Conifer trees (including pines, firs, cedars, and their relatives) are not flowering plants, and nearly all are pollinated by the wind. The problem with wind pollination is that it requires many individuals *of the same species in the same area* to be effective. That's a strong deterrent to diversity. More than about thirty distinct species of wind-pollinated trees simply cannot share the same forest. With that many species, the chance of receiving pollen from a member of your own species—via the wind—becomes hopelessly improbable. However, if it's tall trees and lots of lumber you want, you cannot beat the ancient conifer forests of America's Pacific Northwest. In this corner of the world mild winters and dry summers favor the evergreen cone-bearing trees that don't bother much about growing sideways. No other forests in the world pack so many tree trunks into the same area. But don't let that fool you. These are not diverse forests; they support fewer than two dozen prominent tree species. Also, these are not dynamic forests; old growth in these forests is graced by trees over a thousand years in age. Hurricane-force winds are not known to

occur along this region of the Pacific Coast. Moreover, with ample rainfall, fires are rare. Here is a moist forest environment that supports fewer species than does a more severely stressed alpine meadow or a Midwestern prairie. Wind pollination simply cannot sustain the same kind of species-rich vegetation that flowering plants have created.

With the help of insects, birds, and bats—who can seek and find trees of the same species—the flowering plants can pack as many as 300 species of large trees into a few acres of tropical evergreen forest. In turn, those 300 different trees support a mob of hungry herbivores that have their own predators and parasites. Here is the most significant way in which flowering plants have changed the world. By developing loosely symbiotic relationships with a variety of pollinating animals, flowering plants can survive and propagate as rare individuals or sparsely distributed populations. In this way, they can prosper on the floor of the deep forest where winds are weak or absent. Despite the high costs of fancy petals, enticing aromas, and sugary nectar, the investment has paid off: plant species with pretty flowers outnumber all other land plant species several times over. Using animals for pollination, flowering plants have been able to create ecosystems far richer in species than any that have ever come before. Though some might not think of this as progressive, I'm convinced it is. And there's more.

If most of today's tropical wet forests are highly dynamic, we can understand why flowering plants are dominant in these habitats as well. Most flowering plants grow fast and reproduce quickly. Few flowering plants live beyond 300 years, whereas many gymnosperms can exceed 1,000 years in age. Also, flow-

ering trees rot more rapidly than conifers do, and, for that reason, return their nutrients to the soil more quickly. By growing fast, dying young, and getting animals to help in pollination and in dispersal, flowering plants have brought profound changes to the world's land surfaces. Whether in the rain forest, in deciduous woodland, or on open savanna and steppe, flowering plants have created high-turnover, faster-paced ecosystems, amply supporting many more species. And with their widely spreading upper branches, angiosperm trees did more than provide an arena for swinging apes. In wet forests these widely spreading branches became festooned with many kinds of smaller plants (*epiphytes*). Growing high on tree limbs, epiphytes have provided housing to a specialized fauna, further expanding species richness.

Not so long ago, conifer forests formed a conspicuous stratum in many tropical and temperate mountains at higher elevation. Unfortunately, many of these forests have been cleared by human activities over the last few centuries. Conifers are also dominant in huge areas of seasonally frigid boreal forest (*taiga*) in Alaska, Canada, and northern Russia. But such forests are not—and never were—rich in species. Flowering plants have created a richer world through more rapid vegetational cycling, a greater variety of reproductive strategies, and by becoming much more diverse themselves.

There probably were two major ways in which flowering plants increased their numbers. The first was by moving into new habitats. They did this as they expanded out of the tropics, into cooler regions, drier habitats, and new niches. For example, learning how to live on the branch of a tree opened new "arenas"

for flowering plants, either as parasites (the mistletoe family) or as true epiphytes (many species of orchids, bromeliads, and others). The second way of increasing numbers was to divide the habitat more finely. We've already seen how insect pollination helped create tropical forests with hundreds of *different* flowering tree species. But intense competition has done more. Visit a Midwestern prairie in late May and check out the brightly flowering species. Do the same in late June, late July, late August, and late September. Each time you visit, there will be a new cast of characters putting on their own flowery show. Most prairie species have a precise period during which they flower—usually less than three weeks long. And how might that have come about? The only reasonable answer is that competition for pollinator services has resulted in this nonstop parade of nicely staggered flowering times. One similarly sees precise flowering schedules in a great many tropical species, even in the evergreen tropics. Making more flowering plant species has inevitably helped many other species to diversify as well. And since species are the measure of biodiversity, angiosperms have been a primary engine of biodiversity expanding over time.

Finally, we should remember that the fundamental advantage of animal pollination was maintaining gene flow to isolated plants or distant populations. Only a sentient animal, following a remembered path, could provide this valuable service over long distances. This same mechanism, in turn, allowed for the evolution of many new species represented by only a few populations or isolated individuals. Such "fragmentation" allowed a rich supply of new species to survive, where before most would have been unable to persist.

The ability of angiosperms to produce a multitude of species is really quite marvelous. Flowering plants include a few genera that number over a thousand species. *Senecio* with about 1,500 species and *Vernonia* with about 1,000 are members of the composite family. *Acacia* with 1,200 and *Astragalus* with 2,000 are genera of legumes. *Euphorbia* includes over 1,500 species, and *Psychotria* of the coffee family numbers 1,400. Little Costa Rica is home to about 100 species of *Psychotria*, with 14 found nowhere else in the world. Unfortunately, specialists are dividing many of these large genera into smaller groupings in what has been called *taxonomic inflation* (I think it also inflates the egos of the experts). But back to big genera. We really do not know why one genus will have only a few species when a related genus can have hundreds. In some cases it may be a special set of adaptations, as in the figs and their wasps. In other cases, it may be the ability to "bud off" peripheral founder populations into new and different habitats. We still suffer from an inadequate understanding of speciation, and why flowering plants are so good at it.

Flowering Plants and the Expansion of Biodiversity

The primary importance of angiosperms, however, is that they did much more than simply increase their own numbers. Earlier we discussed the wide array of growth forms found in flowering plants. Such variety is evident within the tropical rain forest itself, ranging from moss-sized orchids growing on tree

limbs, to the tall trees on which the orchids grow. In addition, giant banana-like "herbs" on the forest floor, shrubs and small trees in the understory, thickly sinuous woody vines, and tall but slender palms add to the rain forest's structural richness. And since they're all busy photosynthesizing, these many flowering plants transform the energy of sunlight into materials that help support the monkeys, birds, insects, and all the other inhabitants of the forest. The structural variety of angiosperms provides a richer and more complex environment for all the other members of the forest.

Flowering plants play a similarly important role in seasonally dry tropical woodlands and savannas, as well as in our northern deciduous forests. We've already seen how the deciduous leaf and the invention of grass allowed angiosperms to dominate many seasonally dry habitats, but we haven't focused on how these "inventions" changed the world. Fundamentally, flowering plants created ecosystems that are more nutritious and vastly richer in species. Individually, a coniferous tree simply does not feed as many herbivores as does a flowering tree. As many as 163 different species of beetles and 43 different species of ants have been counted from *a single angiosperm tree* in the rain forest of Amazonia. For animals living in the rain forest, young leaves, sap, nectar, seeds, and especially succulent fruit are major food sources. No other group of plants provides as many calories to the environment as do the flowering plants.[3]

There's also chemistry; flowering plants produce a huge variety of chemical substances. We discussed this in regard to plants defending themselves. Our own species has made use of this cornucopia in many ways. Again, the primary sources of medicinal, herbal, and flavoring agents for all indigenous societies are flowering plants. Our continuing search for potential new medicines in tropical forests is a reflection of the huge diversity of chemical substances constructed by flowering plants. Thus the forest is also a complex mosaic of many different chemicals, created by plants to defend themselves against a hoard of herbivores. Recent research has substantiated these inferences. Using newly developed "knock-out genes," experimenters were able to produce plants lacking key features that normal plants possess. In this research, plants of the genus *Nicotiana* were deprived of a set of their defensive compounds (specifically the "jasmonate cascade"). These relatives of tobacco were then planted in their natural habitat and monitored. Sure enough, their specialized insect foes feasted in far greater abundance on these plants with lowered chemical defenses. More important, other insect pests also joined the party, pests that normally avoided this species.[4] Clearly, these chemicals play the defensive roles that had been ascribed to them.

Such chemical defenses have, in turn, led to a diversification of herbivores, especially among insects. As we noted in the case of milkweeds, different lineages of insects have adapted to dealing with different chemical toxins, becoming more specialized regarding their host plants. These "arms races" between plants and herbivores have helped build a world of much greater variety. Here's another example: *Physalis*, like *Nicotiana*, is a genus in the potato family that includes the ornamental "Chinese lanterns" with about eighty species. One of these species (*Physalis angulata*) produces fruit completely lacking linolenic acid, a constituent of many plant tissues. What's important about this unusual feature is that insects need linolenic acid to develop properly. Nevertheless, the larvae of a small moth (*Heliothis subflexa*) have figured out a way of compensating for this deficiency. This gives these little larvae two distinct advantages. First of all, they're busy eating a plant resource other insects cannot use, and so they've got their own private dining table. The second advantage is that because these larvae contain little linolenic acid, they do not produce some of the volatiles that parasitic wasps use to find these larvae. In addition, when a wasp does find one of these larvae and places her egg on the victim, the wasp's larvae cannot develop properly because of the lack of linolenic acid in its host. Thanks to these unusual biochemical factors, both the plant (*Physalis angulata*) and the moth (*Heliothis subflexa*) have their own special niche.[5] Here's another nice example of how biochemical specialization—starting with a plant—has added diversity to the living world.[6]

Though we can't see the vast chemical arsenal with which flowering plants protect themselves, we can appreciate the results. Also we can investigate their diverse forms. A few of these special morphologies have created their own special worlds. In the American tropics "tank plants" or bromeliads (Bromeliaceae family) are conspicuous members of the epiphyte community. Because the leaves of these plants hold water in their broad leaf bases, they create unusual microhabitats high in the trees.

Here insects and little amphibians live in a world all their own. The bromeliads, perched high in the branches of moist forests, and with their miniature reservoirs, have greatly enriched the American tropics.

Just as the bromeliads provided a new element to the structural diversity of wet forests in the New World, flowering plants appear to have done the same—over the entire world—as they expanded in numbers and forms, beginning more than 100 million years ago. We know that in today's world, forests dominated by angiosperms have many more species of other plants and animals than do forests dominated by gymnosperms. Surely the expansion of angiosperms 100 million years ago must have had the same effect worldwide. A combination of greater diversity of growth forms, greater variety of fruits and flowering elements, and a more open, widely branching, upper canopy all contributed to make a forest dominated by flowering plants a more diverse forest.

A recent study of fern evolution supports this idea. An ancient lineage, ferns were thought to have declined in diversity as the flowering plants expanded during the Cretaceous period, 145 to 65 million years ago. But fossil evidence in combination with recent DNA studies makes clear that the more modern ferns multiplied shortly after the angiosperms themselves were beginning their dramatic expansion. These more modern ferns (the polypody ferns) comprise more than 60 percent of living ferns. Though this group of ferns *originated before* the initial diversification of flowering plants, they *proliferated after* that diversification. In other words, between about 130 and 100 million years ago, the angiosperms had begun diversifying; then, a bit later, between 110 and 65 million years ago, the polypody ferns joined in the festivities.[7] Mosses and liverworts increased in a similar manner, as flowering plants produced a more complex forest of widely branching trees. If the fossil record were better, I suspect we would see insects and many other terrestrial lineages expanding their numbers in exactly the same way—all thanks to the flowering plants. In addition, there is evidence that the expansion of flowering plants allowed communities in the oceans to diversify. The evolution of more complex multilayered rain forests, as well as more productive grasslands, should have resulted in additional organic nutrients being carried into the sea. This correlates with estimates of deep-sea marine communities growing richer in species over the last 80 million years (this is not just a statistical artifact of having more fossil deposits in more recent times).

Clearly, flowering plants brought profound changes to the world. Not only did they build more diverse forests and woodlands, they also enriched drier open vegetation. Perhaps the most significant transformation of recent times was the creation of grasslands. Growing quickly after fire and drought, grasslands sustain herds of large herbivores with an efficiency that earlier dryland vegetation could not. Recall the unusual ability of grass leaves: they can continue growing from near their base. This means that when the top of a grass leaf is destroyed by fire or eaten by cattle, the leaf can keep right on growing. (Most plants need to form an entirely new leaf before continuing to grow again.) Even today, grasslands support our most important food animals; then, too, grasses provided us the grains that helped build great civilizations. But is it possible that grasslands contributed to global climate change as well?

Did Prairies Change the Weather?

The first great success of land plants was in creating the giant coal forests of the Carboniferous period (290–363 million years ago). Such forests were confined to low-lying tropical sites, especially along river bottoms and in deltas. These are the forests that produced the great coal deposits that are one of our most significant sources of industrial energy. All that ancient plant growth, however, appears to have had profound negative consequences for the climates of those ancient times. Many paleontologists believe that these great forests pulled so much carbon dioxide out of the atmosphere that they effectively ended up lowering temperatures around the world. Carbon dioxide has the unusual property of being transparent to high-energy ultraviolet light *while absorbing* lower-energy infrared radiation. Visible and ultraviolet light from the Sun penetrates the atmosphere and warms the Earth. That warmth produces lower-energy infrared that is reradiated back into outer space. And this escaping energy is what a thin blanket of carbon dioxide helps contain (together with other greenhouse gases such as methane and water vapor). In effect, these gases form an insulating blanket that keeps our planet warm. With reduced levels of carbon dioxide, more warmth escapes into space, and the surface of our planet inevitably cools. Permian glaciations, about 280 million years ago, were the longest period of colder weather during the last 500 million years—probably caused by the sequestration of carbon in the great coal forests.

In more recent times, starting about 40 million years ago, global temperatures seem to have begun declining steadily; this led gradually to the so-called Ice Age (Pleistocene) beginning about 1.8 million years ago. Is it possible that the spread of flowering plants over the last 60 million years has played a similar role in bringing on a gradual decline in temperatures? Have flowering plants been sequestering so much carbon dioxide that they, too, have cooled the Earth? Stars like our Sun are hypothesized to get warmer as they age—not colder. Why has Earth been getting so much cooler during these more recent times?

As mentioned earlier, grasslands seem to have begun about 30 million years ago, and they've been expanding ever since. Unfortunately, the fact that herbaceous plants rarely fossilize means that we have to study the expansion of grasslands indirectly. Recall that it is the deep-rooted teeth of fossil herbivores that grazed abrasive grasses that provide our evidence for grassland expansion. Fortunately, there are additional sources of evidence; paleobotanist Gregory Retallack has explored grassland expansion by studying fossil soils.

Analyzing both contemporary and fossil soils in the American Midwest, Oregon, East Africa, and elsewhere, Retallack has found intriguing patterns. He notes that grassland soils have *smaller particle sizes* than do the soils of forest and woodland. These smaller particles give grassland soils much more surface area than the soils of woodlands. (Think of how you can increase the surfaces of a cubic foot of wood by simply cutting it into lots of cubic inches.) By having greater surface area, grassland soils can bind more organic materials, an important factor in making grassland soils so rich.

Earlier, soil scientists had discovered that soils were much deeper under moist grasslands than under the drier grasslands of steppe and semideserts. Thus, soil depth gives paleontologists a clue as to how dry a prehistoric grassland may have been. What Retallack discovered was that grassland soils were shallow and apparently confined to drier regions *before* about six million years ago. Since then, he found, grasslands have been expanding into moister regions—as evidenced by their *deeper* soils. Grasslands with deeper soils appear for the first time around six million years ago, and have been *spreading* into regions of higher rainfall since that time. Based on these findings, Retallack sets forth a bold hypothesis; he suggests that expanding prairies helped bring on the Ice Age!

Because the soils of moist prairies are deeper, and because they have smaller granularity, such soils provide far more surface area than did the open woodland soils that prairies replaced—and bind far more organic matter than was held by the earlier woodland soils. Plants of the prairie store large amounts of organic matter in the soil for a reason. It is this underground biomass that allows them to recover quickly after long droughts, severe fires, and intense grazing. The organic matter, in turn, holds a lot of carbon derived from the carbon dioxide pulled out of the air by photosynthesis. Some of the sequestered carbon is not returned to the atmosphere, with the consequent loss of carbon dioxide from the air. And with less carbon dioxide to insulate the Earth, Retallack claims that the prairies therefore contributed to the onset of our recent ice age—beginning about two million years ago. But why might grasslands have expanded into areas earlier dominated by woodlands?

Working in concert with fire, grasses can "erode" the edges of woodlands. Grasses promote fires; once dry, they burn easily, and many of them burn hot. Nearby woody plants, unable to withstand such hot fires, succumb and the grassland advances. In addition, grasslands are not as dark green as is the forest canopy; grasslands reflect more light back into space than do forests. Additional reflected light heats the air column above, making it warmer. Since warmer air holds more moisture than cooler air, grasslands produce a drying effect *and lowered rainfall*—increasing the incidence of fire. Then, too, animal browsing destroys young tree seedlings that might germinate at the forest's edge. Retallack thinks that the combination of grassland, fire, and grazing animals produced a new dynamic, allowing grasslands to expand their range—and change the climate! In his words: "Unidirectional, stepwise, long-term climatic cooling, drying, and climatic instability may have been driven . . . by the coevolution of grasses and grazers."[8]

Bold hypotheses are supposed to be a hallmark of science. The original evidence supporting a new hypothesis may be weak or unconvincing, but it shouldn't take too long to find new and critical ways of testing Retallack's hypothesis. American geologists rejected the hypothesis of continental drift for half a century. However, when deep ocean rifts and ocean-floor spreading were discovered, the mechanism of continental movements was made clear, and the belief in fixed continental positions was demolished. Can Retallack's notion of prairie-induced global cooling stand up to further analysis? Did the expansion of the prairies really change the world's climate? Stay tuned to this business we call science, as expanding knowledge helps us learn more about our world and its history.

Flowering Plants and People: A New Symbiosis

We have already seen how agriculture transformed the human condition. Thanks to the close interaction of humans and a few species of plants and animals, many landscapes could now support more people than ever before. Whether settled agriculturalists or nomadic herders, humans were now harvesting more of the products of the Sun's energy. Along the floodplains of great rivers, agriculture could support cities populated by thousands of our kind. And with the rise of these early civilizations came the development of ceramics, the invention of metallurgy, the blossoming of metaphysics, and the beginnings of science.[9]

Today, with most foods available in every month of the year, it is difficult to appreciate the marvelous transformation that agriculture created. Today's agriculture sustains well over six billion people. Much of this triumph is based on the same traditional agricultural practices that have been in use for thousands of years. But an increasing percentage of our food is now the product of intensive fertilization and aggressive use of biocides, which has been made possible only because of modern industrial chemistry. In fact, we humans are fixing about as much nitrogen today, mostly for fertilizers, as is the rest of the natural biosphere! Needless to say, without that extra fixed nitrogen (thanks to energy-devouring industrial processes), we could not be feeding so many people. As we mentioned earlier, 25 species of flowering plants provide us with about 90 percent of our vegetarian energy. That's 25 species out of more than 260,000 flowering plant species alive in the world at large! If it weren't for that huge diversity to choose from, our larders might be rather bare. Except for fish, seafood, and a few fungi and algae, it is the flowering plants that fill our "gas tanks" and serve as the fodder for the animals that provide us nourishment. Clearly, without the flowering plants we humans would not be the dominant force we have become. And though many species of flowering plants have transformed the world over the last 50 million years, it is we, a single species, who have made astounding changes across the planet in these last 5,000 years.

Looking out of an airplane window in most regions of the nonwatery world, one can see the impact of our species. Whether in the American Midwest, the highlands of Ethiopia, or across southern Asia, the landscapes we see are primarily *agricultural*. What were once savannas, steppes, and woodlands are now fields of wheat, rice, maize, barley, sugarcane, soybeans, oil palm, and a host of other crops. Additional landscapes provide grazing for cattle, sheep, and goats, while others provide us with lumber and fiber crops.

Our modern industrialized society is itself largely powered by plant energy—energy captured by plants that lived long before the angiosperms appeared. We now live in a civilization that feasts on fossil fuel. Having discovered the energy contained in coal, oil, and gas—stored within our planet's surface—we have hugely enriched our lives and lifestyles. Fossil energy has become the primary motive force in the modern world, and it affects all aspects of our lives. Whether waging war, refrigerating food, or driving to work, oil plays a major role in the modern world. Beginning with steam power, and now with reciprocating engines and turbines, we have created a society with an unquenchable thirst for energy. In the late 1930s about two million people traveled by plane each year, now that number is over 200 million. With expanding food production, improved sanitation, better control of infectious diseases, and the use of antibiotics, our numbers reached six billion people at the end of the twentieth century. Only an estimated 600 million people were alive in 1800; we have increased our numbers tenfold in two centuries! Though the rate of population growth is declining, we are continuing to add more than 70 million people to our planet each year. Cities all over the world are increasing in population as an expanding wave of suburban development spreads out from these centers. Highways and byways interconnect communities of all sizes in a network of human activity that has begun to change the biosphere itself.

Clearing forests to create pastures for the beef cattle we love to eat changes the weather—just as the expansion of grasslands by setting fires may have done in the past. Open pastures and agricultural lands absorb less light than does a dark green forest canopy. Increased reflected light above pasture and farmland warms the air, making the landscape both warmer and drier. Also, when it does rain, there is more erosional runoff from a cornfield or pasture than from under a multilayered forest or a dense prairie. Worse yet, the removal of both livestock and plant produce from the land where they originated is a form of "mining the land"—taking valuable soil nutrients and sending them off to distant markets. All the mineral nutrients embedded in rice, beans, and beef, after being digested in the great population centers, mostly end up being flushed into the oceans. That's a one-way street to ever-greater dependence on factory-produced chemical soil additives. And that means more energy demand to keep producing more fertilizers. It doesn't take a rocket scientist to figure out that, sooner or later, such a system is going to develop serious problems.[10] All the while, we are continuing to diminish the world's natural vegetation, increasing our use of fresh water, belching ever more carbon dioxide into the atmosphere, and, in the wealthy countries, making our lives more extravagant. All this is the most recent installment in a three-and-a-half-billion-year odyssey of life on a planet about four and a half billion years old. Finally, let's take a broader view of how we got to where we are today.

The Big Picture in Ten Major Stages

Flowers, we can all conclude, have played a major role in creating the modern world. But how should we view them from a broader perspective? How important were they in the long history of life, a history that, according to recent scientific thinking, began more than three billion years ago? This unfolding story, often referred to as the "evolutionary epic," is one of the major achievements of modern science. It is a coherent historical narrative, constructed out of many scientific disciplines, from planetary astronomy to geology, and from microbiology to botany and zoology. Based on this long history, I believe that ten major advances or stages can be identified, leading to the world we know today.

Certainly, the first major advance in the history of life was the origin of simple life itself, represented in the modern world by simple bacterial-grade cells. The ability to self-construct, to gather and use energy, to replicate, and to carry the information for all these activities from generation to generation are the fundamental hallmarks of life. For many who have considered the question, this may have been the most difficult—and most unlikely—step in the history of life. A few years ago, carbon-isotope ratios similar to ratios produced by living things were claimed to be present in 3.8-billion-year-old rocks from Greenland. Though often cited, these claims have not been corroborated; it now seems that it took life a little longer to get itself organized here on planet Earth. That makes sense. The "late heavy bombardment" of asteroids and cometary material that peppered the Moon and our planet in those early times is thought to have ended around 3.8 billion years ago—not a propitious time for early life on Earth. But by 3.5 billion years ago there is better geological evidence: stromatolites. These laminated structures, often forming rounded columns a few feet tall, are created by thin layers of bacteria and detritus.

With time, new layers are built over earlier layers forming the laminated cross sections that characterize stromatolites. We know of nothing else that might form such structures. They are strong evidence for early life on Earth.

The second major step—still at the bacterial level of cellular organization—was the invention of photosynthesis: using sunlight to split the water molecules as a source of hydrogen. Combining hydrogen atoms with carbon dioxide, this process transforms solar energy into energy-rich carbohydrates, the primary fuel for biological activities. Chlorophyll and its associated pigments are the major players in this important set of reactions. Today, blue-green bacteria appear to be the living descendants of the earliest photosynthesizers, and they continue to play an important role in the modern world. Once life on Earth had learned how to convert solar energy into chemical energy, organic evolution was able to begin its progressive elaboration.

The third major step in the history of life was the development of the larger, much more complex eukaryotic cell. This cell is characterized by having a nucleus within which the chromosomes are confined and replicated. These cells, about one thousand times larger in volume than the average bacterial cell, contain mitochondria that break down simple carbohydrates to meet the energy demands of the larger cell. Recent research has provided strong support for the notion that the mitochondria were once independent bacterial-grade microbes themselves, becoming integral elements of the eukaryotic cell through a process called *endosymbiosis*. Another major step during these early stages of evolution was when one lineage of eukaryotic cells incorporated a second bacterial partner or endosymbiont—

only in this case a photosynthesizing bacterium. This new union gave rise to the chloroplast, the cell organelle in which photosynthesis takes place in algae and higher plants.

The fourth major stage in our continuing epic of increasing biological complexity was building larger, many-celled organisms. Developing independently in the plant, fungal, and animal lineages, this process gave us the three great kingdoms of larger living things that adorn our planet. Evidence for the independent development of larger forms in each kingdom is that plants, fungi, and animals are constructed in completely different ways, despite the fact that all begin their lives as single cells. Incidentally, no larger organism is built of bacterial-grade cells. Apparently, the greater energy-utilizing ability and information-carrying capacity of the eukaryotic cell was a necessary first step in building complex multicellular organisms.

The fifth major stage in the grand pageant of life was the development of larger and more active animals during what has been called the Cambrian explosion, about 540 million years ago. Suddenly, within less than ten million years, the world's oceans became populated by a zoo of *different* lineages of creepy crawlers, shell-bearers, and wormlike animals. Many of these creatures developed eyes to see where they were going, and teeth to eat each other. Why this happened so suddenly is still an open question. DNA evidence suggests that these different lineages had evolved much earlier, but the absence of even trace fossils (such as worm trails) indicates that they remained minute and insignificant before the Cambrian. Perhaps oxygen in the atmosphere finally reached a high enough concentration so that animals could now grow to a larger size and begin actively pursuing each other.

The sixth major event in the history of life, I would propose, was the colonization of the land by larger green plants. As seen in chapter 6, this was a difficult transition, successfully traversed by only a single plant lineage; it began over 450 million years ago. Once land plants became larger and more complex, warm, moist parts of the world were embellished with a rich vegetation. Building on this important stage was a further development: the pollen grain of the seed plant. With pollination, fertilization was no longer restricted to thin films of water. Seed plants could survive and reproduce in much drier habitats than the plants that preceded them.

The seventh major step, especially from our own point of view, was the origin of the land vertebrates. Like the land plants, four-legged vertebrates arose in freshwater environments and moved onto the land about 360 million years ago. Here was a lineage of animals that could defy gravity, become large, move quickly, and build more complex brains. A terrestrial flora, augmented by insects, spiders, worms, and snails, made it possible for the early land vertebrates to find food and prosper. Our four limbs and many-parted backbone are a clear legacy of our "tetrapod" vertebrate heritage.

Then came one of the major setbacks in the history of complex life. The Permo-Triassic extinction, at about 250 million years ago, devastated animal life, both on the land and in the sea. (Perhaps because their fossil record is so poor, land plants show less evidence of having been decimated at this time.) It is still not clear what caused that massive die-off, but extensive volcanic magma eruptions at this time surely played a role. Nevertheless, those plants and animals that survived quickly proliferated anew.

The eighth grand advance, I insist, was the development of the flowering plants. Of their many clever adaptations, getting animals to carry pollen grains from plant to plant was probably the single most important innovation. Not only did the flowering plants use animals as pollinators, they supported the further diversification of these animals, especially the insects. Together, insects and angiosperms helped produce the incredibly rich world we have today.

Then came another major setback in the history of life, as a large asteroid slammed into the Earth about 65 million years ago: the end-Cretaceous extinction. Throwing up a cloud of dust, this event terminated the lives of many species on both land and in the sea. This event wiped out one of the most impressive lineages of land animals the planet has ever seen, the dinosaurs. Dominating the world's landscapes for more than one hundred million years, every dinosaur became extinct—except for the birds. That was the bad news; the good news was that smaller furry creatures suddenly found themselves in a friendlier environment. With the demise of the dinosaurs, mammals diversified explosively. Meanwhile, the flowering plants, unfazed by the grand extinction, continued enriching the vegetation of the world.

As the pageant of life continued, flowering trees promoted the evolution of monkeys and helped give rise to the swinging apes. Open grasslands, a new landscape in seasonally dry habitats, promoted the proliferation of large mammalian herbivores. These nutrient-rich resources in turn helped a two-legged primate triple its brain volume in only three million years. And once that two-legged primate developed clever cultural symbioses with a bunch of tasty plants and a few animals, we humans became the most successful large animal in the history of life. This, I believe, was the ninth grand advance in the history of life on our planet: the rise of agriculturally based human societies.

And then came modern times. The tenth major stage in the history of life, as I see it, was when we humans united the techniques of scientific inquiry and industrial innovation with the firepower of fossil fuels. With that clever combination we not only began to understand the world in which we live, but also to control it more fully, and to increase our numbers explosively. This most recent advance has made us the dominant life form on planet Earth.

Clearly, angiosperms have changed the world over the last 100 million years. It is also apparent that without the flowering plants, primates would not have diversified, long-armed apes would not have swung through the canopies of tropical forests, and herbivore-rich grasslands would not have come into existence. Not only did flowering plants play a crucial role in the *origin* of the human lineage, but, through agriculture, flowering plants helped humans gain dominion over our planet.[11] By empowering the elaboration of human cultures, flowering plants have changed the world more profoundly than in any earlier ecological transformation. It is now a *single species* that is changing the world so quickly and in so many different ways that the biosphere itself is reeling under our assault. Though there are "futurists" who foresee a glorious utopia for genetically transformed humans voyaging to the stars, many others of us are fearful of an ominous future. But whether you are a utopian optimist or an ecological pessimist, we can all conclude that the flowering plants have, indeed, changed the world. In fact, it is the flowering plants that allowed human beings, for better and for worse, to have become the masters of our planet.

epilogue

Unfortunately, our species' newly acquired dominion over the planet brings with it many awesome responsibilities. Let's consider those burdens from the point of view of our long horticultural heritage.

Our early ancestors probably began their agricultural efforts with loosely organized plots of food plants, flavorful herbs, and medicinal species. As populations expanded, the basic grains, pulses, and root crops had to be grown more intensively over larger acreage. Now it was necessary to grow our food plants in carefully prepared fields. By means of irrigation, agriculture could feed thousands, becoming the foundation of city-states. Egyptian civilization—watered by the Nile, fertilized by the silt of yearly flooding, and protected from invasion by surrounding desert—has flourished over more than five thousand years. But many other high civilizations were not so fortunate, as soil deterioration and climate change brought them crashing down. All the while, and in villages throughout the world, small gardens continued providing variety to the diet and enriching human lives.

Ornamental gardens developed only after societies became wealthier, more complex, and more stratified. Whether surrounding a religious shrine or a quiet refuge within the compounds of the aristocracy, a landscaped garden became a distinctive part of cultural life in many different societies. The Spanish conquistadors were greatly impressed by the Aztec gardens they encountered in Mexico, and also by those in the highlands of Peru. Gardens have also had a long history in China, and later in Japan. Our oldest graphic depictions of gardenlike settings are those of the Egyptians, from about 3,500 years ago. These show a rectangular pond, planted with lotus and papyrus, populated by ducks and surrounded by a walkway, itself bordered by plantings of herbs and small trees. The trees, providing shade, included figs, pomegranates, date palms, and arbors for grapes. All of this in turn was contained within an enclosing wall. Such a wall emphasized the special nature of the space, served to isolate the garden from the bustle of the city, and protected the plants from hungry goats and naughty children.

The word *paradise* is derived from the ancient Persian word for "enclosure." The hunting park of the Persian king was enclosed to exclude poachers and the common folk. Such an enclosed park gave rise to the Greek concept of *paradeisos*: the sumptuous garden. Christian tradition adopted this concept in imagining a celestial paradise, as does the Koran. In this sense, the garden became a very special space—isolated and protected from the chaos and strife of the world outside. As the centuries progressed, the idea of the garden was retained as a special sanctuary for a special few. The monastery garden of the early medieval period often had three areas. The orchard grew fruit trees and served as the monk's cemetery. A kitchen garden produced vegetables while the infirmary garden grew medicinal herbs. In addition, these gardens served as experimental plots in which to make new introductions and select the best varieties. By the seventeenth century homes of wealthy Europeans were embellished with landscaped grounds and carefully tended gardens. Later, as a more prosperous middle class developed, they, too, sought to have small gardens.[1]

Finally, with the advent of the Industrial Revolution and intense crowding within large cities, the need for open public space became ever more apparent. City planning now included tree-lined avenues, larger plazas, and public gardens. Today, landscape design includes the careful arrangement of trees, shrubs, and open expanses of lawn and walkways. City gardens stand in bold contrast to the concrete and steel of the modern urban center; they offer a contrast to and respite from the hectic commercial life of the metropolis. It is difficult to imagine New York City without its Central Park, Washington, DC, without its mall, or Chicago without its public lakefront. Creating such large public spaces took energy, money, and foresight. Montgomery Ward fought for more than two decades to keep Chicago's lakefront public. His opponents claimed that *development* (that is, factories) along the lakefront would provide additional taxes for better schools and hospitals. Preserving open land and building public gardens took much determination.

A recent estimate of how much money Americans spend on leisure-time activities put the figure for "flowers and plants" at nearly nineteen *billion* dollars in 2002.[2] That's almost as much as we spent on sports attendance and movies *combined*. Needless to say, most of this money is going into planting our gardens and beautifying our communities. Similarly, flowers are big business around the world. Commercial orchids alone, sold as pot plants and for ornamental use, have an annual value of around two billion dollars worldwide.

Unfortunately, as urban populations are growing and suburban sprawl expands relentlessly, natural vegetation is retreating. Not only is the world's native vegetation being reduced, it is also being invaded. Second only to the loss of habitat in impact, invasive foreign species are attacking native vegetation and extirpating indigenous species of plants and animals. Whether it's kudzu in our southeast, garlic mustard in our Midwestern woodlands, or the zebra mussels in the Great Lakes, the introduction of foreign species can radically alter local ecology. Arriving in a new territory *without* their pests and pathogens, a few of these newly arriving species proliferate with abandon. The introduction of pests and pathogens can wreck havoc as well. An invasion of Eurasian diseases wiped out the American chestnut and continues to ravage our elms. Clearly, it takes only a few aggressive species to effect major mayhem.

Thanks to the way we humans traverse the planet and ship things around, invasive species are becoming a major headache in all the habitable corners of the world. Whether purchasing exotic wildlife, cultivating ornamentals from around the globe, or simply trekking dirty boots through international air terminals, the movement of invasive species is not likely to diminish anytime soon. When we add the deterioration of agricultural land by salinization, accumulation of biocides, and erosion, the emerging future takes on a very troubling appearance. Meanwhile, sport utility vehicles are selling briskly in China as that populous nation follows the same track taken by every other affluent society. Most of this consumptive exuberance is being propelled by the photosynthesis of many millions of years ago. Though more than a third of the world may be living in abject poverty, the rest of us are having a great time.

Surely one of the most profound ways in which the flowering plants have changed the world was to have helped create a master species—a species so clever, so fertile, and with such an appetite for material resources and sensuous diversions that the biosphere itself is suffering from the effects of our expansion. Though our rate of growth may be much diminished in a few countries, we are continuing to expand our numbers by over seventy million additional people each year—mostly in countries that are already impoverished. Our numbers are now expected to reach nine billion by midcentury. Such growth will make even greater demands on our water resources, our agricultural soils, our urban environments, and the world's fisheries (which are nearly all in serious decline). More than half the world's people are malnourished—the highest number in human history—and yet our numbers keep growing. Clearly, human beings are not living in sustainable harmony with the biosphere that supports them.

Our planet, some people have suggested, has a "global metabolism," an interwoven network involving microbes, plants, animals, rainfall, erosion, ocean currents, and even volcanic activity. Some have argued that these interacting factors are lifelike, working together to keep the whole system in balance; it's called the Gaia hypothesis.[3] I disagree. Fossil evidence for the extirpation of entire faunas during the great geological extinction events, and the turbulence of repetitive ice advances in the last million years, suggest that the planet isn't the quiet and predictable, or self-regulating, place we would like to imagine. Not only are the biosphere and the atmosphere highly dynamic, we humans are beginning to change many of the equations. Global warming is no longer a hypothesis. Now the only questions are exactly how much our own species is contributing to this warming, and how severe the warming will become. Currently, the world appears to be in a period of dynamic flux, with human activities themselves responsible for unprecedented changes.

Perhaps we can think of human history as having had three major stages—all related to our energy consumption. The first stage changed little over hundreds of thousands of years, as small bands of foragers searched the landscape for sustenance. They are referred to as hunters and gatherers, and these people probably lived in groups rarely exceeding fifty mature individuals. Foraging simply could not sustain larger numbers of people. Tasty tubers and energy-rich animals are scarce in most of the world's ecosystems; starvation was a constant threat. The second historical advance came with the invention of agriculture. Suddenly a number of habitats could support ten times as many people as had lived there in the past. By supplying more food, more reliably, agriculture represented our first major advance in energy procurement. The third and final stage was the Industrial Revolution and a complex technology based on fossil fuels; first with coal, then oil, and more recently gas. Worldwide, we are currently turning more than 80 million barrels of oil into carbon dioxide *each and every day*—carbon dioxide that will only further increase the warming of the globe. Anybody who thinks that this kind of activity can be sustained over the long haul lives in a very special world of hopes and dreams. Nevertheless, most people are optimistic, thinking, "If technology created these problems, technology will help us solve them." And then there are those who are unconcerned because they believe the biblical "end-times" are coming, and coming soon. Either way, most people are ignoring the negative consequences of ever-expanding industrial growth and our spiraling numbers.[4] Unfortunately, what seems more likely is the very real possibility of a global calamity of unprecedented enormity. (But then, with population densities increasing and high-speed international air travel around the world, a lethal pathogen might do for planet Earth what the Black Plague did for medieval Europe; that is to save us from a headlong trajectory into ecological self-destruction.)

Like no other heavenly body we know of, this is the blue, white, and green planet. It is, in fact, a planet whose surface was and still is adorned by a multitude of gardens.[5] These many dynamically interactive environments help give us our sustenance.

Like it or not, we humans are now in charge of Earth, the planet that gave us birth. Unless we invest the same intensive efforts to preserve natural environments as we have done for our homes and gardens, the welfare of future generations is at risk. We humans must diminish our appetites and reduce our numbers. If we wish to provide our grandchildren with a better world, we'll have to do better. At present, no political pundits, economists, or journalists seem aware of the huge challenges facing our future. It will take a gargantuan effort to maintain our planet as the lovely home it's been over so many millions of years. From where I sit, it looks like the sixth major extinction over these last 500 million years is now well underway, the first grand extinction to have been caused by a single biological species. Here on planet Earth, we human beings may be nature's supreme intellectual achievement, but we have also become its most profound threat. We must change our ways; we must become the master gardeners of our biosphere, the stewards of planet Earth.

notes

Acknowledgments

1. Karl J. Niklas, *The Evolutionary Biology of Plants* (Chicago: University of Chicago Press, 1997), p. xiii.

Introduction

1. Loren Eiseley, *The Immense Journey* (New York: Random House, 1957), p. 77.

Chapter 1: What, Exactly, Is a Flower?

1. A fine discussion of flowers can be found in Peter Bernhardt, *The Rose's Kiss: A Natural History of Flowers* (Washington, DC: Island Press, 1999). Another very readable book is Sharman Apt Russell, *Anatomy of a Rose: Exploring the Secret Life of Flowers* (New York: Perseus Books, 2001).

2. The "inferior ovary" of the rose is actually quite complex. It is a pouchlike development of the floral axis that encloses the separate ovaries of the pistils clustered in the center of the flower. This is very different from most inferior ovaries, which are the basal part of a single pistil.

3. For more information about flower structure and botany in general, an introductory college-level textbook is your best bet. Skip over the technical areas of physiology and genetics and go to the structure and development sections; then find the parts specifically about flowers. A superbly illustrated introduction to plant identification is James L. Castner, *Photographic Atlas of Botany and Guide to Plant Identification* (Gainesville, FL: Feline Press, 2004). A more demanding and much more detailed text is Peter H. Raven, Ray F. Evert, and Susan E. Eichorn, *Biology of Plants*, 6th ed. (New York: W. H. Freeman & Co., 1999).

4. There is a huge literature regarding both wild orchids and their many cultivated varieties. A good general introduction is Robert Dressler, *The Orchids: Natural History and Classification* (paperback ed.; Cambridge, MA: Harvard University Press, 1990). A larger book with colorful illustrations is William Cullina, *Understanding Orchids* (New York: Houghton Mifflin, 2004).

A short overview of the family is Kenneth Cameron, "Age and Beauty," in *Natural History* 113, no. 4 (June 2004): 27–32.

5. For more about grasses, see Richard W. Pohl, *How to Know the Grasses*, 3rd ed. (Dubuque, IA: Wm. C. Brown, 1978), and Rick Darke, ed., *Manual of Grasses* (Portland, OR: Timber Press, 1995).

6. A good introduction to the composite or Asteraceae family for Midwesterners is found in Thomas M. Antonio and Susanne Masi, *The Sunflower Family in the Upper Midwest* (Indianapolis: Indiana Academy of Science, 2001).

7. For a survey of the aroid family, see Deni Brown, *Plants of the Arum Family*, 2nd ed. (Portland, OR: Timber Press, 2000).

Chapter 2: What Are Flowers For?

1. The original article, "Rescue of a Severely Bottlenecked Wolf (*Canis lupus*) Population by a Single Immigrant," was published by Carles Vila et al. in *Proceedings of the Royal Society of London* B 270 (2003): 91–97.

2. The "Red Queen's Hypothesis" was first published by Leigh Van Valen as "A New Evolutionary Law," *Evolutionary Theory* 1 (1973): 1–30.

3. A recent technical review of self-incompatibility is S. J. Hiscock and S. M. McInnis, "The Diversity of Self-Incompatibility Systems in Flowering Plants," *Plant Biology* 5 (2003): 23–32.

4. For a bit more about when to flower, see Ruth Bastow and

Caroline Dean, "Deciding When to Flower," *Science* 302 (2003): 1695–97.

Chapter 3: Flowers and Their Friends

1. For more about fungi and their importance in making soil nutrients available to plants, see M.-A. Selosse and F. Le Tacon, "The Land Flora: A Phototroph-fungus Partnership?" *Trends in Ecology and Evolution* 13 (1998): 15–20; and Elizabeth Pennisi's short summary of recent work, "The Secret Life of Fungi," *Science* 304 (2004): 1620–22.

2. Insects are the primary pollinators of colorful flowers. An excellent review is Friedrich G. Barth, *Insects and Flowers: The Biology of a Partnership* (Princeton, NJ: Princeton University Press, 1985). A more technical review of pollination is Michael Proctor, Peter Yeo, and Andrew Lack, *The Natural History of Pollination* (Portland, OR: Timber Press, 1996).

3. A recent review of plant scents is Eran Pichersky, "Plant Scents: What We Perceive as a Fragrance Is Actually a Sophisticated Tool Used by Plants to Entice Pollinators, Discourage Microbes and Fend Off Predators," *American Scientist* 92 (2004): 514–21.

4. The life of the honeybee and the analysis of its "dance language" are reviewed in detail by James L. Gould and Carol G. Gould in *The Honeybee* (1988; New York: Scientific American Library, 1995). A popular account of beekeeping is Hattie Ellis, *Sweetness and Light: The Mysterious History of the Honeybee* (New York:

Harmony Books, 2004). For more about the bee family, see Christopher O'Toole and Anthony Raw, *Bees of the World* (London: Blamford Press, 1991); and Dave Goulson, *Bumblebees: Their Behaviour and Ecology* (Oxford and New York: Oxford University Press, 2003).

5. The article about honeybee heads was authored by Victor B. Meyer-Rochow and Olli Vakkuri and titled "Honeybee Heads Weigh Less in Winter than in Summer: A Possible Explanation," *Ethology, Ecology & Evolution* 14 (2002): 69–71.

6. Though technical, this book has proven to be a classic: Bernd Heinrich, *Bumblebee Economics* (1979; Cambridge: Harvard University Press, 2004).

7. The article critical of Darwin's early interpretations is L. A. Nilsson et al., "Monophily and Pollination Mechanisms in *Angraecum arachnites* Schltr. (Orchidaceae) in a Guild of Long-tongued Hawk-moths (Sphingidae) in Madagascar," *Biological Journal of the Linnean Society* 26 (1985): 1–19. A more positive interpretation of this phenomenon in the iris family is given by Ronny Alexandersson and Steven B. Johnson, "Pollinator-mediated Selection on Flower-tube Length in a Hawkmoth-pollinated *Gladiolus* (Iridaceae)," *Proceedings of the Royal Society, London* B 269 (2002): 631–36.

8. The long-tongued pollinating flies of South Africa are discussed by Laura A. Sessions and Steven D. Johnson in their article "The Flower and the Fly," *Natural History* 114, no. 2 (March 2005): 58–63.

9. "A sucker is born every minute" comes from page 191 in Peter Bernhardt, *The Rose's Kiss: A Natural History of Flowers* (Washington, DC: Island Press, 1999).

10. The estimates of seed numbers produced by some orchids are from Edward Ayensu, "Beautiful Gamblers of the Biosphere," *Natural History* 86, no. 8 (October 1977): 37–44.

11. Close chemical mimicry in the odor of an orchid and the odor of a female insect is reported by F. Schiestl et al., "Orchid Pollination by Sexual Swindle," *Nature* 399 (1999): 421–22. A similar report is that of Manfred Ayasse et al., "Pollinator Attraction in a Sexually Deceptive Orchid by Means of Unconventional Chemicals," *Proceedings of the Royal Society, London* B 270 (2003): 517–22.

12. For more about fig trees and fig wasps, see James M. Cook and Jean-Yves Rasplus, "Mutualists with Attitude: Coevolving Fig Wasps and Figs," *Trends in Ecology & Evolution* 18 (2003): 241–48.

13. For a technical report on an unusual disperser, see Anthony C. Nchanji and Andrew J. Plumptre, "Seed Germination and Early Seedling Establishment of Some Elephant-dispersed Species in Banyang-Mbo Wildlife Sanctuary, South-western Cameroon," *Journal of Tropical Ecology* 19 (2003): 229–37. See also D. W. Snow, "Tropical Frugivorous Birds and Their Food Plants: A World Survey," *Biotropica* 13 (1981): 1–14.

14. For the story of plant regeneration after the reintroduction of wolves at Yellowstone National Park, see William J. Ribble and Robert L. Beschta, "Wolves and the Ecology of Fear: Can Predation-risk Structure Ecosystems?" *Bioscience* 54 (2004): 755–66.

15. Fungi helping keep back other fungi in the leaves of the cacao tree are discussed in Keith Clay, "Fungi and the Food of the Gods," *Nature* 427 (2004): 401–402.

Chapter 4: Flowers and Their Enemies

1. Herbivores and how plants have responded to them are discussed in a variety of ecological texts. Two popular books are William Agosta's *Bombadier Beetles and Fever Trees: A Close-up Look at Chemical Warfare and Signals in Animals and Plants* (New York: Addison-Wesley, 1996), and the same author's *Thieves, Deceivers and Killers: Tales of Chemistry in Nature* (Princeton, NJ: Princeton University Press, 2001). A recent one-page summary with a few references is Jack C. Schultz, "How Plants Fight Dirty," *Nature* 416 (2002): 267.

2. If you want to know more about cacti, there is the large and comprehensive *The Cactus Family* by Edward F. Anderson (Portland, OR: Timber Press, 2001).

3. Chemical plant defenses are covered in a broad literature. For starters, see William Agosta's books cited above. This topic is also discussed in several chapters of John King, *Reaching for the Sun: How Plants Work* (Cambridge and New York: Cambridge University Press, 1997).

4. For recent detailed work with the monarch butterfly, see Karen S. Oberhauser and Michelle J. Solensky, eds., *The Monarch Butterfly: Biology and Conservation* (Ithaca, NY: Comstock Press, 2004).

5. The study of "false butterfly eggs" was published by A. M. Shapiro, and titled "Egg-mimics of *Strepanthus* (Cruciferae) Deter Oviposition by *Pieris sisymbrii* (Lepidoptera: Pieridae)," *Oecologia* 48 (1981): 142–45.

6. Two books that discuss ant-plant mutualisms are: Andrew

J. Beattie, *The Evolutionary Ecology of Ant-plant Mutualism* (Cambridge and New York: Cambridge University Press, 1985); and Pierre Jolivet, *Ants and Plants: An Example of Coevolution* (Champaign, IL: Backhuys Publishers, 1996).

7. Here are a few references for a somewhat speculative area of inquiry. The corn story is discussed by H. T. Alborn et al. in "An Elicitor of Plant Volatiles from Beet Armyworm Oral Secretion," *Science* 276 (1997): 945–49. In this case corn plants diffuse volatiles that attract wasp parasites of armyworm. Clarence A. Ryan and Daniel S. Moura discuss "Systemic Wound Signaling in Plants: A New Perception," *Proceedings of the National Academy of Sciences USA* 99 (2002): 6519–20. Going much further out on a limb is Anthony Trewavas, "Aspects of Plant Intelligence," *Annals of Botany* 92 (2003): 1–20.

Chapter 5: How Are the Flowering Plants Distinguished?

1. For a college-level survey of paleobotany highlighting changes in the world's land flora over 400 million years, see K. J. Willis and J. C. McElwain, *The Evolution of Plants* (Oxford and New York: Oxford University Press, 2002). A shorter article (based on their more detailed book) is Paul Kenrick and Peter Crane, "Origin and Early Evolution of Plants on Land," *Nature* 389 (1997): 33–34. Another advanced text that looks at the structural evolution of land plants over time is Karl Niklas, *The Evolutionary Biology of Plants* (Chicago: University of Chicago Press, 1997).

2. For a short review of early land plants, see J. B.

Richardson, "Origin and Evolution of the Earliest Land Plants," in *Major Events in the History of Life*, ed. J. William Schopf (Sudbury, ON: Jones & Bartlett Publishers, 1993), pp. 95–118. A more recent technical paper of algal ancestors is Louise A. Lewis and Richard M. McCourt, "Green Algae and the Origin of Land Plants," *American Journal of Botany* 91 (2004): 1535–56.

3. Will H. Blackwell discusses the origin of land plants' life cycle in "Two Theories of Origin of the Land-plant Sporophyte: Which Is Left Standing?" *Botanical Review* 69 (2003): 125–48.

4. For a better understanding of "plumbing systems" in plants, check the discussions of plant anatomy in the botanical textbooks cited earlier, such as Peter H. Raven, Ray F. Evert, and Susan E. Eichorn, *Biology of Plants*, 6th ed. (New York: W. H. Freeman & Co., 1999).

5. A superbly illustrated introduction to plant identification and important plant families is James L. Castner, *Photographic Atlas of Botany and Guide to Plant Identification* (Gainesville, FL: Feline Press, 2004). A good text that reviews plant classification and terminology, as well as flowering plant families, is Dennis Woodland, *Contemporary Plant Systematics* (Berrien Springs, MI: Andrews University Press, 1997). A review from a more modern perspective is Walter S. Judd et al., *Plant Systematics: A Phylogenetic Approach* (Sunderland, MA: Sinauer Associates, 1999).

6. The angiosperm carpel is a somewhat abstract concept. Here again, a college-level textbook can give you additional insights into the nature and diversity of the gynoecium in flowering plants. Many botanists have thought that both the primitive stamen and primitive carpel were leaflike in nature in early

angiosperms. I think that's right for the carpel but wrong for the stamen; see my paper "Are Stamens and Carpels Homologous?" in *The Anther: Form, Function and Phylogeny*, ed. W. G. D'Arcy and R. C. Keating (Cambridge and New York: Cambridge University Press, 1996), pp. 111–17.

7. The important new fossil find from China was described by Ge Sun et al. in "Archaefructaceae, A New Basal Angiosperm Family," *Science* 296 (2002): 899–903.

8. Again, most botany texts will discuss the salient differences between the monocots and dicots. I've even gone so far as to suggest that the seed leaves or cotyledons of monocots and dicots are not homologous; see "The Question of Cotyledon Homology in Angiosperms," *Botanical Review* 64 (1998): 356–71.

9. Modern ways of dividing up the angiosperms have been discussed in a large number of recent technical papers that will not be accessible to the general reader. However, for those with the proper background, here are a few important titles. Recent molecular studies are discussed by Pamela S. Soltis and Douglas E. Soltis in "The Origin and Diversification of Angiosperms," *American Journal of Botany* 91 (2004): 1614–26. A useful morphological review of what are thought to be ancient lineages of living angiosperms is given by Peter K. Endress in "Structure and Relationships of Basal Relictual Angiosperms," *Australian Journal of Botany* 17 (2004): 343–66. Finally, the Angiosperm Phylogeny Group (Mark Chase et al.) gives an update of the Angiosperm Phylogeny Group classification for the orders and families of flowering plants: APG II. *Botanical Journal of the Linnean Society London* 141 (2003): 399–436.

Chapter 6: What Makes the Flowering Plants So Special?

1. Norman F. Hughes, *Paleobiology of Angiosperm Origins* (Cambridge and New York: Cambridge University Press, 1976), p. 36.

2. Francis Hallé discusses plant form and structure in his elegantly illustrated *In Praise of Plants* (Portland, OR: Timber Press, 2002). For a technical review of the genetic background that has allowed flowering plants to form so many structures, see Annette Becker and Günter Theissen, "The Major Clades of MADS-box Genes and Their Role in the Development and Evolution of Flowering Plants," *Molecular Phylogeny and Evolution* 29 (2003): 464–59.

3. G. Ledyard Stebbins and I discussed the question of "Why are there so many kinds of flowering plants?" in a pair of articles in *Bioscience* 31 (1981): 572–81. For a discussion of diversity in moist tropical environments, see John Terborgh, *Diversity and the Tropical Rain Forest* (New York: Scientific American Library, 1992).

4. For a good general discussion of plant physiology and photosynthesis, see John King, *Reaching for the Sun: How Plants Work* (Cambridge and New York: Cambridge University Press, 1997).

Chapter 7: Primates, People, and the Flowering Plants

1. R. W. Sussman discusses the angiosperm-primate relationship in "Primate Origins and the Evolution of Angiosperms,"

American Journal of Primatology 23 (1991): 209–23, and also in "How Primates Invented the Rainforest and Vice Versa," in *Creatures of the Dark: The Nocturnal Prosimians*, ed. L. Alterman et al. (New York: Plenum Press, 1995), pp. 1–9.

2. I suggested that flowering trees played a key role for the evolution of larger "swinging" apes in chapter 7 of *Perfect Planet, Clever Species: How Unique Are We?* (Amherst, NY: Prometheus Books, 2003).

3. For an up-to-date and readable discussion of how we achieved our bipedal gate, see Craig Stanford, *Upright: The Evolutionary Key to Becoming Human* (New York: Houghton Mifflin, 2003).

4. Human-mediated extinctions at the end of the ice ages are still a very controversial topic in some quarters. But see Todd Surovell, Nicole Waguespack, and P. J. Brantingham, "Global Archaeological Evidence for Proboscidean Overkill," *Proceedings National Academy of Science, USA* 102 (2005): 6231–36; and David A. Burney and Timothy F. Flannery, "Fifty Millennia of Catastrophic Extinctions after Human Contact," *Trends in Ecology and Evolution* 20 (2005): 395–400.

5. For a technical article on ant agriculture, see Ulrich G. Mueller, S. A. Rehner, and T. R. Schultz, "The Evolution of Agriculture in Ants," *Science* 281 (1998): 2034–38.

6. A fine overview of early agriculture is Bruce D. Smith, *The Emergence of Agriculture* (New York: Scientific American Library, 1995). Another important summary is Richard S. MacNeish, *The Origins of Agriculture and Settled Life* (Norman: University of Oklahoma Press, 1992). A more general review is Jack R. Harlan, *The*

Living Fields: Our Agricultural Heritage (Cambridge and New York: Cambridge University Press, 1995). A symposium volume covering this topic is: C. Wesley Cowan and P. J. Watson, eds. *The Origins of Agriculture: An International Perspective* (Washington, DC: Smithsonian Institution Press, 1992).

7. The earliest record of rice growing is presented by Zhao Zhijun, "The Middle Yangtze Region in China Is One Place Where Rice Was Domesticated: Phytolith Evidence from the Diaotonghuan Cave, Northern Jiangxi," *Antiquity* 72 (1998): 885–97.

8. For a review of the idea of "agricultural centers," see D. R. Harris, "Vavilov's Concept of Centers of Origin of Cultivated Plants: Its Genesis and Its Influence on the Study of Agricultural Origins," *Biological Journal of the Linnean Society* 39 (1990): 7–16. For the Western Hemisphere, see Bruce D. Smith, "The Origins of Agriculture in the Americas," *Evolutionary Anthropology* 3 (1944–45): 174–84.

9. For a very negative view of agriculture, see Richard Manning, *Against the Grain: How Agriculture Has Hijacked Civilization* (New York: North Point Press, 2004). The subtitle itself is an oxymoron; without agriculture there are no grand civilizations. But his book does make some telling points regarding modern government-subsidized agriculture.

10. A short technical reference to legume nodules is Jean Marx, "The Roots of Plant-microbe Collaborations," *Science* 304 (2004): 234–36.

11. Not only did humans "domesticate" some plants and animals, many anthropologists believe that in the process we our-

selves became domesticated. See Helen M. Leach's article, "Human Domestication Reconsidered," *Current Anthropology* 44 (2003): 349–68.

12. This story is discussed by Benjamin Orlove, John Chang, and Mark A. Cane, "Ethnoclimatology in the Andes," *American Scientist* 90 (2002): 428–35.

13. For a good review of recent developments in plant breeding, see Paul F. Lurquin, *High Tech Harvest: Understanding Genetically Modified Food Plants* (New York: Perseus Books, 2002). And for a fine review of the science of genetics, focusing on our own species, see Henry Gee, *Jacob's Ladder: The History of the Human Genome* (New York: W. W. Norton, 2004).

14. The figures for domestic animal numbers come from J. R. McNeill, *Something New Under the Sun: An Environmental History of the Twentieth Century* (New York: W. W. Norton, 2000).

Chapter 8: How Flowers Changed the World

1. The beginning quote is from page 108 in Michael Pollon's delightful *The Botany of Desire: A Plant's Eye View of the World* (New York: Random House, 2001). He delves into the history and biology of four cultivated plants that have had strong interactions with our own species: apples, tulips, the marijuana plant, and potatoes.

2. The concept of progress in evolution is discussed in greater detail in chapter 6 of my *Perfect Planet, Clever Species* (Amherst, NY: Prometheus Books, 2003). The idea of progress also brings up the question of "design" in nature; check out how philosopher

Michael Ruse deals with this issue in *Darwin and Design: Does Evolution Have a Purpose?* (Cambridge and New York: Harvard University Press, 2003).

3. The numbers of insects supported by temperate trees was reviewed by T. Southwood in "The Number of Species of Insect Associated with Various Trees," *Journal of Animal Ecology* 30 (1961): 1–8.

4. Evidence for the importance of chemicals in resistance to insects in *Nicotiana* was reported by André Kessler, Rayko Halitschke, and Ian T. Baldwin, in "Silencing the Jasmonate Cascade: Induced Plant Defenses and Insect Populations," *Science* 305 (2004): 665–68.

5. Consuelo M. De Moraes and Mark C. Mescher, "Biochemical Crypsis in the Avoidance of Natural Enemies by an Insect Herbivore," *Proceedings National Academy of Science, USA* 101 (2004): 8993–97.

6. Beetles, especially, have increased their numbers thanks to flowering plants; see B. D. Farrell, "'Inordinate Fondness' Explained: Why Are There So Many Beetles?" *Science* 281 (1998): 555–59.

7. Harald Schneider et al., "Ferns Diversified in the Shadow of Angiosperms," *Nature* 428 (2004): 553–56. For a fine overview of ferns, see Robbin Moran, *The Natural History of Ferns* (Portland, OR: Timber Press, 2004).

8. Gregory J. Retallack, "Cenozoic Expansion of Grasslands and Climatic Cooling," *Journal of Geology* 109 (2001): 407–26.

9. For more about the impact of agriculture, see the references about agriculture in the last chapter.

10. Mining the land is a concept that has escaped the notice of most commentators on our planet's future. In contrast, take a good, long look at Daniel Hillel, *Out of the Earth: Civilization and the Life of the Soil* (Berkeley: University of California Press, 1991). For the demise of forests over the centuries, see Michael Williams, *Deforesting the Earth: From Prehistory to Global Crisis* (Chicago: University of Chicago Press, 2003).

11. The idea that we humans simply wouldn't be here without the flowering plants is part of a series of "lucky breaks" that I argued may make our radio telescopes unique in the Milky Way galaxy. This is the central theme of *Perfect Planet, Clever Species.*

Epilogue

1. For a general history, see Christopher Thacker, *The History of Gardens* (Berkeley: University of California Press, 1979). For a beautiful view into the Japanese garden, see *The Hidden Gardens of Kyoto*, by Katsuhiko Mizuno (photographs) and Masaki Ono (text), trans. L. E. Riggs and C. Imoto (Tokyo: Kodansha International, 2004). A comprehensive survey of the use of flowers in many human societies is Jack Goody's *The Culture of Flowers* (Cambridge and New York: Cambridge University Press, 1993).

2. These numbers come from Christiansen Capital Advisors, quoted in George Vecsey's column, "Forget Pete Rose; The Country Has a Betting Habit," *New York Times*, January 18, 2004, Sports, p. 6.

3. For more about the Gaia hypothesis, see Lynn Margulis,

Symbiotic Planet: A New View of Evolution (New York: Basic Books, 1998), chap. 8; and also Elisabet Sahtouris, *Gaia: The Human Journey from Chaos to Cosmos* (New York: Pocket Books, 1989). For short reviews, see C. Barlow and T. Volk, "Gaia and Evolutionary Biology," *Bioscience* 42 (1992): 686–92; S. Schneider, "A Goddess of Earth or the Imagination of a Man?" *Science* 291 (2001): 1906–1907; and "The Living Earth" by the originator of the Gaia hypothesis, James Lovelock, in *Nature* 426 (2003): 769–70.

4. For references on the human population explosion, you might start with Joel E. Cohen's *How Many People Can the Earth Support?* (New York: W. W. Norton, 1995); and a more recent article by the same author, "Human Population: The Next Half Century," *Science* 302 (2003): 1172–75. For a very optimistic view of feeding all the world's people, see Norman Borlaug, "Feeding a World of 10 Billion People: The Miracle Ahead," in *Global Warming and Other Eco-Myths*, ed. Ronald Bailey (New York: Prima Publisher, 2002), pp. 29–59.

5. Ecotourism is one way to sustain nature's original gardens. Daniel Janzen sees the ecotourist as "a better kind of cow" visiting "pastures" that are maintained as conservation areas. See "How to Grow a Wildland: The Gardenification of Nature," in *The Biodiversity Crisis*, ed. Michael J. Novacek (New York: New Press, 2001), pp. 156–62.

glossary

Abscission: The controlled separation of leaves, flowers, and fruits from the stems that bear them, usually by the formation of a special layer of cells.

Accrescent: Increase in size of parts with age, as in the sepals of some fruits.

Achene: A simple dry, one-locular, one-seeded fruit that does not split open.

Actinomorphic flower: A flower that is radially symmetrical.

ACULEATE: Armed with spines; thorny or prickly.

ADAXIAL: The side toward the stem axis (especially during development, hence the upper side of a leaf).

ADVENTITIOUS ROOT: A root developing in an unusual part of the plant, as from a stem.

AERIAL ROOT: A root arising from above ground level.

AGGREGATE FRUIT: A fruit arising from the close coalescence of several ovaries within the flower.

AIR PLANTS: Epiphytes; growing in the air above the ground.

ALGAE: Eukaryotic; mostly simple aquatic plants.

ALKALOID: A compound with one or more carbon rings, and including nitrogen. Important in plants' defensive chemistry.

ALLELE: One of two or more variants of the same gene. Members of the same species have the same genes but differ in their alleles.

ALLELOPATHY: Where chemicals released by a plant inhibit the growth of other nearby plants.

ALPINE: High mountain vegetation above the timberline.

ALTERNATION OF GENERATIONS: Two life stages in the life cycle of a plant; from haploid plant to diploid plant, to haploid, etc.

Androecium: The male elements of a flower.

Anemophilous: Technical term for pollination by the wind.

Angiosperm: A lineage of seed plants characterized by ovules borne within the locules of ovaries; the division Magnoliophyta.

Annual: A plant that completes its life cycle and produces seed within a single growing season.

Annual ring: A distinctive ring in a cross section of wood, produced during a single growing season.

Anther: The saclike structure at the apex of the stamen where pollen grains are formed; see figure 2B (page 25).

Anthesis: The time at which a flower is in full bloom.

Anthocyanin: An important group of water-soluble plant pigments, varying from red to blue in color.

Apical bud: The growing tip at the end of shoots, often protected by bud scales.

Aquatic: Growing naturally on or under water.

Arborescent: Treelike, resembling a tree.

Archegonium: Flasklike female sex organ of nonseed plants in which the egg cell resides.

ARMED: Protected by thorns, spines, prickles, etc.

AROID: A member of the Araceae family.

AUTOGAMY: Self-fertilization.

AUTOTROPHIC NUTRITION: Nutrition produced by the plant itself through the process of photosynthesis.

AXIL: The angle between leaf attachment and the stem, in the direction of the stem apex. Where axillary buds develop.

BARK: The outer tissues surrounding the cambium helping protect the stem or trunk.

BERRY: A fruit in which the ovary wall and inner parts are enlarged and become juicy, as in grapes, tomatoes, etc.

BIENNIAL: A plant completing its life cycle in two growing seasons; producing fruits and seeds in the second year.

BIOSPHERE: Our all-encompassing ecosystems, including land, soil, oceans, and the atmosphere. Ecosphere.

BISEXUAL: Producing both male and female gametes. (Note that some plants are bisexual while bearing unisexual flowers.)

BLADE: The flattened part of a leaf or perianth part.

BOREAL FORESTS: Cold northern forests, usually dominated by evergreen conifers such as spruce, pine, and fir.

BRACT: A scalelike part or much-reduced leaf that subtends and protects a flower or inflorescence in early stages. A modified leaf enlarged in some flowers for show, as in *Poinsettia*.

BROMELIAD: A member of the Bromeliaceae family.

BRYOPHYTA: A major division of simple, nonvascular plants, including mosses, liverworts, and hornworts.

BUD: An undeveloped shoot with embryonic (meristem) tissues that can develop into stems, leaves, or flowers.

BUD SCALE: A modified and much-reduced leaf that protects the bud in early stages of development.

BULB: A shortened compact stem bearing modified leaves that store energy reserves, as in an onion.

CALLUS: A corky tissue covering wounds in woody tissues.

CALYX: Collective term for all the sepals of a flower. Plural: calyces. See figure 3A (p. 26).

CAMBIUM: A tubular meristem in stems where cells divide continually to produce new xylem toward the interior and new phloem toward the exterior.

Capitulum: An inflorescence of densely compacted flowers. Also called a head, as in the composite family.

Carpel: A simple pistil with one locule, one style, and one stigma. Also, the hypothetical leaf that became folded to enclose and protect the ovules in early angiosperm history.

Caryopsis: Technical term for the grain of a grass, a dry nondehiscent fruit with seed coat and ovary wall united.

Catkin: An elongate, slender, pendulous inflorescence, usually bearing wind-pollinated flowers, as in oaks, mulberry, etc.

Cell: The smallest functional unit in complex plants and animals. Plant cells have walls made of cellulose.

Cellulose: A complex carbohydrate built of glucose units; the chief constituent of plant cell walls.

Chlorophyll: A green pigment molecule central to photosynthesis, located in the chloroplast in cells of land plants.

Chloroplast: An organelle in photosynthesizing eukaryotic cells where chlorophyll resides and photosynthesis takes place.

Chromosome: A slender strand of DNA carrying the genes in a linear order; confined within the nucleus in plant cells.

Circadian rhythms: Regular rhythms of growth and activity in plants and animals, usually in step with the twenty-four-hour day.

CLADISTIC ANALYSIS: Method of using unique and derived characteristics to determine patterns of relationships.

CLASS: A rank in classification above the order, and included within the division.

CLASSIFICATION: The science of arranging organisms into groups based on their characteristics. See **Systematics** and **Taxonomy**.

CLEISTOGAMOUS FLOWERS: Flowers that do not open and are self-pollinated; many violets have such flowers in addition to their colorful normal flowers.

CLONES: Genetically identical plants or animals produced by asexual reproduction from the same individual parent.

COEVOLUTION: When two species interact closely and have affected each other's characteristics over time.

COMPLETE FLOWER: A flower that has all the normal flower parts.

COMPOSITAE: An older name for the sunflower family, Asteraceae.

COMPOSITE HEAD: The compact inflorescence of tightly compacted flowers in the composite or Asteraceae family; capitulum.

COMPOUND LEAF: A leaf with separate stalked leaflets. Leaves usually have axillary buds in their axils; leaflets do not.

Conifer: A general term for the Coniferinae, a class of cone-bearing gymnosperms, including pines, firs, junipers, cypress.

Conspecific: Within or belonging to the same species.

Cork: The outer, protective, tissue of bark.

Corm: A short, thickened, underground stem or base of the stem where the plant stores food.

Corolla: The petals of the flower considered together as a unit; see figures 1 (p. 24) and 3 (p. 26).

Cotyledon: The first leaf or leaves of the embryo, modified in some species to serve for food storage. The first seed leaf.

Cross-pollination: The transfer of pollen to another plant.

Cryptogams: Plants reproducing by spores (not seeds) such as ferns, lycopods, mosses, and liverworts. See **Phanerogams**.

Cultivar: A cultivated plant variety produced by horticultural or agricultural techniques.

Cuticle: A waxy layer on the surface of leaves, stems, and other parts, protecting the surface against water loss.

Cyathium: The pseudoflower of the genus *Euphorbia*; see color insert I.

Cycad: A member of cycadinae, a class of gymnosperms; usually with thick stems, large cones, and palmlike pinnate leaves.

Cytoplasm: The living protoplasm that surrounds the nucleus within the eukaryotic cell.

Deciduous: (1) Shedding parts at the end of the growing season or at the end of flowering, as in deciduous leaves and fallen petals. (2) Trees and shrubs that drop their leaves at the end of the growing season.

Dehiscent: Fruits or other parts that split open at maturity are said to be dehiscent.

Derived: Said of a structure or organism believed to be descended from an older or more primitive precursor.

Determinate growth: Limited growth, ending at maturity; often used in inflorescences. See **Indeterminate growth**.

Development: The changes and growth of an individual from the fertilized zygote to the mature plant or animal.

Dicot: A member of the Dicotyledoneae; formerly the largest class of angiosperms, having two cotyledons on the embryo. The class Magnoliopsida.

Diffuse secondary growth: Thickening of stems by scattered cell divisions (rather than from a tubular cambium), as in palms.

DIMORPHISM: Having two different forms, as in juvenile and adult leaf forms, or having two kinds of flowers in the same species.

DIOECIOUS: A species having functional male and female parts on different individual plants; each plant being unisexual.

DIPLOID: Having two sets of chromosomes in the cells of the organism. One from a male gamete, the other from the female. Nearly all larger plants and animals are diploid.

DISC FLOWER: The small tubular and radially symmetrical flowers in the central disc of a composite inflorescence (capitulum).

DISSEMINULE: A plant part that can give rise to a new plant, or be dispersed and then give rise to a new plant.

DIURNAL: Being active during the day. Compare **Nocturnal**.

DIVISION: A rank in classification above the rank of class, and below the rank of kingdom. Zoologists call this rank a *phylum*.

DNA: Deoxyribonucleic acid; the basic chemical unit that forms a long chain on which base pairs are aligned to provide genetic information.

DORMANCY: Period of time during which the plant is inactive.

DOUBLE FERTILIZATION: Where one nucleus from the pollen tube fertilizes the egg cell, and an additional nucleus joins with one or two female nuclei to initiate endosperm.

Drip tip: A long and narrowly pointed leaf tip, common in rain forest species. Thought to help water run off the leaf.

Drupe: A simple fleshy fruit from a single ovary and with a single seed, as in peaches, cherries, olives, etc.

Egg cell: A female gamete in plants, residing within a protected site such as the ovule or archegonium.

Embryo: The earliest organized stage of plant development.

Endemic: Native to a limited region and not found elsewhere.

Endosperm: A food tissue, stored within the seed.

Endosymbiosis: A symbiotic relationship where one partner lives within the other. This is thought to have been how mitochondria and chloroplasts became a part of eukaryotic cells.

Entomophilous: Technical term for pollination by insects.

Enzyme: A protein molecule functioning as a catalyst in biochemical reactions.

Epidermis: The outer layer of cells on most plant parts.

Epipetalous: Borne on the petals, as in epipetalous stamens.

Epiphyte: A plant growing on another plant or plant part for physical support. Not an invasive parasite.

Essential elements: Elements required by plants or animals for proper growth and development.

Eukaryotic cells: Cells with a nucleus and mitochondria; the basis of all higher life forms on Earth.

Evergreen: A woody perennial plant that retains its leaves throughout the year or over a number of seasons.

Evolution: The biological concept of changes in plants and animals having accrued over periods of time. Species and populations (not individuals) evolve over time.

Exotic: Originating from another place; foreign, not native.

Family: A rank in classification above the genus, including genera more closely related to each other than to genera outside the family.

Fascicle: A close cluster of flowers, leaves, stems, etc.

Fermentation: Breakdown of food molecules in the absence of oxygen to produce carbon dioxide, ethyl alcohol, and energy.

Fertilization: (1) Union of male and female sex cells to create the zygote. (2) Adding nutrients to soil to increase productivity.

Fiber: A long, thick-walled cell, dead at maturity and providing woody stems their strength.

Fibrous roots: A branched, spreading root system, lacking a strong taproot. Characteristic of monocotyledonous plants.

Filament: The usually slender stalk of a stamen, bearing the anther; see figure 2B (p. 25).

Fleshy: Plant tissues that are succulent, thick, and moist within.

Flora: (1) The plant life of a region. (2) A book that lists and describes the plants of a region.

Floral tube: Tube or cup formed by union of floral parts.

Floret: A small flower; often used for the small flowers in composite heads (as in figure 4A [p. 27]), and within the spikelets of grasses (see color insert F).

Flower: The reproductive unit in angiosperms. Flowers vary greatly in different lineages and appear to be polyphyletic.

Food: Organic substances rich in energy and mineral nutrition; used by animals and plants for their sustenance. In plants, mineral supplements are often called *plant food*.

Food chain: Pathway of chemical energy in an ecosystem, from photosynthetic producers to herbivores to carnivores and decomposers.

Frugivore: Animal that eats fruits.

Fruit: The product of a fully mature ovary, usually with seeds.

Gall: A local abnormal growth on a plant, usually caused by insects, bacteria, or fungi.

Gamete: A sex cell with haploid number of chromosomes; the sperm or egg cell.

Gametophyte: A haploid plant that produces gametes among nonseed plants. See **Alternation of generations**.

Gene: A unit of genetic inheritance borne as a sequence of DNA units on the chromosome; often named and defined by the effects caused by its mutant forms.

Genotype: The basic genetic makeup of an organism. The effects of the genotype can be modified by environmental and other effects during development; see **Phenotype**.

Genus: A rank above the species level, including all those species more closely related to each other than to any other species.

Geotropism: Growth of a plant part in response to gravity.

Germination: The initial growth of a seed, spore, or pollen grain.

Graft: The union of one plant stem to that of another in order to propagate the grafted stems. Important in horticulture.

GRAIN: The single-seeded fruit of a grass; see **Caryopsis**.

GUARD CELL: One of a pair of cells that surround the stomate.

GYMNOSPERM: A member of the division Gymnospermae whose seeds often develop on bracts that are part of a cone.

HABIT: The general form or appearance of a plant.

HABITAT: The natural environment of a species or organism.

HAPLOID: Having only one set of chromosomes; having half the number of chromosomes most organisms possess; see **Gamete**.

HAUSTORIUM: A rootlike organ produced by parasitic plants that penetrates the tissue of a host for the extraction of nutrition.

HAWK MOTHS: An important group of pollinators, the Sphingidae; their fast wing beats allow them to hover like hummingbirds.

HEARTWOOD: The darker-colored wood in the center of most woody tree trunks.

HERBACEOUS: Usually smaller plants that grow for one season and then die back completely to the ground.

HERBARIUM: A collection of many dried plant specimens, systematically filed, used for study and for comparisons.

HERBIVORE: An animal that consumes only plants.

HETEROTROPHIC NUTRITION: Using organic materials for nutrition; from meat and potatoes, to rotting logs. See **Autotrophic nutrition**.

HETEROZYGOUS: Having both dominant and recessive alleles for a particular gene. Also used to indicate mixed parentage and having greater genetic diversity. See **Homozygous**.

HOMOZYGOUS: Having identical alleles for a particular gene, or groups of genes in an individual plant or animal.

HORMONE: Mobile messenger molecules, operating in small amounts, that control growth and development within the plant.

HORTICULTURE: The science of garden plants, ornamentals, and orchard crops.

HOST: A plant harboring a parasite, or subject to being fed upon or infected.

HUMMINGBIRDS: An important guild of pollinators in the Americas that do not need to perch while feeding from flowers.

HUMUS: Organic matter in the soil from the decomposition of plants and animals. A major source of soil nutrients.

HYBRID: A plant that is the offspring of two different species or different varieties of the same species.

Inbreeding: Self-pollination (within the same plant) or pollination among the plants of the same small population. Mating of closely related individuals.

Incomplete flower: Old term for a flower lacking one or more of the normal floral whorls.

Indeterminate growth: Unlimited continuing growth. Also used for a shoot apex that can produce many lateral shoots.

Inflorescence: A shoot bearing a number of flowers; the arrangement of flowers along an expanded axis or series of stems.

Insectivorous plants: Plants that capture and hold insects, digesting them to obtain nitrogen and other nutrients.

Internode: The section of a stem between two leaves or two nodes.

Introduced: A plant brought into one area from another.

Involucre: One or several whorls of bracts subtending a flower or inflorescence.

Kingdom: The largest, most inclusive, taxonomic rank.

Labiate: Having a lip, as in flowers with a large lower corolla lobe. Such flowers are characteristic of the mint family (as in fig. 3A [p. 26]), orchids (see color insert N), and others.

LAMINA: The flat expanded part of a leaf; the blade.

LATERAL: Borne on the side of an organ.

LATERAL BUD: Same as axillary bud, from the axil of a leaf, or from the axil of an undeveloped or fallen leaf.

LATERAL MERISTEM: An active tissue of dividing cells, along the length of the stem or root. See **Cambium**.

LATEX: A usually colored sap in the outer tissues of many plants; a source of rubber from a few species.

LEAF: A usually flattened outgrowth of a stem; the principal site of photosynthesis in most plants.

LEAFLET: A distinct leaflike part of a compound leaf, with its own stalk (petiolule).

LIANA: A long-stemmed woody climbing plant, growing from the ground up into the trees.

LICHEN: A fungus harboring algae to form distinctive leafy structures or thin crustose films. Now called *lichenized fungi*.

LIFE CYCLE: The sequence from fertilization, early growth, development, maturity, and senescence of an organism.

LIGNIN: A complex organic substance, found around the cellulose fibers of cells in wood.

Locule: A chamber within the ovary; see figure 2B (p. 25).

Lycopods: Seedless vascular plants, traditionally including the club mosses (*Lycopodium*), spike mosses (*Selaginella*), and the quillworts (*Isoetes*); often referred to as fern allies.

Magnoliophyta: A major division of seed plants including all the flowering plants: angiosperms.

Marine: Living in the ocean.

Megaspore: A spore that develops into a female gametophyte.

Meiosis: The process of cell division producing the sex cells; cells with half the normal complement of chromosomes.

Meristem: A tissue of embryo-like cells capable of continuing cell division and differentiation over long time periods.

Metabolism: The sum of biochemical processes that keep a cell or organism alive.

Micronutrient: Any element required for successful growth and reproduction by a plant or animal in very small amounts. For plants iron, manganese, zinc, chlorine, etc., are micronutrients.

Microspore: A spore that develops into a male gametophyte.

Mitochondria: Cell organelles in which respiration takes place.

Mitosis: Normal cell division, producing two cells with the same chromosome number as the parent cell.

Monocot: Short for Monocotyledonae, once considered a class of flowering plants (Liliopsida), with only one seed leaf on the embryo.

Monophyletic: A group of organisms postulated to have had a single common ancestor.

Morphology: The study of form, structure, and development in plants.

Multiple fruit: A cluster of mature ovaries united together in fruit by a common axis, such as a pineapple.

Mutation: A heritable change in a gene or chromosome, whether natural or induced by experimentation.

Mutualism: A form of symbiosis in which the two partners, living together, both benefit.

Mycorrhiza: A symbiotic, nonpathogenic association between a fungus and the roots of a higher plant.

Natural selection: A process of population change over time, as those organisms more successful at reproduction replace the less successful individuals.

Nectar: A secretion from nectaries (usually within flowers) that contains sugars and other nutrients.

Nectar guides: Color patterns and grooves within the flower that help guide the pollinator to the nectar.

Nectary: A gland or tissue secreting nectar in plants.

Nitrogen-fixation: The conversion of atmospheric nitrogen into organic nitrogen compounds by a few lineages of bacteria.

Nocturnal: Active at night, as in flowers that open at night.

Node: The point on the stem where leaves are attached and axillary buds arise.

Nomenclature: In biology, the naming of species and higher ranks; the correct usage of scientific names in taxonomy.

Nucleus: The organelle within the eukaryotic cell that contains the chromosomes and their normal functioning.

Nutrient: Any substance that provides energy or promotes growth in living organisms.

Omnivore: An animal that eats both plants and animals.

Order: A rank in classification above the family, including families more closely related to each other than to families outside the order.

Organelle: A separate, smaller structure within the cell that performs a special function.

Outbreeding: The pollination of plants unlike each other or less closely related. The opposite of inbreeding.

Ovary: The basal part of a pistil that encloses the ovules and will develop into the fruit.

Ovule: A special structure that houses the female sex cell and will, with fertilization, become the seed. The evolutionarily reduced female gametophyte of seed plants. See figure 2B (p. 25).

Palmate: A pattern radiating from a basal point, as in many palm leaves and in palmate leaf venation.

Panicle: A highly branched indeterminate inflorescence, as in many grasses.

Pappus: The scales, hairs, or bristles ringing the base of the corolla tube in the composite family; see figure 4B (p. 27).

Parallel venation: Where the veins run parallel with each other for most of the length of the leaf; common in monocots.

Parasite: An organism obtaining nutrition from the living tissue of a host, usually to the detriment of the host.

Pedicel: The stalk of an individual flower in an inflorescence. Compare **Peduncle**.

Peduncle: The stalk of an inflorescence or of a solitary flower.

Perennial: A plant living through several growing seasons. Note that the aboveground parts may die back in perennial herbs.

Perianth: All the sepals and petals (or tepals) of a flower.

Pericarp: The outer fruit wall derived from the ovary wall.

Petal: A broad and visible part of the flower, usually arising in a whorl from within the sepals; see figure 1 (p. 24).

Petiole: The stalk of a leaf; the long, narrowed leaf base.

Phanerogams: Plants reproducing by seeds, also called *spermatophytes*. Compare **Cryptogams**.

Phenology: The study of seasonality in plants and animals.

Phenotype: The form of an organism after it has been affected by its development and surroundings. Compare **Genotype**.

Phloem: The food-conducting vascular tissue of plants.

Photoperiodism: The response of organisms to the length of day or night, controlling many aspects of plant growth and reproduction.

Photosynthesis: The process of transforming light energy into chemical energy, using carbon dioxide and water to build carbohydrates.

Phototropism: Growth response by a plant to the direction of light, both positive (toward the light) or negative (away).

Phyllotaxy: The arrangement of leaves along the stem.

Phylogeny: The evolutionary history and relationships within a group of organisms.

Physiology: The study of the life processes of organisms.

Phytotoxin: A plant substance having toxic effects on herbivores.

Phytogeography: The study of plant distributions and geography.

Pinnate venation: Leaves with a single midvein from which emerge the lateral secondary veins.

Pistil: The female element of a flower usually comprising ovary, style, and stigma; see figure 2 (p. 25). A pistil may be constructed of one or more carpels.

Pollen grain: The minute rounded structure that germinates on the stigmatic surface. A microspore carrying a reduced male gametophyte in seed plants. (*Pollen* is also a collective term.)

Pollination: The transfer of pollen to stigma (in angiosperms) or ovule (in gymnosperms).

Pollinium: A coherent mass of pollen grains, as in orchids.

Polyphyletic: A group of similar organisms but that are believed to lack a single common ancestor; compare **Monophyletic**.

Polyploid: An organism or tissue with more than two sets of chromosomes.

Primary growth: Growth arising from the apical meristems, as in shoots and roots. Compare **Secondary growth**.

Primitive: The ancestral condition (often hypothetical).

Primordium: A plant part in its first stages of development.

Propagule: Any plant part that becomes separated from the plant and can give rise to a new plant.

Protoplasm: The living material of the cell, including cytoplasm, nucleus, and organelles.

Pseudoflowers: Inflorescences structured so as to resemble a true flower. See figure 4 (p. 27) and color inserts I and J.

Raceme: Inflorescence with a single erect axis bearing many stalked (pedicellate) flowers.

Ray flower: Floret with a long strap-shaped petal circling the disc flowers in many composite heads.

Receptacle: The base of the flower from which floral parts arise; the basal center of the flower.

Reproduction: The formation of new individuals by sexual or asexual means.

Respiration: The process of deriving energy from the breakdown of carbohydrates by using oxygen.

Rhizome: An underground horizontal stem.

Root: Usually the underground portions of a plant that anchor the plant, and absorb water and nutrients from the soil.

Saprophyte: A plant that derives energy from dead organic matter.

Secondary growth: Growth resulting from the lateral meristems, increasing the girth of woody tissues and bark.

Seed: The product of a fertilized ovule, containing the embryo and stored food for the growth of the seedling.

Seed coat: The protective outer layer of the seed.

Self-incompatible: A plant that is self-sterile.

Self-pollination: Transfer of pollen from the anther to a stigma of the same flower or another flower of the same plant.

Sepal: Outer part of the perianth, usually protecting the flower bud in early stages; see figure 1A (p. 24).

Sessile: Sitting on a base; without a stalk.

Sexual reproduction: Union of sex cells (gametes) produced by meiosis to produce a new diploid organism.

Shoot: A stem bearing leaves.

Shrub: A woody plant with multiple stems from the base, lacking a single trunk.

Spadix: A thick floral axis bearing flowers; often enclosed within a leaflike spathe in aroids.

Spathe: A leaflike or petal-like structure that encloses or subtends the inflorescence as in aroids (see color insert H).

Species: A population or series of populations capable of interbreeding, and (usually) not interbreeding with any other species. The basic unit in biological classification, designated by a Latin binomial name (the specific epithet).

Sperm: The male sex cell; the male gamete.

Spike: Inflorescence with a single erect axis bearing many sessile flowers; see figure 3A (p. 26).

Spikelet: A small condensed spike, characteristic of grass and sedge inflorescences; see color insert F.

Spore: A disseminule of primitive plants that grows into a new plant. Usually a single cell that may be haploid or diploid.

Sporophyte: A diploid plant that produces haploid spores after meiosis. See **Alternation of generations**.

Spur: (1) A tubular projection at the rear of a flower holding nectar; see color insert D. (2) A greatly shortened lateral branch, a short shoot.

Stamen: The male, pollen-producing part of a flower, usually made up of slender filament and pouchlike anther; see figures 1A (p. 24) and 2B (p. 25).

Staminode: A sterile stamen, often making the flower showier.

Starch: The primary food-storage carbohydrate of higher plants, made up of many glucose units.

Stem: The aboveground axis on which leaves and flowers are borne. Also, highly modified underground stems such as bulbs, corms, and rhizomes.

STERILE: Lacking functional reproductive structures.

STIGMA: The part of the pistil or style that receives pollen; the surface on which pollen germinates.

STIPULE: A lateral outgrowth at the base of the petiole, often protecting the shoot apex in early development, usually paired.

STOMATE: A minute pore on leaf surfaces that can open for gas exchange and close to prevent water loss; stoma (pl. stomata).

STYLE: The narrow portion of a pistil above the ovary that bears the stigmas. (Some pistils lack styles, having sessile stigmas.)

SUBSPECIES: A distinct subdivision of a species, often geographically separate from other subspecies of that species.

SUCCESSION: The ecological sequence as bare earth is colonized and, with new species entering, becomes a larger stable community.

SUCCULENT: A plant with thick water-storing stems; fleshy.

SYMBIOSIS: The living together of two different organisms for mutual benefit, or for the benefit of one if the other is not harmed. Compare **Mutualism** and **Parasite**.

SYSTEMATICS: The science of determining the relationships of living things, including the naming, classifying, and analysis of their evolutionary diversification over time.

TANNIN: A complex astringent substance found in many plant tissues; a defense against herbivores and decay.

TAP ROOT: A prominent primary root derived from the base of the embryo; characteristic of dicots and gymnosperms.

TAXONOMY: The science of classification; giving names to plants and animals. Now included under systematics.

TENDRIL: A slender modified stem or leaf that usually coils and serves as a means of attachment for climbing plants.

TEPAL: A perianth part in flowers where sepals and petals are very similar (as in the lilies of chapter 1).

TERMINAL BUD: The apical bud of the individual shoot.

TIMBERLINE: The upper limit of tree growth on mountains.

TISSUE: A group of cells of the same type, forming a unit with common function.

TRACHEOPHYTES: Plants with vascular tissue; ferns, seed plants.

TRANSPIRATION: The loss of water as water vapor from a plant.

TREE: A large long-lived plant with distinctive woody trunk.

TRICOLPATE: With three grooves; as in the pollen of most dicots.

TROPISM: The growth response of a plant to an external stimulus, such as light or gravity.

TUBER: A thick dense or fleshy, usually underground, stem.

TUNDRA: A low-growing, treeless vegetation in very cold areas.

TURGOR PRESSURE: The pressure within a cell when it is filled with water. Lack of turgor results in wilting.

TYPE SPECIMEN: A specimen permanently attached to a scientific species name, and the physical reference for that name.

UMBEL: An inflorescence where all the flower stalks (pedicels) arise from the same point at the apex of the stem; see color insert K.

UNDERSTORY: Smaller herbs, shrubs, and trees growing under the canopy of the forest.

VASCULAR BUNDLE: A strand of specialized cells containing both xylem and phloem; the water- and food-conducting tissue.

VASCULAR PLANTS: Plants with water- and food-conducting tissues in their stems; ranging from ferns through seed plants.

VEIN: A strand of vascular tissue (a vascular bundle) in the leaf blade or floral parts.

VENATION: The pattern of veins in leaves and other structures; see **Parallel** and **Pinnate venation**.

VESSEL: A water-conducting xylem cell in flowering plants.

WEED: (1) A general term for plants that colonize newly disturbed open soil. (2) A plant growing where we don't want it.

WHORLED: Arrangements of leaves or flowers arising at the same level around the stem.

WOOD: The dense tissue composed of secondary xylem in stems, trunks, and roots.

XANTHOPHYLL: A pigment associated with photosynthesis, yellow or brown in color; related to carotene.

XYLEM: The water- and mineral-conducting tissue of plants. Dead and empty when mature, xylem cells also provide strength.

ZYGOMORPHIC FLOWER: A bilaterally symmetrical flower.

ZYGOTE: The result of fertilization of an egg cell by sperm. The diploid zygote begins to divide and will form the embryo.

index

References to figures are in **boldface**.

lonely planet
D0052004
LONDON
ENCOUNTER
JOE BINDLOSS

London Encounter

Published by Lonely Planet Publications Pty Ltd
ABN 36 005 607 983

Australia Head Office, Locked Bag 1, Footscray, Victoria 3011
☎ 03 8379 8000 fax 03 8379 8111
talk2us@lonelyplanet.com.au

USA 150 Linden St, Oakland, CA 94607
☎ 510 250 6400
toll free 800 275 8555
fax 510 893 8572
info@lonelyplanet.com

UK 2nd fl, 186 City Rd, London EC1V 2NT
☎ 020 7106 2100 fax 020 7106 2101
go@lonelyplanet.co.uk

This title was commissioned in Lonely Planet's London office and produced by: **Commissioning Editor** Clifton Wilkinson **Coordinating Editors** Kristin Odijk, Pat Kinsella **Coordinating Cartographer** Jolyon Philcox **Assisting Cartographer** Erin McManus **Layout Designer** Indra Kilfoyle **Senior Editors** Helen Christinis, Katie Lynch **Managing Cartographer** Mark Griffiths **Project Manager** Rachel Imeson **Managing Layout Designer** Celia Wood **Thanks to** Chris Girdler, Lisa Knights, Charity Mackinnon, Katie O'Connell, Trent Paton, Amanda Sierp

ISBN 978 1 74179 047 4

Printed through Colorcraft Ltd, Hong Kong.
Printed in China.

Acknowledgements London Underground Map © Transport for London 2006; Central London Bus Map and Tourist Attractions Map © Transport for London 2006

HOW TO USE THIS BOOK

Colour-Coding & Maps

Colour-coding is used for symbols on maps and in the text that they relate to (eg all eating venues on the maps and in the text are given a green knife and fork symbol). Each neighbourhood also gets its own colour, and this is used down the edge of the page and throughout that neighbourhood section.

Shaded yellow areas on the maps denote 'areas of interest' – for their historical significance, their attractive architecture or their great bars and restaurants. We encourage you to head to these areas and just start exploring!

Prices

Multiple prices listed with reviews (eg £10/5 or £10/5/20) indicate adult/child, adult/concession or adult/child/family.

Send us your feedback We love to hear from readers – your comments help make our books better. We read every word you send us, and we always guarantee that your feedback goes straight to the appropriate authors. The most useful submissions are rewarded with a free book. To send us your updates and find out about Lonely Planet events, newsletters and travel news visit our award-winning website: ***lonelyplanet.com/contact***.

Note: We may edit, reproduce and incorporate your comments in Lonely Planet products such as guidebooks, websites and digital products, so let us know if you don't want your comments reproduced or your name acknowledged. For a copy of our privacy policy visit ***lonelyplanet.com/privacy***.

JOE BINDLOSS

Joe Bindloss lived briefly on London's Caledonian Rd as a child and he headed back here as soon as he was old enough to work out the tube map. He's lived in north London ever since, in between stints working overseas. Joe has written for more than 30 Lonely Planet guidebooks, from *India* and *Nepal* to Lonely Planet's *Gap Year* book. When not travelling for Lonely Planet, he writes about travel, restaurants and life in London for the *Independent*, the *Guardian, Wanderlust* and other publications. For more information, see www.bindloss.co.uk.

The 1st edition of *London Encounter* was written by Sarah Johnstone.

JOE'S THANKS

First and foremost, my thanks to my partner Linda for providing a second opinion on trips to restaurants, bars and sights. Thanks also to the friends who provided dining tips – Gar Powell Evans, I owe you one – and the helpful Londoners who gave interviews for this book.

Our readers Many thanks to the travellers who wrote to us with helpful hints, useful advice and interesting anecdotes. Erica Mccreath, Jill Murphy, Michel Porro, Chris Rickwood, Caroline Winter, Rob Woodburn.

Cover photograph Millennium Bridge by Erica Rodriguez Sanz. This image was selected as one of the winners of a competition to find four unique travellers' photos that really convey the experience of using an Encounter guide. Entrants were challenged to submit eye-catching photos that get right to the heart of the city and give the impression of seeing and experiencing it for themselves. Check out the other winning images on *Barcelona*, *Paris* and *Istanbul Encounter* covers.
Internal photographs p21 Alex Segre/Alamy, p21, p77, p105, p141, p149 Joe Bindloss. All other photographs by Lonely Planet Images and by Adina Tovy Amsel p57, p90; Alain Evrard p27, p142; Barbara Van Zanten p14; Charlotte Hindle p20; Doug McKinlay p6 (bottom), p10, p11, p12, p17, p29, p69, p73, p95, p132, p161, p168, p178, p188, p191, p202, p203; Gavin Gough p28, p33; Jonathan Smith p23, p81, p110, p181, p201; Juliet Coombe p136; Krzysztof Dydynski p25; Lawrence Worcester p32; Lou Jones p63; Manfred Gottschalk p64; Mark Daffey p34 (bottom); Martin Moos p24; Neil Setchfield p6 (top left), p8, p13, p15, p16, p19, p22, p30, p48, p78, p100, p106, p114, p116, p124, p128, p151, p170, p193, p195, p196, p197, p198, p199, p200; Orien Harvey p4, p8, p18, p26, p36, p184; Rachel Lewis p34 (top); Travis Drever p6 (top right), p38, p189.

Familiar faces – Parliament's Big Ben clocktower (p14) seen through the London Eye (p10)

CONTENTS

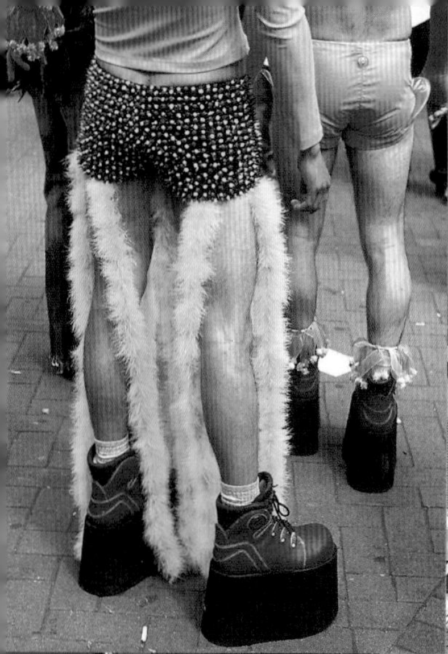
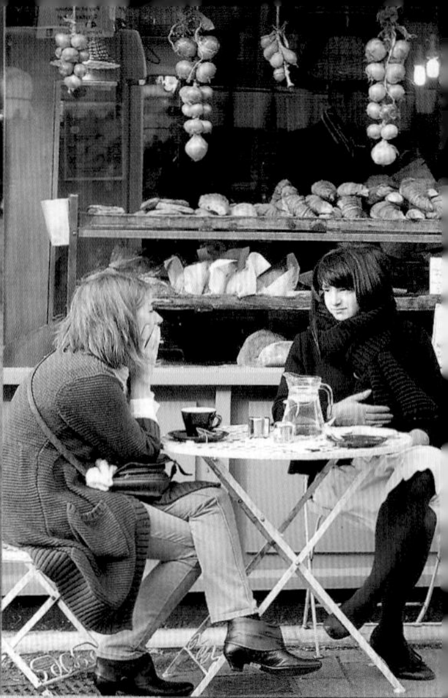
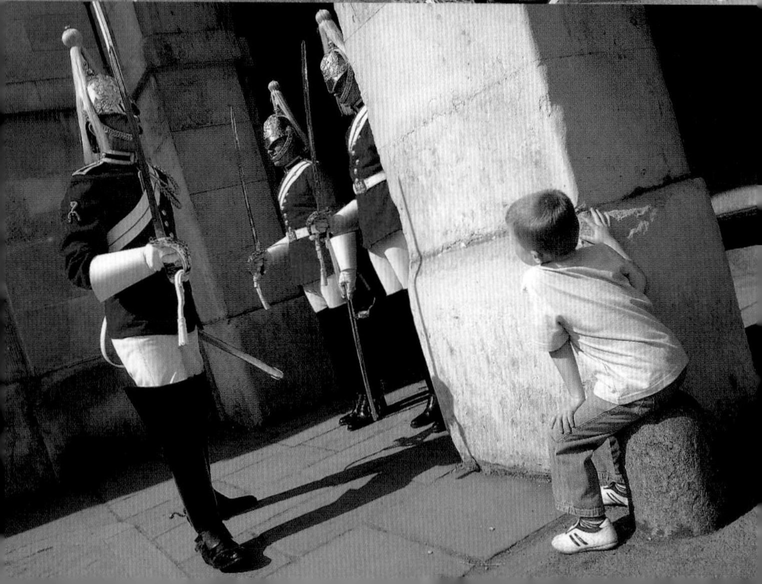

>THIS IS LONDON

'When a man is tired of London, he is tired of life,' claimed the great lexicographer Samuel Johnson, and we're inclined to agree. Few world cities can compete with the English capital for variety, vitality and pace.

More than seven million people, from six continents, are crammed into this heaving metropolis, creating one of the largest cultural mixing pots on earth. This is where the money that drives the British economy is made and spent, where the Queen reigns and parliament governs, and where trends in music, fashion and the arts are made and discarded, sometimes in the space of a single evening.

London is the fashion capital of Mary Quant and Stella McCartney, the music capital of the Rolling Stones and Coldplay, the arts capital of the Tate Modern and Damien Hirst, the shopping capital of Harrods and Harvey Nichols, the culture capital of the BBC Proms and the Victoria & Albert Museum. Almost 2000 years of history are writ large in the streets, from the medieval spires of Westminster Abbey to the soaring dome of St Paul's Cathedral and the phallic exclamation mark of Norman Foster's 'Gherkin'.

London obviously has its faults – it's expensive, overcrowded and often run-down around the edges – but Londoners love it with a giddy passion. This urban love affair is fuelled by romantic dinners at restaurants run by uberchefs like Gordon Ramsay and Marcus Wareing, by neon-coloured cocktails in the sleek bars of Shoreditch and Notting Hill, and by bass-charged all-nighters at super clubs like Fabric and the Ministry of Sound.

London is changing, and many would say improving, in the run-up to the 2012 Olympics. You'll still see the London clichés – red buses, black taxis and bobbies on the beat – but you'll also find improved infrastructure, reinvigorated public spaces and some of the best nightlife, shopping, theatre, music and dining in the world. Catch it now before they start charging Olympic prices.

Top left Sole brothers – participants on a high during a London Pride parade (p30) **Top right** A coffee break from the Big Smoke **Bottom** Life changing – a young boy watches the Changing of the Guard (p87)

>HIGHLIGHTS

Soaking up the sun outside the National Gallery (p46) in Trafalgar Sq (p47)

>1 LONDON EYE & SOUTH BANK

UP, UP AND OVER THE CITY ON THE WORLD'S BIGGEST FERRIS WHEEL

The London Eye (p76) was the world's largest observation wheel when it opened in 2000, soaring 135m above the River Thames. Cities around the world are now copying London's futuristic Ferris wheel, but this was the original. The podlike gondolas offer a sweeping vista over central London, from the Houses of Parliament and Westminster to the distant towers of the City and the smokestacks of Battersea Power Station.

The wheel was originally planned as a temporary exhibit, but it was such a massive hit with visitors and locals that the owners made it a permanent feature. It has even seeped into popular culture, appearing in *Harry Potter & the Order of the Phoenix* and an episode of *Dr Who* as a communication device for the sinister Nestene Consciousness. Writer Will Self dubbed it 'God's bicycle wheel'.

Many visitors combine a spin on the Eye with a stroll through the Southbank Centre (p78), and a walk east to the similarly eye-catching Tate Modern (p79).

>2 BRITISH MUSEUM

VIEW TREASURE AFTER TREASURE IN BRITAIN'S GREATEST MUSEUM

Who said museums have to be fusty? The curators at Britain's largest museum (p104) have done a fantastic job of reinvigorating the displays in this centuries-old powerhouse of history. Founded in 1753 to house the personal collection of Sir Hans Sloane, the museum was dramatically extended in the Georgian period, and the central Grand Court was covered by a soaring geometric canopy by Sir Norman Foster in 2000.

Inside you can see such historical wonders as the Rosetta Stone (room 4), which enabled archaeologists to translate Egyptian hieroglyphs, and the controversial Elgin (Parthenon) Marbles (room 18), swiped from the Acropolis by Britain's ambassador to the Ottoman Empire. The collection of Egyptian mummies in rooms 62 to 63 is legendary, but don't overlook English relics like the Mildenhall Treasure and the Snettisham Hoard (rooms 49 to 50).

Funky new additions include the basement Africa gallery (room 25) and the Wellcome Trust's stunning Living & Dying exhibition in room 24. Entry is free, except for temporary exhibitions.

>3 ST PAUL'S CATHEDRAL

THE HISTORY OF LONDON CAPTURED IN ONE ARCHITECTURAL MASTERPIECE

If we had to pick one building that summed up the spirit of London it would be St Paul's Cathedral (p63). Designed by Sir Christopher Wren as London emerged from the ashes of the Great Fire, the building even survived the firestorm of the Blitz thanks to the efforts of volunteer fire fighters. Nowhere else in London conjures up such a sense of British grit.

Today, stripped of the grime deposited by the coal-fired power station that now houses the Tate Modern, the cathedral gleams inside and out. Most visitors head straight to the vertiginous Golden Gallery atop St Paul's famous dome, reached by 530 winding steps. En route, explore the open-air Stone Gallery (378 steps) and the acoustically brilliant Whispering Gallery (259 steps) inside the dome.

In the crypt are tombs and memorials to the greatest heroes of the British Empire – Nelson, Wellington and Churchill, and Wren himself, who lies buried under a plain stone with the epitaph: 'Reader, if you seek his monument, look around you'.

>4 TATE MODERN

GREAT WORKS OF ART IN A WORK-OF-ART BUILDING

Housed in a stunning conversion of the power station that once supplied the City with electricity, the world's most popular modern art gallery is going from strength to strength. After years of focusing on high-profile temporary exhibitions in the hanger-like Turbine Hall, Tate Modern (p79) is getting serious about its permanent collection, rehanging major works in chronological – or at least logical – order and bringing unseen masterpieces out of storage.

The big news on the horizon is that the Tate will soon be expanding. Work is underway on a £215 million extension at the rear of the main building to house a dramatically expanded collection (see boxed text, p79) and facilities for new media.

For the time being the focus is still on well-known modern names, like Picasso, Mondrian, Pollock, Lichtenstein and Mark Rothko. Our favourite works are Francis Bacon's *Triptych,* Max Ernst's *Celebes*, Alberto Giacometti's spindly figurines and the excellent collection of Russian propaganda posters.

>5 WESTMINSTER

LONDON'S SEAT OF GOVERNMENT AND A POWERHOUSE OF ARCHITECTURE

Since the English civil war, politicians have been filing into the Houses of Parliament (p90) to thrash out the issues of the day, accompanied by raucous catcalls of 'Hear, hear!' and 'rhubarb rhubarb rhubarb'. The seat of British politics is still big on pomp and circumstance. During the state opening of Parliament each December, the doors are slammed in the face of the Queen's representative to symbolise the independence of the state.

For many visitors, the history of Westminster matters less than the monuments of state. Thousands pose for photos in front of the Houses of Parliament and their famous clocktower, Big Ben, named for the giant bell inside. Across St Margaret's St is Westminster Abbey (p92), where British monarchs have been crowned ever since 1066, and where a dozen former kings and queens are buried.

Close by is the famous black door of Number 10 Downing St (p91), currently occupied by Prime Minister Gordon Brown (though for how long remains to be seen).

>6 THE ROYAL PARKS

IDYLLIC SUMMERS IN THE GREEN LUNGS OF LONDON

In 1808 the parliamentarian William Windham declared that Hyde Park (p121) should be preserved at all costs as the 'lungs of London'. Good thing, too. Along with London's other Royal Parks (www.royalparks.gov.uk), Hyde Park is a national treasure. Together, these open spaces make London (surprisingly) the greenest city in Europe, and Londoners love the parks as places to picnic, sunbathe, play sport and even demonstrate against the government.

In alphabetical order, the eight Royal Parks are Bushy Park, Green Park, Greenwich Park, Hyde Park, Kensington Gardens, St James's Park, Regent's Park and Richmond Park, but the best and busiest parks are the four in the centre – St James's, Regent's, and joined-at-the-hip Hyde Park and Kensington Gardens – which form a crescent around the West End.

Visit St James's Park (p91) for rose gardens, tame squirrels and pet pelicans, Regent's Park (p115) for city views (from nearby Primrose Hill) and London Zoo, and Hyde Park (pictured above) and Kensington Gardens (p121) for deck chairs, concerts, and the cultural delights of the Serpentine Gallery.

>7 THE KENSINGTON MUSEUMS

A DOSE OF SCIENCE, HISTORY AND THE ARTS AT LONDON'S VICTORIAN MUSEUMS

Thank the Victorians for this wonderful cluster of fascinating museums. The Great Exhibition of 1851 provided the inspiration for the Victoria & Albert Museum (V&A; p126), while the Victorian fascination with science led to the creation of the Science Museum (p125) in 1857 and the Natural History Museum (p124) in 1881. Today this trio of museums covers all the bases of human knowledge – what can't be learned here probably isn't worth learning.

Most locals feel a strong affection for the Natural History Museum (pictured above), not least for the friendly diplodocus skeleton that welcomes visitors into the Great Hall. Just behind is the Science Museum, with vast rooms full of boy's toys and excellent exhibits on genetics and future medicine provided by the Wellcome Trust.

This leaves the spectacular V&A to cover the decorative arts. Ongoing renovations, and the inclusion of quirky modern items like the heels that made Naomi Campbell trip on the catwalk, are opening the V&A up to a whole new generation.

>8 TOWER OF LONDON

OFF WITH THEIR HEADS AT LONDON'S FIRST ROYAL PALACE

From palace to prison – it's funny how things turn out. The castle (p64) founded by William the Conqueror in 1078 was the original seat of the English monarchy, but it found a new use as a prison for traitors during the backstabbing years of the Tudors. Dozens of heroes and rogues were locked up or beheaded inside the Tower, from Sir Thomas More and Sir Walter Raleigh to Guy Fawkes and Ann Boleyn.

Today the castle is mobbed, but the highlights outweigh the inconvenience of the queues. Don't miss the Crown Jewels, perhaps the most ostentatious collection of baubles in existence, or the gruesome armoury of medieval weapons in the White Tower.

Live medieval re-enactments and interactive displays add a splash of colour to famous sights like the Bloody Tower, which stills contains graffiti scratched into the walls by former prisoners. And, of course, there are the cheerful Yeoman Warders (Beefeaters) who lead free tours around the castle daily.

>9 NATIONAL GALLERY & NPG

OLD MASTERS AND FAMOUS FACES AT THE NATIONAL AND NATIONAL PORTRAIT GALLERIES

Facing onto Trafalgar Sq, the National Gallery (p46) and neighbouring National Portrait Gallery (NPG; p46), just around the corner, fill in all the gaps in the history of painting that aren't covered by the Tate Modern and Tate Britain. These grand neoclassical monuments house some of the country's finest paintings and, refreshingly, not all are stuffy 17th-century oils – there are wacky modernist works and cartoon caricatures as well.

The National (pictured above) is the official gallery of the British Isles and the collection has a predictably UK-centric focus. High-profile paintings include Constable's *Hay Wain* and Monet's *The Thames below Westminster*, but there's plenty to see so grab a free floor plan and decide on a handful of works you wouldn't want to miss.

Next door, the more whimsical NPG contains paintings and photographic portraits of British celebrities going back to the time of Henry VIII, some flattering, some abstract and some openly mocking, such as Gerald Scarfe's withering caricatures of Margaret Thatcher.

>10 CRUISING THE THAMES

SEE A DIFFERENT SIDE OF LONDON ON A RIVERBOAT CRUISE

With all the new skyscrapers planned for central London (see boxed text, p65), the views of the skyline from the River Thames are only going to get better. Passengers on the river already get a grandstand view of St Paul's, Tower Bridge, the Tower of London, the London Eye, the Houses of Parliament, the maritime precinct at Greenwich and the garden of office towers at Canary Wharf.

Half a dozen companies offer transfers along the river, from chugging commuter boats to the zippy shuttle service between the Tate Modern (p79) and Tate Britain (p126). With time to spare, you can even ride the river all the way to Kew Gardens or Hampton Court Palace (see boxed text, p92).

Advance booking is rarely necessary – to save money on commuter boats, buy an all-day rover ticket and pay with your Oyster card. The main central piers are at Westminster, Waterloo, Embankment, Blackfriars, London Bridge, Tower of London, Canary Wharf and Greenwich – for boat operators, see boxed text, p222.

>11 GREENWICH

EAST ALONG THE RIVER TO THE MARITIME CAPITAL OF LONDON

Green and pleasant Greenwich (p176) is the precise location from which global time and points of distance around the east–west axis of the earth are measured. These things aren't so important in the digital age, but back when Britain depended on its navy for its wealth and its power, a few degrees out here or there could spell disaster – as happened in 1707, when the British fleet sailed too close to the Scilly Isles and 1400 perished.

The Royal Observatory (p180) still marks the Prime Meridian with a metal line, but the most impressive thing is the setting, on a hilltop surrounded by parkland with panoramic views over the Old Royal Naval College (p179) and the National Maritime Museum (p179), with the River Thames and Docklands' futuristic skyscrapers beyond.

As well as the historical attractions, people come to Greenwich for the feeling of escape from the big city. Greenwich was – and still feels like – a village, and the backstreets are lined with interesting and quirky shops.

>12 HAMPSTEAD HEATH

PACK A PICNIC AND HEAD FOR LONDON'S MOST RELAXING OPEN SPACE

The Royal Parks (p15) are one thing, but to really get away from it all, you should head to the suburbs. Sprawling up the eastern side of 98m-high Parliament Hill, Hampstead Heath (p168) is that rare thing: a public park where nature prevails. In among the artificial water features and pedestrian walkways, you'll find areas of overgrown woodland and swishing tall grasses, perfect for a secluded picnic, or a secret romantic liaison.

People come from all over London to picnic, splash around in the outdoor swimming ponds, fly kites, enjoy outdoor concerts and soak up the sweeping views from the top of Parliament Hill. Across North End Rd, the West Heath is a popular gay meeting point – there's even a special contingent of police officers who protect participants from harassment.

Many visitors combine a trip to the heath with lunch at one of the surrounding pubs – the Hollybush (p173) in Hampstead is our top pick.

>13 NIGHTLIFE

SEE LONDON COME ALIVE AS THE LIGHTS GO DOWN

With the relaxing of Britain's strict licensing laws, going out in London has become a much more laid-back affair. No longer are expensive clubs the only places to drink after 11pm. Of course, relaxed doesn't have to mean boring – the city is awash with edgy nightclubs, cocktail bars, party pubs, comedy clubs and live-music venues.

If bands are your thing, head to Camden (p173) or Hoxton (p156) to get an earful of the latest up-and-coming indie talent. Fans of synth sounds and booming basslines can dance their socks off at iconic London DJ clubs such as Ministry of Sound (p83) and Heaven (see boxed text, p56).

Currently there's a massive buzz about cabaret, with venues like the Pigalle Club (p101) and Volupté Lounge (p70) offering a cheeky, almost Victorian take on striptease. And Londoners are also going wild for swing – the dance not the lifestyle – with dozens of small venues offering jive, boogie-woogie and Lindy-hop nights (see boxed text, p163).

Also see the 'Play' sections of this book for suggestions.

>14 EATING OUT

THERE'S MORE TO BRITISH FOOD THAN FISH AND CHIPS

OK, we admit it – British food doesn't have the best reputation. Put that down to the fact thate most people who moan about 'British food' have never eaten in one of London's modern British restaurants. London has become one of the great food capitals of Europe – American gastro mag *Gourmet* even recently described London as 'the best place in the world' to eat.

In recent years pioneering chefs like Gordon Ramsay (see boxed text, p98) and Tom Aikens have revived the traditional British cooking that existed before WWI and WWII, with an emphasis on ingredients from small, artisan producers. And London has attracted chefs from all over the world, creating a cooking pot of culinary cultures. Among other things, the capital has some of the best Indian, Thai, Chinese, Japanese, Vietnamese and Korean food outside of Asia.

However, all this divine dining doesn't come cheap. To escape penury, look out for set-price menus or head to London's excellent ethnic canteens.

>15 THEATRE

IF THE WORLD IS A STAGE, WELCOME TO THE FRONT ROW

Londoners first got the theatre bug in the 17th century when Shakespeare and his contemporaries starting performing comedies and dramas for the masses on the stage of the original Globe Theatre (p84). Four hundred years later and Londoners are still obsessed with the stage. Theatreland in the West End boasts more than 50 theatres, staging everything from period romps and big-name musicals to occasionally cringe-worthy shows based on banging dustbin lids and rap poetry.

Soho and Covent Garden are the London equivalent of New York's Broadway, with all-singing, all-dancing extravaganzas that seem to run and run (*The Phantom of the Opera* has been running since 1984). At the other end of the spectrum are small theatres like the Almeida (p164), Royal Court (p133) and Young Vic (p85), which showcase cutting-edge drama and famous actors on small, intimate stages.

For a breakdown of all the shows currently taking place in London, and a map of London theatres, visit www.officiallondontheatre.co.uk.

>16 PUBS & BARS

MAKE MINE A PINT, NO WAIT, A PINEAPPLE MOJITO

Londoners have turned going out for a drink into an artform. Many of the capital's historic public houses still swim with Victorian grandeur, and boozers like the Old Cheshire Cheese (p69) and George Inn (p82) have been pulling pints since the time of the Great Fire of 1666.

London's pubs also provide a vital forum for live music, with Camden's Dublin Castle (p174) and Hoxton's Old Blue Last (p157) leading the scene. Then there are the gastropubs – public houses with gourmet kitchens that are giving established restaurants a run for their money.

At the other end of the spectrum are London's ubercool cocktail bars, such as Notting Hill's Lonsdale (p140) and Hoxton's Hawksmoor (p155). Recently mixologists have been pulling out all the stops, inventing surprising new creations like the pineapple and sage mojito and the wasabi and basil kiss.

Bars open late most nights, but many pubs still close at 11pm (10.30pm Sunday). If you're a pub connoisseur, check out websites like www.fancyapint.co.uk or www.beerintheevening.com for more recommendations.

>17 RETAIL THERAPY

LONDON IS A SHOPAHOLIC'S HEAVEN

Few people come to London expecting bargain prices, but the range and quality of London's shops is truly world-class. You want high street fashions? Join the crowds on Oxford St (p47). You want space-edge gadgets? Look no further than Tottenham Court Rd (p106). You want designer clothes with the London look? Head to glamorous Mayfair (p93).

Every district in London has its own shopping theme. Knightsbridge, Kensington and Chelsea (p127) are dominated by glitzy brand names, while fans of edgy streetwear and retro fashion buy their glad rags in trendy Shoreditch (p147). Bookworms flock to the second-hand bookshops on Charing Cross Rd (p47), and vinyl junkies trawl the record shops of Soho (see boxed text, p50).

Then there are London's bustling markets. Borough Market (p79) is foodie heaven, and antique collectors go gaga for Portobello Rd (p136). Camden Markets (p169) is an open-air department store for punks, goths and indie kids, and Columbia Road Flower Market (see boxed text, p150) is a riot of pot plants and bouquets. If you'd rather shop indoors, the department stores of Knightsbridge, Chelsea and the West End (see boxed text, p116) are cathedrals to the consumer instinct, selling everything from iPods to trampolines.

See the 'Shop' sections of this book for more suggestions.

>LONDON DIARY

There's always something going on in London, and a couple of times a year there is something really big going on. The only problem is that over seven million locals want to get in on the action as well. London's festivals can attract millions (really!) of visitors, so it pays to book tickets early and arrive early to secure a spot.
For more listings of festivals and events, go to www.visitlondon.com, and www.bbc.co.uk/london, or sign up to the mailing lists at Urban Junkies (www.urbanjunkies.com/london), Daily Candy (www.dailycandy.com/london) and Flavour Pill (www.flavourpill.com/London).

Rhythm and hues at the Notting Hill Carnival **(p30)**

JANUARY & FEBRUARY

London Parade, New Year's Day

www.londonparade.co.uk

After the official fireworks on New Year's Eve, the Lord Mayor of Westminster leads a parade of musicians and street performers from Parliament Sq to Berkeley Sq.

Chinese New Year

www.chinatownchinese.co.uk

In late January/early February Chinatown explodes with fireworks as the Chinese community ushers in its new year.

MARCH & APRIL

Oxford & Cambridge Boat Race

www.theboatrace.org

Crowds line the Thames west of Putney to watch Britain's poshest universities go oar-

Enter the dragon – Trafalgar Sq during Chinese New Year celebrations

to-oar in a battle to prove who has the stiffest upper lip.

London Marathon

www.london-marathon.co.uk

Some 35,000 fitness freaks, costume fanatics, charity fundraisers and masochists join the world's biggest road race from Greenwich Park to the Mall.

Camden Crawl

www.thecamdencrawl.com

London's most rocking borough pulls out all the stops for 130 gigs spread over two sweat-filled nights.

MAY

Chelsea Flower Show

www.rhs.org.uk/chelsea

The world's most famous horticultural show features celebrity gardeners and outrageous floral displays at the Royal Hospital Chelsea.

FA Cup Final

www.thefa.com/TheFACup

The domestic football season climaxes with this epic cup match at the revamped Wembley Stadium.

JUNE

Trooping the Colour

The Queen was born in April, but her official birthday is celebrated in June at Horse Guards Parade, with much flag waving and pageantry.

Pounding the streets in the London Marathon

Wimbledon Lawn Tennis Championships

www.wimbledon.com

Two glorious weeks of tennis, strawberries and cream, and finally a player for British supporters to get behind in the form of Andrew Murray.

Red Bull Flugtag

www.redbullflugtag.co.uk

Every summer, madcap inventors send their flying machines crashing into the Serpentine in Hyde Park, watched by crowds of thousands.

JULY

Pride

www.pridelondon.org

London's gay community go into overdrive for this annual extravaganza around Soho and Trafalgar Sq.

Proms (BBC Promenade Concerts)

www.bbc.co.uk/proms

The finest orchestras and musicians play for two months, finishing with a patriotic extravaganza at Kensington's Royal Albert Hall.

Loud and proud – a Pride participant

V&A Village Fete

www.vam.ac.uk

A weekend of fairground fun in this arty take on the traditional village fete, held in the Victoria & Albert Museum's John Madejski Gardens.

AUGUST

Notting Hill Carnival

www.nottinghillcarnival.biz

More than a million people show up for Europe's biggest outdoor carnival, an uninhibited celebration of Afro-Caribbean culture.

Innocent Village Fete

www.innocentvillagefete.com

An August weekender with bands, DJs, kids' events and stage performances, sponsored by the smoothie company.

SEPTEMBER

Thames Festival

www.thamesfestival.org

A cosmopolitan weekend festival of fairs, street theatre, music, fireworks and river races, finishing with a Sunday-night carnival at Victoria Embankment.

OUTDOOR SUMMER MUSIC

You don't have to sit in a muddy field to enjoy the best of British music. London has plenty of its own outdoor music festivals, including the following:

> Hard Rock Calling (June; Hyde Park; www.hardrockcalling.co.uk) – a weekend rocker at Hyde Park, with huge names like Eric Clapton and the Police.

> Kenwood Picnic Concerts (June–August; Hampstead Heath; www.picnicconcerts.com) – two months of classical and acoustic events targeting 30-somethings and older.

> Lovebox Weekender (July; Victoria Park, Hackney; www.lovebox.net) – Groove Armada's out-there weekend extravaganza attracted Manu Chao and the Flaming Lips in 2008.

> Rise (July; Finsbury Park; www.risefestival.org) – free rock-against-racism festival, with urban, African, Asian and Middle Eastern music and stage shows.

> Somerset House Summer Series (July; Somerset House; www.somersethouse.org.uk) – grown-up pop and soul and film premieres in the wonderful courtyard at Somerset House.

> Wireless (July; Hyde Park; www.o2wirelessfestival.co.uk) – a proper outdoor rockathon; 2008 acts included Morrissey, Sam Sparro and Jay Z.

> Loaded in the Park (August; Clapham Common; www.getloadedinthepark.com) – lots of indie action south of the river, with the likes of Supergrass and the Hives.

> South West Four (August; Clapham Common; www.southwestfour.com) – one-day hardcore dancefloor party with veteran DJs like Carl Cox and Sven Vath.

London Open House

www.londonopenhouse.org

For one day only, Londoners and visitors get to peek inside the doors of some 500 buildings that are normally off-limits, including Norman Fosters' Gherkin.

OCTOBER

BBC Electric Proms

www.bbc.co.uk/electricproms

Camden rocks for five days of concerts covering every musical genre, held at venues like the Roundhouse, Barfly and Koko.

Dance Umbrella

www.danceumbrella.co.uk

Five weeks of cutting-edge contemporary dance from around the world; venues include the British Film Institute (BFI), the Royal Opera House and the Southbank Centre.

London Film Festival

www.bfi.org.uk/whatson/lff

A two-week-long city-wide celebration of the moving image centred on the BFI, with special screenings and Q&A sessions with leading actors and directors.

Fireworks over the River Thames illuminate Westminster Abbey (p92)

NOVEMBER

Guy Fawkes Night (Bonfire Night)

Londoners celebrate Guy Fawkes' failed attempt to blow up the Houses of Parliament with bonfires, fireworks, burning effigies and general merriment on 5 November. Top spots to marvel at the pyrotechnics include Blackheath, Alexandra Palace, Battersea Park and Hackney's Victoria Park.

Lord Mayor's Show

www.lordmayorsshow.org

Expect plenty of medieval pageantry as the new Lord Mayor of the City of London travels in a horse-drawn carriage to the Royal Courts of Justice.

DECEMBER

Carols in Trafalgar Sq

www.london.gov.uk/trafalgarsquare

Every winter the festive lights on Oxford, Regent and Bonds Sts are switched on by a minor celeb, then carol singers warble on till Christmas at Trafalgar Sq.

>ITINERARIES

Top brass – the Queen's Life Guards in Horse Guards Parade (p87)

ITINERARIES

You could spend a lifetime in London and still not see all the city has to offer. Fortunately, most of the major sights are clustered together in a central nucleus between the City, the South Bank, Westminster and Marylebone. If time is tight, consider the following as a guide. For details on organised tours, see p224.

ONE DAY

Days are short, so start early with a visit to Trafalgar Sq (p47) and the National Gallery (p46), before heading south to Westminster for a peek at Westminster Abbey (p92) and the Houses of Parliament (p90). Stroll over Westminster Bridge for a spin on the London Eye (p76), then head east along the river past the attractions of the South Bank (p73) before crossing over to Covent Garden (see p42) and walking north to the magnificent British Museum (p104). After hours, drink and dine in the hedonistic streets of Soho (p52) – Andrew Edmunds (p52) or Arbutus (p52) are top picks for dinner.

TWO DAYS

Follow the one-day itinerary and get up early the next day for some serious retail therapy in the backstreets of Marylebone (see p115) and Mayfair (see p93). Head over to St James's for a peak at Buckingham Palace (p87), then enjoy a picnic lunch in Hyde Park (p121). In the afternoon, head to the City to explore St Paul's Cathedral (p63), then cross the Millennium Bridge (p76) to Tate Modern (p79). If it's a Thursday, Friday or Saturday, visit Borough Market (p79) and cross Tower Bridge (p64) to the Tower of London (p64). In the evening, let your hair down at Loungelover (p155) or one of the other Hoxton bars (p154).

THREE DAYS

If the first two days haven't totally exhausted you, spend the morning of day three at the Kensington Museums (p124, p125, p126) followed by lunch at Pétrus (p129) or one of the other exclusive restaurants in Kensington (make a reservation ahead of time; see p128). In the afternoon, recharge your batteries in Regent's Park (p115). This should leave

Top A couple of park-life fans relax in Regent's Park (p112) **Bottom** Children chase Trafalgar Sq's once-ubiquitous pigeons in front of the National Gallery (p46)

you with enough energy for a night of culture at the Royal Opera House (p57), the National Theatre (p83) or Sadler's Wells (p70), or a riotous West End show (see p24).

FOUR DAYS

Depending on the day of the week, you can spend the morning of day four browsing the market stalls of Portobello Rd Market (p136) or Camden Markets (p169). In the afternoon, head north to Hampstead Heath (p168) for the soothing sound of the wind in the trees and the views from Parliament Hill. Alternatively, take the boat downriver to Greenwich (p176) to explore the maritime relics and bijou shops. On your final evening, catch a live band in trendy Hoxton (see p156) or Camden (see p173).

RAINY DAYS

Use rainy days to 'do' the city's numerous galleries and museums by tube, or spend some time in London's showy department stores (see boxed text, p116). Alternatively, sneak off for a matinee performance in the West End (see p24), or visit one of London's excellent alternative cinemas (see boxed text, p83). Rainy days are also a prime opportunity for a lingering meal at one of London's many gourmet restaurants (see p23).

Capital dining at a restaurant on the Thames near Southwark Bridge

FORWARD PLANNING

The trick in London is either to book very early, or to try at the last minute and hope you get lucky.

Three to six months before you go You may need to book this far ahead for big-name restaurants, such as the Ivy (p54) and Gordon Ramsay (see boxed text, p98), and Saturday-night performances of big West End shows.

Two to three months before you go Check out websites like www.ticketmaster.co.uk and www.seetickets.com for tickets to big-name rock acts at venues like Koko (p174) and Brixton Academy (p83) or any theatrical performances with high-profile stars (see p24).

Three to four weeks before you go Check for any 'general sale' tickets for football or cricket matches at London's big stadiums.

Two weeks before you go Sign up for an email newsletter, such as *Urban Junkies* at www.urbanjunkies.com, and double-check review sites, such as www.londonnet.co.uk. Two weeks is usually plenty of time to get a reservation for trendy, interesting restaurants, such as Bistrotheque (p150), and venues like the Pigalle Club (p101).

A few days before you go Tickets for blockbuster exhibitions at the Royal Academy of Arts (p91), Tate Modern (p79) and Tate Britain (p126), or the Victoria & Albert Museum (p126) can usually be booked a few days beforehand or even on the day.

LONDON FOR FREE

London is one of the world's most expensive cities, but many of the most impressive sights won't cost you a penny. There's no charge to visit the permanent collections at any of the Kensington Museums (p126, p125 and p124) or the British Museum (p104), National Gallery (p46), National Portrait Gallery (p46), Tate Britain (p126) or Tate Modern (p79). Alternatively, why not spend a day relaxing in one of the wonderful Royal Parks (p15), or the green open spaces of Greenwich (p176) and Hampstead Heath (p168). You can also pass some free time exploring the old churches of the City (see boxed text, p70) or the modernist monuments of the South Bank (see p73). And there's no charge for looking at fabulous buildings like St Paul's (p63), Westminster Abbey (p92) and the Houses of Parliament (p90) from the outside.

>NEIGHBOURHOODS

Primrose Hill panoramas (p166)

NEIGHBOURHOODS

London has grown organically over two millennia, swallowing up parks and villages on all sides. As a result, the city is a mixed-up jigsaw of small neighbourhoods, and working out how these little pieces fit together can take time.

First up, you'll find it a lot easier to fit London into geographical space if you travel overland. The famous tube map designed by Harry Beck is schematic and not geographically accurate, so distances are misleading. It takes two stops and a change of lines to get to Covent Garden from Tottenham Court Rd by tube, compared to 10 minutes on foot.

Fortunately, the centre of London is surprisingly compact. The City covers a single square mile, and the attractions of Soho, Covent Garden, Mayfair, Bloomsbury, Westminster, Marylebone and St James's are squeezed into an area just 2 miles across. The most useful landmark is the River Thames, which coils through the city centre like a python with indigestion.

Tall buildings are another useful navigational aid – the Centre Point and BT towers are invaluable markers for the West End; Big Ben points the way to Westminster; the London Eye marks the South Bank; and St Paul's and the Gherkin act as lodestones to the City. Then there are the Royal Parks, arranged in a crescent around the West End.

Surrounding this inner core are a series of suburbs that once existed as separate villages – we strongly recommend escaping the crush of the city centre for the diverse charms of Camden, Clerkenwell, Islington, Hampstead, Knightsbridge, Kensington, Chelsea and Notting Hill.

Wherever you go, the best way to get to grips with London is to break it down into manageable chunks. Set aside one day for the City, another for the South Bank, another for Soho, Mayfair and Marylebone and so on.

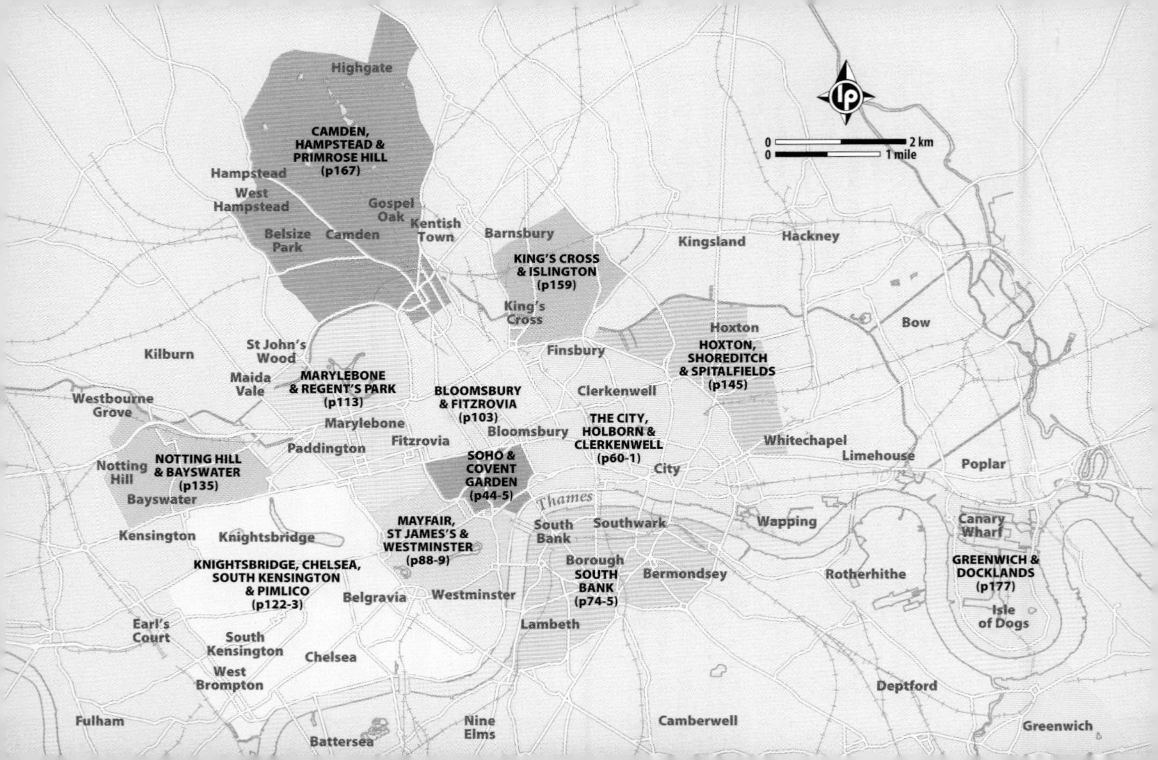
0 2 km
0 1 mile
Highgate
CAMDEN, HAMPSTEAD & PRIMROSE HILL (p167)
Hampstead
West Hampstead
Gospel Oak
Kentish Town
Belsize Park
Camden
Barnsbury
KING'S CROSS & ISLINGTON (p159)
King's Cross
Kingsland
Hackney
Bow
Hoxton
HOXTON, SHOREDITCH & SPITALFIELDS (p145)
Kilburn
St John's Wood
Maida Vale
MARYLEBONE & REGENT'S PARK (p113)
BLOOMSBURY & FITZROVIA (p103)
Finsbury
Clerkenwell
Westbourne Grove
Marylebone
Bloomsbury
THE CITY, HOLBORN & CLERKENWELL (p60-1)
Whitechapel
Limehouse
Poplar
Paddington
Fitzrovia
Notting Hill
NOTTING HILL & BAYSWATER (p135)
Bayswater
SOHO & COVENT GARDEN (p44-5)
City
Thames
Kensington
Knightsbridge
MAYFAIR, ST JAMES'S & WESTMINSTER (p88-9)
South Bank
Southwark
Wapping
Canary Wharf
KNIGHTSBRIDGE, CHELSEA, SOUTH KENSINGTON & PIMLICO (p122-3)
Belgravia
Westminster
Borough
SOUTH BANK (p74-5)
Bermondsey
Rotherhithe
GREENWICH & DOCKLANDS (p177)
Isle of Dogs
Lambeth
Earl's Court
South Kensington
Chelsea
West Brompton
Deptford
Fulham
Battersea
Nine Elms
Camberwell
Greenwich

>SOHO & COVENT GARDEN

Perhaps the most famous neighbourhood in London, Soho hovers somewhere between sophisticated and seedy. The winding backstreets between overcrowded Oxford St and Shaftesbury Ave are lined with cool shops, famous theatres, trendy bars and some very expensive restaurants, yet this is also London's main red-light district, complete with peep shows, sex shops and men in dirty macs. Londoners are prepared to overlook Soho's grubby side for its anti-establishment air and the libertarian vibe of the gay scene along Old Compton St. Incidentally, the name 'Soho' comes from a medieval hunting cry – there's no connection to the Soho (South Houston) in New York.

The bohemian atmosphere extends to neighbouring Covent Garden, once you escape the tourist trap of the main piazza. Radiating north and south from Long Acre, the bustling lanes are dotted with gourmet restaurants, oddball shops and more London theatres staging all-singing, all-dancing spectaculars. Shopping highlights in Covent Garden include the fashion boutiques on Floral St, the specialist bookshops on Charing Cross Rd and the GoreTex-filled adventure shops on Southampton St.

Between these two neighbourhoods is a diamond of land containing London's small but colourful Chinatown – a tight tangle of Chinese supermarkets, Cantonese roast-duck houses and dim sum canteens – and touristy Leicester Sq, with its giant multiplex cinemas and tacky opening nights. Nearby Piccadilly Circus also draws a crowd, despite the waning appeal of illuminated advertising billboards. One tourist attraction that everyone seems to agree on is Trafalgar Sq, with its wonderful art galleries and dramatic monuments to heroes of state.

SOHO & COVENT GARDEN

SEE

Courtauld Gallery........... 1 H4
London Transport Museum 2 F4
National Gallery 3 E6
National Portrait Gallery 4 E5
Photographer's Gallery 5 E4
Somerset House 6 H5
Trafalgar Square 7 E6

SHOP

Agent Provocateur 8 C3
Apple Store 9 A3
Coco de Mer 10 E3
Covent Garden Market 11 F4
Ellis Brigham 12 F4
Forbidden Planet 13 E2
Foyles 14 D3
Hamleys 15 A4
Harold Moores Records 16 B3
Jessops 17 D2
Keonig Books 18 D4
Liberty 19 A3
Murder One 20 D4
Neal's Yard Dairy 21 E3
Neal's Yard Remedies ... 22 E3
Paul Smith 23 F4
Poste Mistress 24 E3
Prowler 25 C4
Revival Records 26 B3
Shipley 27 D4
Sister Ray 28 B3
Stanfords 29 E4
Tatty Devine 30 C4
Ted Baker 31 F4
Vintage & Rare Guitars 32 D2

EAT

Andrew Edmunds 33 B4
Arbutus 34 C3
Café de Hong Kong 35 D4
Carluccio's 36 E4
Great Queen Street 37 F2
Haozhan 38 D4
Ivy 39 E4
J Sheekey 40 E4
Maison Bertaux 41 D3
Mildred's 42 B3
Myung Ga 43 B4
Neal's Yard Salad Bar .... (see 22)
Patisserie Valerie 44 D3
Paul 45 F4
Red Fort 46 C3
Yauatcha 47 C3

DRINK

Aka Bar .... (see 57)
Balans 48 C4
Bar Italia 49 D3
LAB 50 D3
Lamb & Flag 51 E4
Rupert St 52 C4
Salisbury 53 E4

PLAY

Comedy Store 54 C5
Curzon Soho 55 D4
Donmar Warehouse 56 E3
End 57 E2
English National Opera 58 E5
Ghetto 59 D2
Heaven 60 F6
Karaoke Box 61 D3
Madame Jo Jo's 62 C4
Prince Charles Cinema 63 D4
Ronnie Scott's 64 D3
Royal Opera House 65 F3
SIN 66 D2
Soho Revue Bar .... (see 62)
TKTS Booth 67 D5

Please see over for map

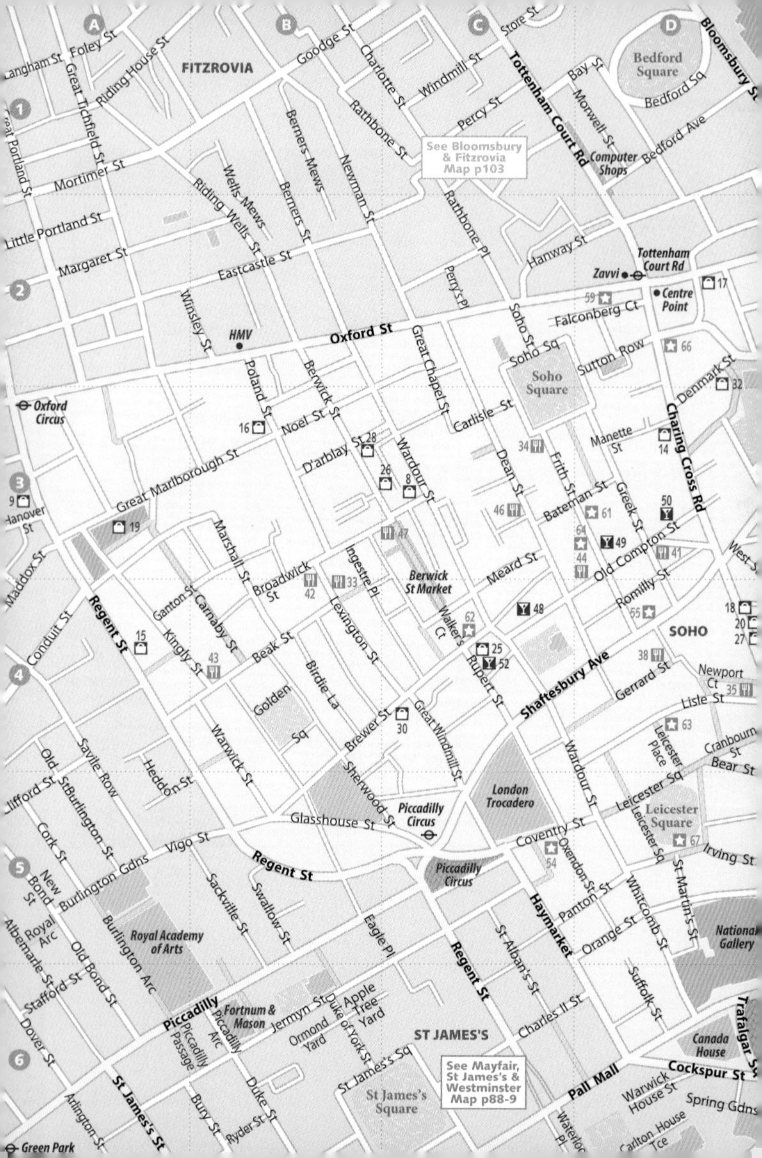
A
B
C
D
1
2
3
4
5
6
FITZROVIA
SOHO
ST JAMES'S
Bedford Square
Soho Square
Golden Sq
Leicester Square
St James's Square
See Bloomsbury & Fitzrovia Map p103
See Mayfair, St James's & Westminster Map p88-9
Oxford Circus
Tottenham Court Rd
Piccadilly Circus
Green Park
Zavvi
HMV
Centre Point
Computer Shops
Berwick St Market
London Trocadero
Royal Academy of Arts
Fortnum & Mason
National Gallery
Canada House
Oxford St
Regent St
Shaftesbury Ave
Charing Cross Rd
Tottenham Court Rd
Piccadilly
Haymarket
Pall Mall
Cockspur St
Goodge St
Foley St
Riding House St
Mortimer St
Margaret St
Eastcastle St
Great Marlborough St
Noel St
D'arblay St
Broadwick St
Beak St
Brewer St
Glasshouse St
Vigo St
Burlington Gdns
Savile Row
Old Bond St
New Bond St
Conduit St
Maddox St
Hanover St
Carnaby St
Kingly St
Ganton St
Marshall St
Poland St
Berwick St
Wardour St
Dean St
Frith St
Greek St
Old Compton St
Bateman St
Carlisle St
Soho Sq
Sutton Row
Falconberg Ct
Denmark St
Manette St
Romilly St
Meard St
Gerrard St
Lisle St
Newport Ct
Leicester Sq
Coventry St
Rupert St
Great Windmill St
Sherwood St
Lexington St
Ingestre Pl
Walker's Ct
Berners St
Berners Mews
Newman St
Rathbone Pl
Rathbone St
Charlotte St
Windmill St
Percy St
Store St
Bay St
Morwell St
Bedford Sq
Bedford Ave
Bloomsbury St
Hanway St
Perry's Pl
Soho St
Great Chapel St
Wells Mews
Riding Wells St
Winsley St
Little Portland St
Great Portland St
Great Titchfield St
Langham St
Warwick St
Heddon St
Birdie La
Clifford St
Cork St
Old Burlington St
Burlington Arc
Royal Arc
Albemarle St
Stafford St
Dover St
Arlington St
St James's St
Piccadilly Passage
Piccadilly Arc
Bury St
Ryder St
Duke St
Jermyn St
Duke of York St
Apple Tree Yard
Ormond Yard
St James's Sq
Sackville St
Swallow St
Eagle Pl
St Alban's St
Charles II St
Oxendon St
Panton St
Orange St
Whitcomb St
Suffolk St
St Martin's St
Irving St
Bear St
Cranbourn St
Leicester Place
Waterloo Pl
Warwick House St
Spring Gdns
Carlton House Tce
Trafalgar Sq
West St

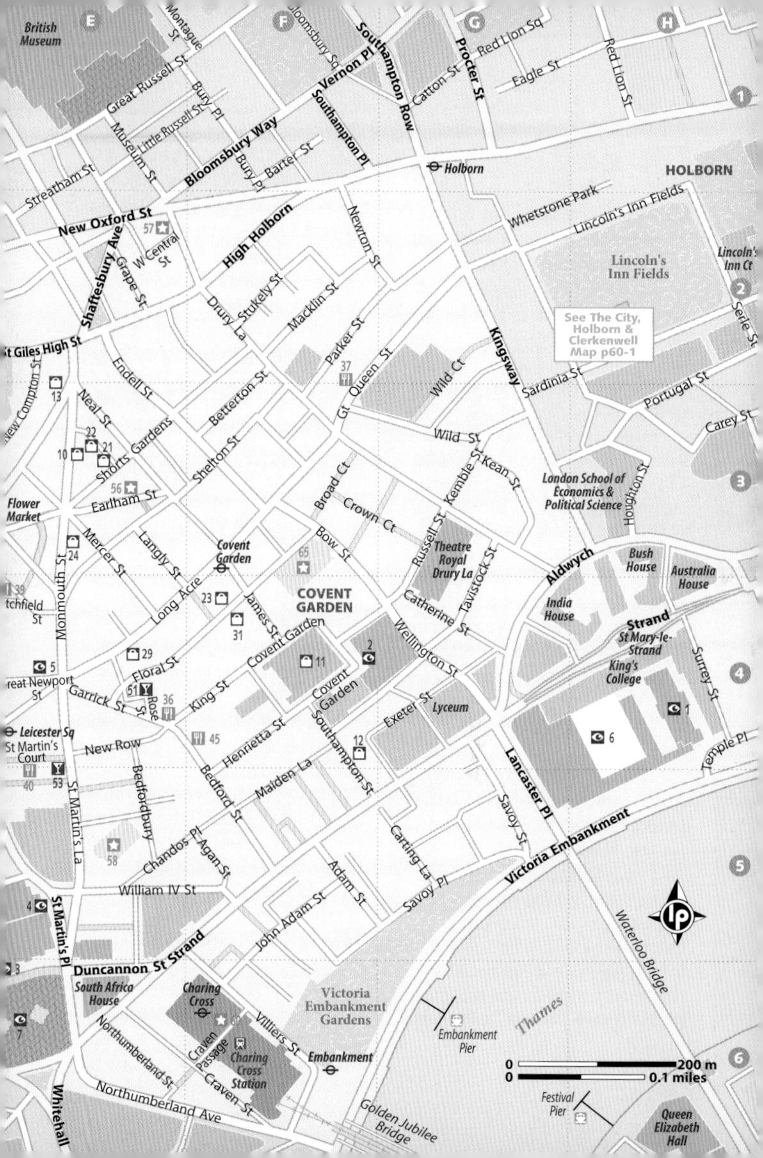

British Museum
Montague St
Bloomsbury Sq
Southampton Row
Procter St
Red Lion Sq
Red Lion St
Eagle St
Catton St
Vernon Pl
Great Russell St
Bury Pl
Little Russell St
Museum St
Bloomsbury Way
Southampton Pl
Barter St
Streatham St
Holborn
HOLBORN
Whetstone Park
Lincoln's Inn Fields
Lincoln's Inn Fields
Lincoln's Inn Ct
Serle St
New Oxford St
57
W Central St
Grape St
Shaftesbury Ave
High Holborn
Newton St
Stukely St
Macklin St
Drury La
Parker St
Kingsway
See The City, Holborn & Clerkenwell Map p60-1
St Giles High St
New Compton St
13
Endell St
Neal St
37
Gt Queen St
Wild Ct
Sardinia St
Portugal St
Carey St
Betterton St
Wild St
22
21
10
Shorts Gardens
Shelton St
Kemble St
Kean St
Houghton St
London School of Economics & Political Science
56
Earlham St
Flower Market
Broad Ct
Crown Ct
Russell St
Theatre Royal Drury La
Tavistock St
Aldwych
Bush House
Australia House
India House
Mercer St
24
Langly St
Covent Garden
65
Bow St
COVENT GARDEN
39
Litchfield St
Monmouth St
Long Acre
23
James St
31
Catherine St
Strand
St Mary-le-Strand
King's College
Surrey St
Wellington St
Covent Garden
2
11
29
Floral St
5
Great Newport St
51
Rose St
36
King St
Covent Garden
Garrick St
Exeter St
Lyceum
1
6
Leicester Sq
St Martin's Court
New Row
45
Henrietta St
Southampton St
12
Lancaster Pl
Temple Pl
40
53
Bedfordbury
Bedford St
Maiden La
Savoy St
Victoria Embankment
St Martin's La
58
Chandos Pl
Agan St
Carting La
Adam St
William IV St
John Adam St
Savoy Pl
4
St Martin's Pl
Waterloo Bridge
Duncannon St
Strand
3
South Africa House
Charing Cross
Victoria Embankment Gardens
60
Thames
Embankment Pier
7
Craven Passage
Charing Cross Station
Villiers St
Embankment
Northumberland St
Craven St
0
200 m
0
0.1 miles
Festival Pier
Northumberland Ave
Golden Jubilee Bridge
Queen Elizabeth Hall
Whitehall
E
F
G
H
1
2
3
4
5
6

SEE

COURTAULD GALLERY

7848 2526; www.courtauld.ac.uk; The Strand; UK students free, adult/concession £5/4, free entry Mon before 2pm; 10am-6pm; ; Temple or Covent Garden

Set inside gorgeous Somerset House (opposite), this upscale gallery is loved as much for its architecture as for the artworks inside. Art buffs come here to browse the collection of impressionist and post-impressionist works, including daubings by Cézanne, Degas, Gauguin, Monet, Matisse, Renoir and Van Gogh. The attached Courtauld Institute is the UK's leading centre for the study of art history and conservation.

LONDON TRANSPORT MUSEUM

7565 7299; www.ltmuseum.uk; Covent Garden; adult/child/concession/senior £10/free/5/8; 10am-6pm Sat-Thu, 11am-9pm Fri; ; Covent Garden or Charing Cross

There's something nostalgic about public transport in London – it must be the old red double-decker buses. Adults will appreciate the history at this cute museum; kids will love the old-fashioned vehicles.

NATIONAL GALLERY

NPG; 7747 2885; www.nationalgallery.org.uk; Trafalgar Sq; admission free, prices vary for special exhibitions; 10am-6pm Thu-Tue, 10am-9pm Wed, tours 11.30am & 2.30pm; ; Charing Cross

Looking out over Trafalgar Sq, this huge gallery displays some of the world's most famous paintings, including Constable's *Hay Wain,* Seurat's *Bathers at Asnières,* Van Gogh's *Sunflowers,* Monet's *Water-Lily Pond* and Holbein's *The Ambassadors* with its hidden anamorphic skull. See also p18.

NATIONAL PORTRAIT GALLERY

NPG; 7306 0055; www.npg.org.uk; St Martin's Pl; admission free, prices vary for exhibitions; 10am-6pm, to 9pm Thu & Fri; ; Charing Cross or Leicester Sq

Almost an annexe to the National Gallery, the NPG concentrates on the job of putting faces to the famous names in British history, from medieval lords and ladies to modern-day celebs. See also p18.

PHOTOGRAPHERS' GALLERY

7831 1772; www.photonet.org.uk; 5 & 8 Great Newport St; admission free; 11am-6pm Mon-Sat, to 8pm Thu, noon-6pm Sun; ; Leicester Sq

THE FOURTH PLINTH

The four plinths surrounding Nelson's Column were built in the early 19th century to display statues of British national heroes, but nobody could agree on a fourth hero, so the last plinth was left empty – until 1999, when the Royal Society of Arts adopted the space for temporary exhibitions of modern art. Since then the fourth plinth has displayed such controversial artworks as Thomas Schütte's gaudy Perspex *Hotel for the Birds* and Marc Quinn's naked statue of the pregnant disabled artist Alison Lapper. The latest proposal is an interactive sculpture by Antony Gormley – 2400 volunteers will each stand on the plinth for one hour, with the option to address the crowd on any issue they choose. There is, however, a rival plan for a permanent statue of Queen Elizabeth II. See www.london.gov.uk/fourthplinth for the latest on the plinth debate.

Split between two buildings on Great Newport St, this tiny gallery is London's leading exhibition space for art photography, though some of the work on display wanders into pretentious territory.

SOMERSET HOUSE

☎ 7845 4600; www.somerset-house.org.uk; The Strand; adult/under 12yr/concession £8/free/6; Embankment Galleries 10am-6pm, to 9pm Thu; ♿ ; ⊖ Temple or Covent Garden

This Palladian masterpiece is loved as much for its courtyard and riverside terrace café as for the art inside. The best works are in the Courtauld Gallery (opposite), but the Embankment Galleries host contemporary art exhibitions. On summer evenings the courtyard is used for trendy film shows and concerts; in winter the courtyard is iced over for skating.

TRAFALGAR SQUARE

♿ ; ⊖ Charing Cross

The grand piazza that separates the West End from Whitehall is the ultimate London cliché. Tourists come here in their hundreds of thousands to pose for photos in front of Nelson's Column – built to celebrate the British victory at Trafalgar in 1805 – and clamber all over the four bronze lions at its base. Fronting onto the square are the National Gallery and Hawksmoor's St Martins-in-the-Fields. See also boxed text, above.

SHOP

APPLE STORE *Computers*

☎ 7153 9000; www.apple.com/uk; 235 Regent St; 10am-9pm Mon-Sat; ⊖ Oxford Circus

Apple's flagship UK store attracts hordes of gadget addicts looking for the latest laptops and iPods (and checking their email for free on the demo computers).

COVENT GARDEN MARKET

Market

☎ 7395 1010; www.coventgardenmarket.co.uk; 3-11 Southampton St; 10am-7pm Mon-Sat, 11am-6pm Sun; ; Covent Garden

Designed by Inigo Jones, Covent Garden piazza has been attracting shoppers since medieval times. Today most of the market stalls sell touristy knickknacks, but the attendant street performers lend a carnival air to proceedings.

ELLIS BRIGHAM

Outdoor Gear

☎ 7395 1010; www.ellis-brigham.com; 3-11 Southampton St; 10am-7pm Mon-Fri, to 8pm Thu, 9.30am-6.30pm Sat, 11.30am-5.30pm Sun; ; Covent Garden or Charing Cross

The ultimate emporium for trekking, camping, backpacking, climbing and snow gear. There's even an 8.5m indoor ice-climbing wall (www.vertical-chill.com).

FORBIDDEN PLANET *Books*

☎ 7420 3666; www.forbiddenplanet.com; 179 Shaftesbury Ave; ; 10am-7pm Mon-Sat, to 8pm Thu, noon-6pm Sun; Tottenham Court Rd

All that glitters – shopping at Covent Garden Market

CENTRAL SHOPPING SPOTS

Shopping is one of the main reasons to visit Covent Garden and Soho by day, and there are more shops than any book could reasonably list. Here's what to expect in the squares and streets of this neighbourhood.

- Berwick St (B3) – second-hand record and fabric shops
- Carnaby St (B4) – what all shopping streets would look like if the chain stores got their way
- Charing Cross Rd (D4) – excellent book shops like Foyle's (below), Murder One and Shipley (see boxed text, p50)
- Covent Garden Piazza (F4) – arty crafty market stalls targeted at tourists
- Denmark St (D3) – more cool guitars than a Rolling Stones tour
- Floral St (F4) – upmarket fashion stores like Paul Smith (p50) and Ted Baker (p51)
- Long Acre (E4) – classier high-street fashions and the wonderful Stanfords (p51)
- Monmouth St (E3) – floaty fashions and funky footwear in fun, cool boutiques
- Neal St (E3) – mainstream fashion stores masquerading as boutiques
- Oxford St (B2) – shocking crowds, chain stores, and HMV and Zavvi
- Regent St (A4) – more crowds and more chain stores, plus Hamleys (below)
- Southampton St (F4) – everything you need for an Everest expedition

Looking for the complete back issues of *Sandman*? Need a framed print of a Stan Lee supervillain? Desperate for a poseable plastic Dr Who figure? Then look no further than this comics and collectables superstore. For more West end comic shops, see boxed text p107.

FOYLES *Books*

☎ 7437 5660; www.foyles.co.uk; Charing Cross Rd; 9.30am-9pm Mon-Sat, noon-6pm Sun; ⊖ Oxford Circus

Established in 1903, Foyles is a huge sprawling bookshop with a vast range of reading and listening material – head to Ray's Jazz Café for jazz and world music CDs. There's a branch on the South Bank.

HAMLEYS *Toys*

☎ 0844 855 2424; www.hamleys.com; 188-196 Regent St; 10am-8pm Mon-Fri, 9am-8pm Sat, noon-6pm Sun; ; ⊖ Oxford Circus

Live demonstrations add an interactive element at London's most famous toy shop. Selling everything from puzzles and pogo-sticks to Lego and Xbox games, the shop gets packed – only the brave or masochistic visit on weekends.

JESSOPS *Cameras*

☎ 0845 458 7201; www.jessops.co.uk; 63-69 New Oxford St; 9am-7pm Mon-Fri, to 8pm Thu, 11am-5pm Sun; ⊖ Tottenham Court Rd

Jessops has two floors of cameras, accessories, and processing and

BOOKWORM-A-GO-GO

Charing Cross Rd is the centre of the London book trade. Superior specialist bookshops include the following:
Koenig Books (☎ 7240 8190; www.koenigbooks.co.uk; 80 Charing Cross Rd) Photography and design.
Murder One (☎ 7539 8820; www.murderone.co.uk; 76-78 Charing Cross Rd) Crime fiction.
Shipley (☎ 7240 1559; www.artbook.co.uk; 72 Charing Cross Rd) The arts.

printing equipment – it's a firm favourite of photography students and pro-snappers.

LIBERTY *Department Store*

☎ 7734 1234; www.liberty.co.uk; 210-220 Regent St; 10am-9pm Mon-Sat, to 8pm Thu, noon-6pm Sun; ⊖ Oxford Circus

Step into another era at this old-fashioned, mock-Tudor department store, best known for its printed fabrics, homewares and exotic rugs. 'Liberty prints' still appear regularly on the pages of *Vogue* and *Elle*.

PAUL SMITH *Clothing*

☎ 7379 7133; www.paulsmith.co.uk; 40-44 Floral St; 10.30am-6pm Mon-Wed, 10.30am-7pm Thu, 10.30am-6.30pm Fri & Sat, noon-5pm Sun; ⊖ Covent Garden

Paul Smith represents the best of British design, with sharply cut shirts, suits, trousers and womenswear for the perfect London look. You'll find branches all over town, but for bargains, head to the sale shop (p96).

POSTE MISTRESS *Shoes*

☎ 7379 4040; www.office.co.uk/postemistress; 61-63 Monmouth St; 10am-7pm Mon-Sat, 11.30am-6pm Sun; ⊖ Covent Garden or Leicester Sq

Shoe shoppers go weak at the knees when confronted by Poste Mistress' pastel-coloured wellies and shoes from the likes of Emma

BACKSTREET VINYL

Oxford St's **HMV** (www.hmv.com) and **Zavvi** (www.zavvi.co.uk) are stacked floor to ceiling with new CDs, DVDs and vinyl, but hardcore music collectors head to Berwick St (accessible from Oxford Circus or Piccadilly Circus tube) for the best in second-hand music. Top stores include **Revival Records** (☎ 7437 4271; www.revivalrecords.uk.com; 30 Berwick St) and **Sister Ray** (☎ 7734 3297; www.sisterray.co.uk; 34-35 Berwick St). Fans of classical music can find some gems at specialist music seller **Harold Moores Records** (☎ 7437 1576; www.hmrecords.co.uk; 2 Great Marlborough St).

NAUGHTY SOHO

Since the rise of the internet, the old-men-in-trench-coats-style sex shops in Soho have gone into rapid decline, but a new kind of adult store has emerged on the scene. Soho's new intimate emporiums are smart, sophisticated and firmly aimed at grown-up couples; try the following:

Agent Provocateur (☎ 7439 0229; www.agentprovocateur.com; 6 Broadwick St; ⊖ Oxford Circus or Piccadilly Circus) Sleek and sexy lingerie emporium, with branches in Selfridges and Harrods.

Coco de Mer (☎ 7836 8882; www.coco-de-mer.co.uk; 23 Monmouth St; ⊖ Tottenham Court Rd) A classy take on silky lingerie, ticklers and spankers – perfect for role playing.

Prowler (☎ 7734 4031; www.prowlerstores.co.uk; 5-7 Brewer St; ⊖ Oxford Circus or Piccadilly Circus) The largest gay sex shop in Britain – not for the faint-hearted.

Hope, Vivienne Westwood, Chloe Sevigny and Eley Kishimoto.

STANFORDS *Books*

☎ 7836 1321; www.stanfords.co.uk; 12-14 Long Acre; 9am-7.30pm Mon-Fri, from 9.30am Tue, to 8pm Thu, 10am-8pm Sat, noon-6pm Sun; ⊖ Covent Garden or Leicester Sq

Travellers have been coming to Stanfords of Long Acre for more than a century. Inside you'll find London's best selection of guidebooks, travel literature, maps and nifty travel gifts.

TATTY DEVINE *Jewellery*

☎ 7434 2257; www.tattydevine.com; 57b Brewer St; 11am-7pm Mon-Sat; ⊖ Oxford Circus

Chintzy, camp and high-spirited, Tatty Devine assembles unique costume jewellery from plastic and found objects – check out the baked-Perspex name pendants and plastic crisp necklaces. There's a branch on Brick Lane (p150).

TED BAKER *Clothing*

☎ 7836 7808; www.tedbaker.com; 9-10 Floral St; 10am-7pm Mon-Sat, to 8pm Thu, noon-6pm Sun; ⊖ Covent Garden

A great British designer, Ted Baker offers stylish retro men's and women's wear. Even if you can't afford the clobber, it's worth coming into see the shop floor – currently styled like a 1950s ballroom.

VINTAGE & RARE GUITARS *Music*

☎ 7240 7500; www.vintageandrareguitars.com; 6 Denmark St; 10am-6pm Mon-Sat, noon-4pm Sun; ⊖ Tottenham Court Rd

Denmark St is lined with guitar shops, but hardcore axe-addicts make a beeline for this Aladdin's cave of vintage Stratocasters, Les Pauls, Gretschs and Rickenbackers.

EAT

ANDREW EDMUNDS

Modern European £££-£££

☎ 7437 5708; 46 Lexington St; 12.30-3pm Mon-Fri, 1-3pm Sat & Sun, 6-10.30pm daily; ⊖ Oxford Circus

Squeezed onto two floors of a cramped Georgian townhouse, this Soho favourite has been at the crest of the modern British cooking wave since 1983. It serves hearty, homey meals prepared with high-quality seasonal British ingredients.

ARBUTUS

Modern European ££-£££

☎ 7734 4545; www.arbutusrestaurant.co.uk; 63-64 Frith St; noon-2.30pm & 5-10.30pm Mon-Sat, noon-3.30pm & 5.30-9.30pm Sun; ⊖ Tottenham Court Rd

The minimalist décor of this sleek modern European restaurant stands in stark contrast to the imaginative Mediterranean-inspired food. Step outside your comfort zone with the braised pig's head and squid and mackerel burgers.

CAFÉ DE HONG KONG

Chinese £

☎ 7534 9898; 47 Charing Cross Rd; 11.30am-11pm Mon-Sat, 11am-10.30pm Sun; ⊖ Leicester Sq

Chinese students gather at this inexpensive canteen for tasty Cantonese food served double fast. Order hawker soup noodles or ho fun with black bean sauce and swish it down with sago pearl tea.

GREAT QUEEN STREET

British ££

☎ 7242 0622; 32 Great Queen St; noon-2.30pm Tue-Sun, 6-10.30pm Mon-Sat; ⊖ Holborn

Owned by the same team as Waterloo's Anchor & Hope (p80),

THE YARD THAT NEAL BUILT

Tucked away behind Monmouth St, Neal's Yard has been transformed into a major international brand. The **Neal's Yard Dairy** (☎ 7240 5700; www.nealsyarddairy.co.uk; 17 Shorts Gardens) is a fabulous emporium of pungent English cheeses (there's a branch at Borough Market; p79), while **Neal's Yard Remedies** (☎ 7379 7222; www.nealsyardremedies.com; 15 Neal's Yard) sells organic skincare products and herbal remedies. While you're here, order a fresh fruit smoothie from the cheerfully Bohemian **Neal's Yard Salad Bar** (☎ 7836 3233; www.nealsyardsaladbar.com; 1, 8/10 Neal's Yard). Covent Garden is the nearest tube.

COOL CHAINS

Londoners frequently mourn the loss of small independent cafés, but not all chains are bad. Look out for branches of the following superior café chains.

Carluccio's (☎ 7836 0990; www.carluccios.com; Garrick St; ⊖ Leicester Sq or Covent Garden) One of London's favourite lunch spots, with fabulously fresh ciabatta and pizza, and top-quality deli ingredients.

Patisserie Valerie (☎ 7437 3466; www.patisserie-valerie.co.uk; 44 Old Compton St; ⊖ Leicester Sq) Founded in Soho in 1926, serving delicious sweet pastries and gelato.

Paul (☎ 7836 5321; www.paul-uk.com; 29 Bedford St; ⊖ Leicester Sq or Covent Garden) Appeals to fans of baguette sandwiches, fruit tarts, Viennoiserie and French glacé.

this down-to-earth eatery serves wine by the carafe and well-prepared modern British dishes, including hearty roasts to share.

HAOZHAN *Chinese* ££

☎ 7434 3838; www.haozhan.co.uk; 8 Gerrard St; noon-11pm Sun-Thu, noon-midnight Fri & Sat; ⊖ Leicester Sq

The chef at Haozhan used to work at the glitzy Hakkasan (p108), ensuring a refreshingly modern take on traditional soup noodles and other Cantonese classics.

J SHEEKEY *Seafood* £££

☎ 7240 2565; www.j-sheekey.co.uk; 28-32 St Martin's Ct; noon-3pm & 5.30pm-midnight Mon-Sat, noon-3.30pm & 6-11pm Sun; ⊖ Leicester Sq

Sister-restaurant to the Ivy (p54), J Sheeky specialises in finely prepared Atlantic salmon, razor shells and other fruits from the seas around Great Britain. You still get the celebrities and the long waiting list, but some say less attitude.

MAISON BERTAUX *French* £

☎ 7437 6007; 28 Greek St; 8.30am-11pm Mon-Sat, 8.30am-7pm Sun; ⊖ Leicester Sq or Tottenham Court Rd

Get the *Amelie* vibe at this 130-year-old Parisian café, decked out with stripy tablecloths. Enjoy sinful pastries and coffee by the demitasse.

MILDRED'S *Vegetarian* ££

☎ 7494 1634; www.mildreds.co.uk; 45 Lexington St; noon-11pm Mon-Sat; V; ⊖ Piccadilly Circus

Central London's most famous veggie restaurant, Mildred's has been serving up hearty, old-fashioned home cooking for more than 17 years. Service can be variable, but you can't fault the food.

MYUNG GA *Korean* ££

☎ 7734 8220; www.myungga.co.uk; 1 Kingly St; noon-3pm Mon-Sat, 5.30-11pm Mon-Sun; ⊖ Piccadilly Circus

Myung Ga is a big hit with the pre-theatre crowd – most diners order several types of meat for the barbecue and back it up with *bibimbap* (rice cooked in a hot stone bowl) and *chigae* (spicy cabbage and chilli soup).

RED FORT *Indian* £££-£££

☎ 7437 2525; www.redfort.co.uk; 77 Dean St; noon-2.30pm Mon-Fri, 5.45-11.30pm Mon-Sat; ⊖ Leicester Sq

The menu at this stylish Indian in the heart of Soho has a North-west Frontier edge, with lots of tender lamb and unusual, tongue-tingling curry dishes that you won't find at any of its competitors. In the basement is the stylish Akbar cocktail bar (yep, we get the pun).

THE IVY *British* £££

☎ 7836 4751; www.the-ivy.co.uk; 1 West St; noon-3pm & 5.30pm-midnight, to 11pm Sun; ⊖ Leicester Sq

You'll have to book one to six months ahead to enjoy the innovative modern British cuisine at this favourite hangout of London celebrities. The main reason to visit is the exquisite food, but there's a definite fascination in seeing how the other half live.

YAUATCHA *Chinese* ££

☎ 7494 8888; 15 Broadwick St; 11am-11.45pm Mon-Sat, 11am-10.45pm Sun, restaurant from noon; ⊖ Oxford Circus

Swooningly elegant, this blue-glass dim-sum restaurant and tearoom is another project from the unstoppable Alan Yau. Soho trendies flock here daily for the excellent dim sum and blue, white, black and green teas.

DRINK

BALANS *Café*

☎ 7439 2183; www.balans.co.uk; 60 Old Compton St; 8am-5am Mon-Thu, 8am-6am Fri & Sat, 8am-2am Sun; ⊖ Leicester Sq

A funky and gay-friendly Art Deco café with street-side tables along Old Compton St. It's a great spot for people-watching.

BAR ITALIA *Café*

☎ 7437 4520; www.baritaliasoho.co.uk; 22 Frith St; 24hr; ⊖ Leicester Sq

This old-school Italian café opened in 1949 and Soho revellers still gather here at the start and tail-end of an evening for coffee, panettone and intense conversations about European football.

LAB *Bar*

☎ 7437 7820; www.lab-townhouse.com; 12 Old Compton St; 4pm-midnight Mon-Thu, 4pm-12.30am Fri & Sat, 4-11pm Sun; ⊖ Tottenham Court Rd or Leicester Sq

Staffed by graduates from the London Academy of Bartending,

this pocket-sized cocktail bar offers cool and colourful drinks mixed to perfection.

LAMB & FLAG *Pub*

☎ 7497 9504; 33 Rose St; ⊖ Leicester Sq or Covent Garden

Most pubs in Covent Garden are touristy as hell, so this historic backstreet alehouse feels like a little piece of heaven. It's compact, cosy and crowded, and it serves draught ales.

SALISBURY *Pub*

☎ 7836 5863; 90 St Martin's Lane; ⊖ Leicester Sq

The Salisbury is a 'quick drink after work' kind of place, but the etched and engraved Victorian windows and lamps shaped like Art Nouveau nudes make it worth a quick look.

PLAY

Soho and Covent Garden form the heart of Theatreland – see p24 for more information.

COMEDY STORE *Comedy*

☎ Ticketmaster 0844 847 1728; www.thecomedystore.co.uk; Haymarket House, 1a Oxendon St; from 8pm, also 12pm show Fri & Sat; ⊖ Piccadilly Circus

Definitely at the funnier end of stand-up comedy in London, this fun-house attracts all the big names. Come on Wednesday for improv, often led by the brilliant Paul Merton.

WEST END TICKETS

There are more than 50 theatres in the West End, but websites such as www.whatsonstage.com can help you choose between the weird and wacky productions. Full-price tickets to most shows can be booked through **Ticketmaster** (☎ 0870 060 2340; www.ticketmaster.co.uk). Alternatively, you can buy discounted seats the same day at the **TKTS Booth** (www.officiallondontheatre.co.uk; 10am-7pm Mon-Sat, noon-3pm Sun; ⊖ Leicester Sq) in Leicester Sq. Also see p196.

CURZON SOHO *Cinema*

☎ information 7292 1686, bookings 0871 703 3988; www.curzoncinemas.com; 99 Shaftesbury Ave; ⊖ Leicester Sq

More than just a cinema, this arty Soho movie house pulls in just as many punters with its window-front Konditor & Cook café (p79) and basement bar.

DONMAR WAREHOUSE *Theatre*

☎ 0870 060 6624; www.donmarwarehouse.com; 41 Earlham St; ⊖ Covent Garden

This is the theatre where Nicole Kidman administered 'theatrical

GAY SOHO

Soho is the cabaret-kicking, cycle-shorting, party-preening capital of the London gay scene, centred on Old Compton St. Following are our favourite venues:

Ghetto (☎ 7287 3726; www.ghetto-london.co.uk; 5-6 Falconberg Ct; 🕑 10.30pm-3am Mon-Wed, 10.30pm-4am Thu & Fri, 9.30pm-5am Sat, 9.30pm-2am Sun; ⊖ Tottenham Court Rd) Noisy, sweaty and fun – where the party people go after Old Compton St winds down for the evening.

Heaven (☎ 7930 2020; www.heaven-london.com; Under the Arches, Villiers St; 🕑 11pm-6am Mon, 11pm-4am Wed, 10.30pm-6am Sat; ⊖ Charing Cross or Embankment) Under Charing Cross railway arches, London's best-known gay club is the showground for the glamorous and beautiful on a Saturday night.

Rupert St (☎ 7494 3059; www.rupertstreet.com; 50 Rupert St; 🕑 noon-11.30pm; ⊖ Piccadilly Circus) A spic-and-span gay bar offering a more relaxed experience than the boisterous party pubs.

SIN (☎ 7240 1900; www.sinlondon.com; 144 Charing Cross Rd; 🕑 event times vary; ⊖ Tottenham Court Rd) The new venue for the famous Popstarz, London's biggest gay indie night every Friday.

Soho Revue Bar (☎ 7734 0377; www.sohorevuebar.co.uk; 11 Walker's Court; 🕑 event times vary; ⊖ Covent Garden) This former strip bar hosts the riotous Circus, London's favourite 'polysexual' Friday night.

Viagra' by stripping in Sam Mendes' *Blue Room*. It still attracts major stars looking for a more intimate forum to practise their art. Shows lined up for 2009 include Dame Judi Dench in *Madame de Sade* and Jude Law as Hamlet.

END *Club*

☎ 7419 9199; www.endclub.com; 18 West Central St; 🕑 10.30pm-3am Mon & Wed, 11pm-6am Fri, 11pm-7am Sat; ⊖ Holborn

Gritty and industrial, the End is the perfect setting for electronic soundscapes from resident DJs like Laurent Garnier, Groove Armada and DJ Fabio. In the same location is the hugely popular **Aka Bar** (www.akalondon.com).

ENGLISH NATIONAL OPERA *Opera*

☎ 0870 145 0200; www.eno.org; Coliseum, St Martin's Lane; ⊖ Leicester Sq or Charing Cross

Operating out of the renovated Coliseum Theatre, the English National Opera offers a very interesting program of classic and modern operas and performances by the **English National Ballet** (www.ballet.org.uk).

KARAOKE BOX *Karaoke*

☎ 7494 3878; www.karaokebox.co.uk; 18 Frith St; ⊖ Covent Garden

This cute, kitsch, six-booth karaoke bar is a favourite hangout of Kate Moss and Sarah Michelle Gellar, when she's in town.

MADAME JO JO'S *Club/Cabaret*

☎ 7734 3040; www.madamejojos.com; 8-10 Brewer St; 🕒 from 8pm Tue-Thu, from 10pm Fri & Sat, from 9.30pm Sun, special event times vary; ⊖ Piccadilly Circus

Downstairs from the revamped Raymond's Revue Bar, Madame Jo Jo's is a mecca for fans of funky beats and cabaret kitsch. Keb Darge's Deep Funk Fridays are a Soho legend.

PRINCE CHARLES CINEMA *Cinema*

☎ 0870 811 2559; www.princecharlescinema.com; 7 Leicester Pl; ⊖ Leicester Sq

This Leicester Sq cinema shows a mixture of old classics and recent releases after they have stopped showing at mainstream cinemas. Look out for screenings of vintage treats like *Touch of Evil*.

RONNIE SCOTT'S *Live Music*

☎ 7439 0747; www.ronniescotts.co.uk; 47 Frith St; 🕒 6pm-3am Mon-Sat, 6pm-3am Sun; ⊖ Leicester Sq

The iconic London nightspot founded by Ronnie Scott is still pulling in the biggest names in jazz from around the world. Ronnie's was taken over by theatre impresario Sally Greene in 2005, but the vibe and the line-up are as good as ever.

ROYAL OPERA HOUSE *Opera*

☎ 7304 4000; www.royaloperahouse.org; Bow St; ⊖ Covent Garden

Since its £210-million makeover in the 1990s, the Royal Opera House has been on a mission to open operatic theatre up to the masses. Popular shows in 2008 included Damon Albarn's cartoon opera *Monkey: Journey to the West*. The **Royal Ballet** (www.royalballet.co.uk) also performs here.

Rejuvenated classic – the Royal Opera House

>THE CITY, HOLBORN & CLERKENWELL

Until the medieval period, only the square mile of land inside the Roman city walls could officially be called 'London'. Even today the 'City of London' is distinct from the rest of the metropolis, with its own mayor, its own local government and the wealth of the nation in its pockets, thanks to the mighty financial institutions that drive the British economy.

Visited by some 350,000 commuters every weekday, the City might not seem an obvious tourist attraction, but the history of London is written large in the streets. The great cathedral built by Sir Christopher Wren after the Great Fire of 1666 still dominates the skyline, accompanied by a crystal garden of modern skyscrapers.

To the north of the City, Clerkenwell is a rapidly developing entertainment quarter, and further west is Holborn (*ho*-bern), the cradle of English justice, worth a stop for the quirky Hunterian and Sir John Soane's museums.

THE CITY, HOLBORN & CLERKENWELL

SEE

30 St Mary Axe **1** F4
Bank of England & Museum **2** E4
Dr Johnson's House **3** B4
Hunterian Museum **4** A4
Inner Temple **5** B4
Leadenhall Market **6** E4
Mansion House............... **7** E4
Monument **8** E5
Museum of London **9** D3
Royal Exchange **10** E4
Saint Bartholomew-the-Great **11** C3
Saint Bride's Church **12** C4
Saint Paul's Cathedral **13** D4
Sir John Soane's Museum **14** A3
Temple Church **15** B4
Tower Bridge **16** G6
Tower of London **17** F5

SHOP

Guildhall Library Bookshop **18** D4
International Magic **19** B2
Lesley Craze Gallery **20** C2
Magma Books **21** B3

EAT

Asadal.......................... **22** A3
Bevis Marks Restaurant **23** F4
Café Spice Namaste...... **24** G4
Club Gascon.................. **25** C3
Coach & Horses............. **26** B2
Eagle............................ **27** B2
Moro **28** B2
Peasant........................ **29** C2
Place Below **30** D4
Saint John..................... **31** C3
Smiths of Smithfield **32** C3
Well **33** C2

DRINK

Café Kick....................... **34** B2
Jerusalem Tavern **35** C3
Vertigo 42.................... **36** E4
Ye Olde Cheshire Cheese **37** C4
Ye Olde Mitre................ **38** B3

PLAY

Barbican **39** D3
Fabric........................... **40** C3
Sadler's Wells **41** C1
Volupté Lounge............ **42** B3

Please see over for map

SEE

BANK OF ENGLAND MUSEUM

☎ 7601 5545; www.bankofengland.co.uk; Bartholomew Lane; admission free; 10am-5pm Mon-Fri; Bank

Parts of the fortress-like Bank of England date back to the original banking house built by Sir John Soane in 1788. Most of the bank is locked up tight, but you can lift up a genuine gold bar under the watchful eye of CCTV in the museum.

HUNTERIAN MUSEUM

☎ 7869 6560; www.rcseng.ac.uk/museums; Royal College of Surgeons of England, 35-43 Lincoln's Inn Fields; admission free; 10am-5pm Tue-Sat; ; Holborn

Named for the pioneering 18th-century surgeon John Hunter, this stylish museum displays Britain's finest collection of medical specimens, human dissections and surgical tools. It's not for the squeamish, but the displays are quite astonishing.

LEADENHALL MARKET

www.leadenhallmarket.co.uk; Whittington Ave; 7am-4pm Mon-Fri; ; Bank

You might almost believe you've been whisked away to Victorian London as you step into this glorious covered arcade. Gorgeous ornamental ironwork curves overhead and City traders talk deals and mergers in the market's restaurants and pubs. It's off Gracechurch St.

L IS FOR LEXICOGRAPHER

Samuel Johnson (1709–84) was one of London's great mould breakers – he spoke his mind, he upset the establishment and he wrote the first ever English dictionary. Published in 1755, the *Dictionary of the English Language* was a nine-year labour of love that included such charming definitions as 'Dull – not exhilarating, as in to make dictionaries is dull work'.

Johnson also gained a reputation for outspoken tirades and nervous tics, leading many modern historians to suspect Tourette's Syndrome. Nevertheless, he was an outspoken campaigner on the issues of the day, writing against slavery a full century before it was abolished. Upon his death, he left his wealth to his black manservant Francis Barber.

Dr Johnson's house (☎ 7353 3745; www.drjohnsonshouse.org; 17 Gough Sq; adult/child/concession £4.50/1.50/3.50; 11.30am-5.30pm Mon-Sat, to 5pm Oct-Apr; Chancery Lane) off Fleet St has been restored as a museum full of Johnson memorabilia and period furnishings.

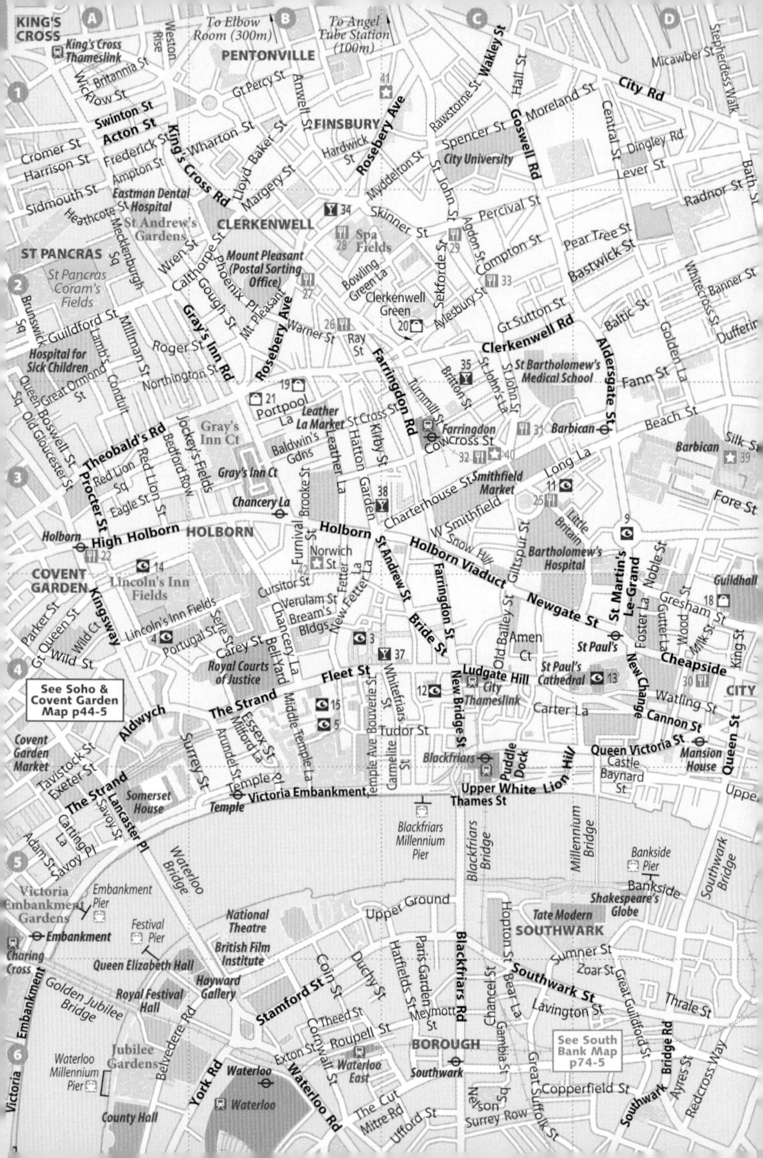

KING'S CROSS
King's Cross Thameslink
To Elbow Room (300m)
To Angel Tube Station (100m)
PENTONVILLE
Weston Rise
Britannia St
Wicklow St
Swinton St
Acton St
Cromer St
Harrison St
Frederick St
Ampton St
Sidmouth St
Heathcote St
Eastman Dental Hospital
St Andrew's Gardens
Mecklenburgh Sq
ST PANCRAS
St Pancras Coram's Fields
Brunswick Sq
Guildford St
Lamb's Conduit
Millman St
Hospital for Sick Children
Great Ormond St
Queen Sq
Boswell St
Old Gloucester St
Theobald's Rd
Red Lion Sq
Red Lion St
Eagle St
Procter St
Bedford Row
Jockey's Fields
Gray's Inn Ct
Gray's Inn Rd
Roger St
Northington St
King's Cross Rd
Wharton St
Lloyd Baker St
Margery St
Gt Percy St
Amwell St
FINSBURY
Hardwick St
Rosebery Ave
Myddelton St
CLERKENWELL
Wren St
Calthorpe St
Gough St
Phoenix Pl
Mount Pleasant (Postal Sorting Office)
Mt Pleasant
Warner St
Spa Fields
Bowling Green La
Clerkenwell Green
Ray St
Skinner St
Rawstorne St
Wakley St
Hall St
Goswell Rd
Spencer St
City University
Moreland St
City Rd
Micawber St
Shepherdess Walk
Central St
Dingley Rd
Lever St
Bath St
Radnor St
St John St
Percival St
Agdon St
Sekforde St
Compton St
Aylesbury St
Pear Tree St
Bastwick St
Whitecross St
Banner St
Baltic St
Dufferin
Gt Sutton St
Clerkenwell Rd
Aldersgate St
Golden La
Fann St
Beach St
St Bartholomew's Medical School
St John's La
Britton St
Turnmill St
Farringdon Rd
Farringdon
Cowcross St
Barbican
Silk St
Fore St
Portpool La
Leather La Market
Cross St
Baldwin's Gdns
Leather La
Hatton Garden
Kirby St
Brooke St
Chancery La
Charterhouse St
Smithfield Market
Long La
Little Britain
W Smithfield
Snow Hill
Giltspur St
Bartholomew's Hospital
Holborn
High Holborn
HOLBORN
Furnival St
Norwich St
St Andrew St
Holborn Viaduct
Farringdon St
COVENT GARDEN
Kingsway
Lincoln's Inn Fields
Cursitor St
Fetter La
New Fetter La
Verulam St
Bream's Bldgs
Newgate St
St Martin's Le-Grand
Noble St
Guildhall
Gresham St
Foster La
Gutter La
Wood St
Milk St
King St
Parker St
Gt Queen St
Wild Ct
Wild St
Portugal St
Serle St
Carey St
Bell Yard
Bride St
Old Bailey St
Amen Ct
St Paul's
New Change
Cheapside
CITY
Royal Courts of Justice
Fleet St
Whitefriars
Ludgate Hill
City Thameslink
St Paul's Cathedral
Watling St
Cannon St
See Soho & Covent Garden Map p44-5
Aldwych
The Strand
Essex St
Milford La
Middle Temple La
Bouverie St
Temple Ave
Carmelite St
Tudor St
New Bridge St
Carter La
Queen Victoria St
Mansion House
Queen St
Castle Baynard St
Covent Garden Market
Tavistock St
Exeter St
Surrey St
Arundel St
Temple Pl
Blackfriars
Puddle Dock
Upper White Lion Hill
Upper Thames St
Temple
Victoria Embankment
Somerset House
Savoy St
Lancaster Pl
Carting La
Adam St
Savoy Pl
Waterloo Bridge
Blackfriars Millennium Pier
Blackfriars Bridge
Millennium Bridge
Bankside Pier
Bankside
Southwark Bridge
Victoria Embankment Gardens
Embankment Pier
Embankment
Festival Pier
Charing Cross
National Theatre
British Film Institute
Upper Ground
Shakespeare's Globe
Tate Modern
SOUTHWARK
Hopton St
Queen Elizabeth Hall
Hayward Gallery
Royal Festival Hall
Golden Jubilee Bridge
Stamford St
Coin St
Duchy St
Hatfields St
Paris Garden
Blackfriars Rd
Chancel St
Bear La
Southwark St
Sumner St
Zoar St
Great Guildford St
Thrale St
Lavington St
Theed St
Meymott St
Roupell St
Cornwall St
Exton St
Waterloo East
BOROUGH
Southwark
Gambia St
Great Suffolk St
See South Bank Map p74-5
Bridge Rd
Ayres St
Redcross Way
Copperfield St
Jubilee Gardens
Belvedere Rd
Waterloo Millennium Pier
York Rd
Waterloo
Waterloo Rd
County Hall
The Cut
Mitre Rd
Ufford St
Nelson Sq
Surrey Row
Southwark Bridge Rd
Victoria Embankment

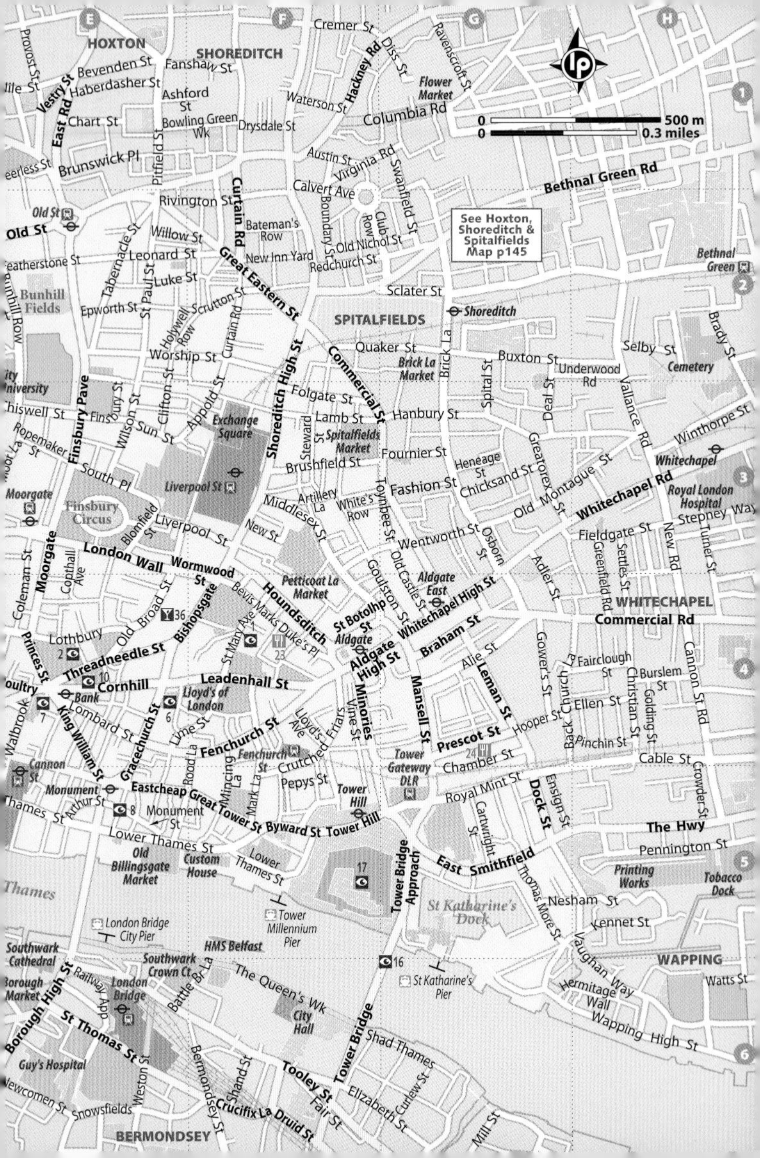

E
F
G
H
1
2
3
4
5
6
HOXTON
SHOREDITCH
SPITALFIELDS
WHITECHAPEL
WAPPING
BERMONDSEY
0 500 m
0 0.3 miles
See Hoxton, Shoreditch & Spitalfields Map p145
Provost St
Bevenden St
Fanshaw St
Haberdasher St
Vestry St
East Rd
Chart St
Ashford St
Bowling Green Wk
Drysdale St
Brunswick Pl
Pitfield St
Cremer St
Hackney Rd
Diss St
Ravenscroft St
Flower Market
Waterson St
Columbia Rd
Austin St
Virginia Rd
Swanfield St
Calvert Ave
Boundary St
Club Row
Bethnal Green Rd
Rivington St
Curtain Rd
Bateman's Row
Old St
Willow St
Old Nichol St
Redchurch St
New Inn Yard
Leonard St
Great Eastern St
Tabernacle St
Luke St
St Paul St
Epworth St
Scrutton St
Holywell Row
Bunhill Fields
Bunhill Row
Sclater St
Shoreditch
Bethnal Green
Quaker St
Brick La
Brick La Market
Buxton St
Selby St
Brady St
Underwood Rd
Cemetery
Worship St
Shoreditch High St
Commercial St
Spital St
Deal St
Vallance Rd
Folgate St
Lamb St
Hanbury St
Appold St
Clifton St
Wilson St
Finsbury St
Chiswell St
Sun St
Exchange Square
Steward St
Spitalfields Market
Fournier St
Greatorex St
Winthorpe St
Whitechapel
Finsbury Pave
Ropemaker St
South Pl
Brushfield St
Heneage St
Chicksand St
Old Montague St
Liverpool St
Fashion St
Toynbee St
Artillery La
White's Row
Whitechapel Rd
Royal London Hospital
Moorgate
Finsbury Circus
Blomfield St
Liverpool St
Middlesex St
New St
Wentworth St
Osborn St
Fieldgate St
Stepney Way
New Rd
Turner St
Settles St
Greenfield Rd
Adler St
London Wall
Wormwood St
Coleman St
Copthall Ave
Old Broad St
Bishopsgate
Petticoat La Market
Goulston St
Old Castle St
Aldgate East
Whitechapel High St
Houndsditch
Bevis Marks
Duke's Pl
St Mary Axe
St Botolph St
Aldgate
Aldgate High St
Braham St
Commercial Rd
Lothbury
Threadneedle St
Princes St
Alie St
Leman St
Gower's St
Fairclough St
Burslem St
Christian St
Cannon St Rd
Poultry
Bank
Cornhill
Leadenhall St
Lloyd's of London
Minories
Vine St
Mansell St
Back Church La
Ellen St
Golding St
Walbrook
King William St
Lombard St
Gracechurch St
Lime St
Lloyd's Ave
Fenchurch St
Crutched Friars
Hooper St
Prescot St
Pinchin St
Cable St
Cannon St
Rood La
Mincing La
Mark La
Fenchurch St
Pepys St
Tower Gateway DLR
Chamber St
Crowder St
Monument
Eastcheap
Great Tower St
Arthur St
Monument St
Tower Hill
Royal Mint St
Cartwright St
Dock St
Ensign St
Thames St
Byward St
Tower Hill
The Hwy
Lower Thames St
Old Billingsgate Market
Custom House
Lower Thames St
Tower Bridge Approach
East Smithfield
Pennington St
Printing Works
Tobacco Dock
Thames
Tower Millennium Pier
St Katharine's Dock
Thomas More St
Nesham St
Kennet St
London Bridge City Pier
HMS Belfast
Southwark Cathedral
Southwark Crown Ct
Battle Br La
St Katharine's Pier
Vaughan Way
Hermitage Wall
Watts St
Borough Market
Railway App
London Bridge
The Queen's Wk
City Hall
Borough High St
St Thomas St
Tower Bridge
Shad Thames
Wapping High St
Guy's Hospital
Weston St
Bermondsey St
Shand St
Tooley St
Curlew St
Newcomen St
Snowsfields
Crucifix La
Druid St
Fair St
Elizabeth St
Mill St
36
1
23
2
10
6
7
8
24
17
16

LONDON'S INNS OF COURT

Since the 13th century, all London barristers have worked from one of London's four Inns of Courts, archaic institutions akin to the Freemasons in their intricate protocol. Entry to the inns is generally reserved for members of the legal profession, but visitors are free to wander the alleyways of the **Inner Temple** (www.innertemple.org.uk) off Fleet St to visit the **Temple Church** (www.templechurch.com), founded by the Knights Templar – predictably, it attracts a fair few *Da Vinci Code* enthusiasts looking for evidence of a papal conspiracy; see the website for current hours. Temple is the nearest tube station.

MANSION HOUSE & AROUND

Mansion House Pl; ⊖ Bank

Facing the Bank of England, the imposing Mansion House is the official residence of the Lord Mayor of London (not to be confused with the Mayor of London, Boris Johnson). The adjacent **Royal Exchange** was the original home of the London Stock Exchange; today it houses swanky boutiques and a swish courtyard café and bar.

MONUMENT

☎ 7626 2717; Monument St; adult/child 5-15yr £2/1; 9.30am-5.30pm; ⊖ Monument

Christopher Wren's memorial to the Great Fire of London stands 202ft tall – the exact distance to the bakery where the Great Fire started in 1666. After a £4.5 million restoration, visitors can once again climb the 311 steps for dizzying City views.

MUSEUM OF LONDON

☎ 0870 444 3582; www.museumoflondon.org.uk; 150 London Wall; admission free; 10am-6pm, last entry 5.30pm, to 9pm 1st Thu of month; ⊖ Barbican

Partner to the Museum in Docklands (p179), this fine museum evocatively recreates the history of the capital, from prehistoric times to the Great Fire of 1666 (more recent galleries are closed for renovations until 2010), using quirky displays, sound effects and hands-on reconstructions. Several medieval bastions and stretches of the old Roman walls are visible nearby.

SIR JOHN SOANE'S MUSEUM

☎ 7405 2107; www.soane.org; 13 Lincoln's Inn Fields; admission free, tours £5; 10am-5pm Tue-Sat, 6-9pm 1st Tue of month, museum tours 11am Sat; ⊖ Holborn

Sir John Soane was the architectural genius behind the Bank of England, and his former home is now a museum containing all the bits of architectural bric-a-brac that Soane accumulated in his lifetime. The candlelit Tuesday evenings are magical.

ST BARTHOLOMEW-THE-GREAT

☎ 7606 5171; www.greatstbarts.com; West Smithfield; 🕑 8.30am-5pm Mon-Fri, 10.30am-4pm Sat, 8am-1pm & 2.30-8pm Sun; ⊖ Barbican or Farringdon

St Bart's was the original priory church for the medieval St Bartholomew's hospital. Henry VIII trashed the monastery during his Reformation rampage, but the church was saved and upgraded by the Elizabethans. Appropriately, it was used as a shooting location for the film *Shakespeare in Love*.

ST BRIDE'S CHURCH

☎ 7427 0133; www.stbrides.com; Fleet St; admission free; 🕑 8am-4.45pm Mon-Fri; ⊖ Blackfriars, 🚆 City Thameslink

The stepped steeple of this Wren masterpiece allegedly inspired the tiered wedding cake, and the first mechanical printing press in England was set up here in the 1500s. The church has been associated with printing ever since – today it houses a memorial to journalists killed in conflicts around the world.

ST PAUL'S CATHEDRAL

☎ 7236 4128; www.stpauls.co.uk; St Paul's Churchyard; adult/under 7yr/7-16yr/concession £10/free/3.50/8.50-9; 🕑 8.30am-4pm (last entry) Mon-Sat; ♿ ; ⊖ St Paul's or Blackfriars

Even the bombs of the Blitz couldn't erase the distinctive dome of London's most famous church. St Paul's was built by Sir Christopher Wren to replace the medieval cathedral destroyed in the Great Fire, but he had to sneak his plans for a dramatic dome

The bomb-dodging dome of St Paul's Cathedral

past the City planners. Head to the crypt for memorials to famous Londoners and the Golden Gallery atop the dome for awesome City views. See p12 for more.

TOWER BRIDGE

☎ 7940 3985; www.towerbridge.org.uk; adult/child 5-15yr/concession/family £6/3/4.50/10-14; 9.30am-6pm Oct-Mar, 10am-6.30pm Apr-Sep; ; ⊖ Tower Hill

According to legend, the German Luftwaffe left this famous bridge intact to act as a navigation aid for bombings raids on London in WWII. True or not, the upper gallery offers a bombadier's view over the river, and there's an engaging exhibition on the steam-powered swing bridge, which still opens periodically for ships.

TOWER OF LONDON

☎ 0870 756 7070; www.hrp.org.uk/toweroflondon; Tower Hill; adult/under 5yr/5-15yr/concession/family £16.50/free/9.50/14/46, discounts online; 9am-5.30pm Tue-Sat, 10am-5.30pm Sun & Mon, to 4.30pm daily Nov-Feb, last admission 1hr before closing; ; ⊖ Tower Hill

The good news is that the City's biggest tourist attraction more

Towering presence – historic Tower Bridge is one of London's most recognisable landmarks

NEXT STOP, THE STRATOSPHERE

For centuries the height of buildings in the centre of London was restricted by medieval laws protecting specific views of St Paul's Cathedral, Westminster and the Tower of London. A few architectural leviathans crept through – most notably John Mowlem's Tower 42 (183m) – but for the most part, the City remained refreshingly low rise. Then along came Norman Foster's **30 St Mary Axe** (www.30stmaryaxe.com; 30 St Mary Axe; ⊖ Bank or Aldgate) at 180m, aka the Swiss Re Tower, aka the Gherkin (a reference to its curving, cucumber-like shape).

The taboo on building high-rises in the City was shattered and permission was granted for a series of monster towers in the heart of the City. The new Bishopsgate Tower (288m) and Heron Tower (246m) will soar above Liverpool St station, the Minerva Tower (216m) will tower over Aldgate, the Leadenhall Building (225m) and 20 Fenchurch St (160m) will rise above Leadenhall Market, and the new Shard of Glass (310m) at London Bridge will alter the protected view of St Paul's Cathedral from Parliament Hill forever.

The City's new zeal for skyscrapers has bought condemnation from the Prince of Wales, London Mayor Boris Johnson and the UN World Heritage Committee among others. For an update on the latest developments, visit the exhibitions at **New London Architecture** (Map p104; ☎ 020 7636 4044; The Building Centre, 26 Store St; admission free; 🕑 9am-6pm Mon-Fri, 10am-5pm Sun; ⊖ Goodge St).

than lives up to the hype. Kids will rave about the Crown Jewels and Tower ravens, gawp at the Yeoman Guards and revel in the gruesome tools for separating human beings from their body parts in the White Tower. See p17 for a more complete description. Only the Jewel House is accessible to wheelchair users.

SHOP

GUILDHALL LIBRARY BOOKSHOP *Books*

☎ 7332 1858; Guildhall, Gresham St; 🕑 9.30am-6pm Mon-Fri; ⊖ Bank or St Paul's

Part of the offices of the Corporation of London, this specialist bookshop carries an excellent selection of books on London and prints of historic maps of the city.

INTERNATIONAL MAGIC *Magic*

☎ 7405 7323; www.internationalmagic.com; 89 Clerkenwell Rd; 🕑 11.30am-6pm Mon-Fri, 11.30am-4pm Sat; ♿; ⊖ 🚆 Farringdon

Last of the old-fashioned magic shops, International Magic has been open for 45 years, selling jokes, novelties and props for the magicians' art.

PAVING THE STREETS WITH NEWSPRINT

In recent years London has become overrun by 'free papers' – lightweight dailies full of advertising and recycled news, handed out for free at tube and train stations. *Metro, London Lite, thelondonpaper* and *City AM* now account for a third of the paper rubbish on the London Underground – do the cleaners a favour and pop yours in the bin when you finish reading.

LESLEY CRAZE GALLERY *Jewellery*

☎ 7608 0393; www.lesleycrazegallery.co.uk; 33-35a Clerkenwell Green; 10am-5.30pm Tue-Sat; ⊖ Farringdon

Delicate gold gauze, anodised titanium and metal-plated twigs and leaves are just some of the materials used to create the stunning contemporary jewellery at this small boutique gallery.

MAGMA BOOKS *Books*

☎ 7242 9503; www.magmabooks.com; 117-119 Clerkenwell Rd; 10am-7pm Mon-Sat; ⊖ Farringdon

Proof of the existence of intelligent design, Magma is piled floor to ceiling with designer tees and books on the latest trends in graphic art.

EAT

As well as the following restaurants, world food stalls set up along Exmouth Market (B2) at lunchtime on weekdays.

ASADAL *Korean* ££

☎ 7430 9006; www.asadal.co.uk; 227 High Holborn; noon-3pm & 6-11pm Mon-Sat, 5-10.20pm Sun; ⊖ Holborn

Sizzling away next to Holborn tube, this stylish basement restaurant specialises in meaty Korean barbecues. It's probably the most sophisticated Korean restaurant in London and a firm favourite with the after-work crowd.

BEVIS MARKS RESTAURANT *Jewish* £££

☎ 7283 2220; www.bevismarkstherestaurant.com; Bevis Marks; noon-3pm & 5.30-10pm Mon-Thu, noon-3pm Fri; ⊖ Aldgate or Liverpool St

Enjoy upmarket kosher cuisine in elegant surroundings next to London's oldest synagogue, built in 1701. All the old Jewish favourites are here – chicken soup with matzos, salt beef and fries – plus some splendid Mediterranean-inspired roasts.

CAFÉ SPICE NAMASTE *Indian* ££

☎ 7488 9242; www.cafespice.co.uk; 16 Prescot St; 🕑 noon-3pm & 6.15-10.30pm Mon-Fri, 6.15-10.30pm Sat; ⊖ Tower Hill

Housed in a former magistrates' court, this superior Parsee (Indian-Persian) and Goan restaurant, oozes Bollywood glam. The house *dhansaak* (hot and sour lamb stew with lentils) swims with the rich flavours of the subcontinent.

CLUB GASCON *French* £££

☎ 7796 0600; www.clubgascon.com; 57 West Smithfield; 🕑 noon-2pm & 7-10pm Mon-Thu, noon-2pm & 7-10.30pm Fri, 7-10.30pm Sat; ⊖ Farringdon or Barbican

Club Gascon has earned a Michelin star for its degustation (read 'fine flavours in small portions') approach to French cuisine. Order four or five plates to make a meal.

EAGLE *Mediterranean* ££

☎ 7837 1353; 159 Farringdon Rd; 🕑 noon-11pm Mon-Sat, noon-5pm Sun; ⊖ Farringdon

Media types have long been fans of the Mediterranean-influenced roasts and salads at this pioneering eatery, the first gastropub in Britain. Freed from the constraints of being a current 'hot spot', it offers good food in convivial surroundings.

MORO *Spanish* £££

☎ 7833 8336; www.moro.co.uk; 34-36 Exmouth Market; 🕑 12.30-2.30pm & 7-10.30pm Mon-Sat; ⊖ Farringdon

There's a definite snooty air to Moro, but the North African influenced tapas menu earns the restaurant some loyal fans. Order a bottle of red and graze on tapas treats like hummus, quails' legs and tortilla.

TWO PINTS & A CONFIT OF OLD SPOT BELLY PORK, PLEASE

Ever since the Eagle (above) came up with the idea of serving gourmet food in a pub setting in 1991, gastropubs have been tickling London's collective palate. Following are some more recommendations (all accessible from Farringdon station or Barbican tube):

Coach & Horses (☎ 7278 8990; www.thecoachandhorses.com; 26-28 Ray St) It's the unpretentious menu that appeals here – roast pork, rabbit pie and a proper ploughman's lunch.

The Peasant (☎ 7336 7726; www.thepeasant.co.uk; 240 St John St) Fine Mediterranean fusion cooking, backed up by a good range of lagers and ales.

The Well (☎ 7251 9363; www.downthewell.co.uk; 180 St John St) A wood-lined gastropub with imported premium beers and quirky Modern European cooking.

SMITHS OF SMITHFIELDS
British £££-£££

☎ 7251 7950; www.smithsofsmithfield.co.uk; 67-77 Charterhouse St; from 7am Mon-Fri, from 10am Sat, from 9.30am Sun; ⊖ Farringdon

This multilayered bistro and booze hall can turn into a bit of a meat market, but it's undeniably popular with the office crowd. Skip the 4th-floor restaurant for the cheaper and more consistent 3rd-floor brasserie.

ST JOHN *British* ££-£££

☎ 7251 4090; www.stjohnrestaurant.co.uk; 26 St John St; noon-3pm & 6-11pm Mon-Fri, 6-11pm Sat; ⊖ Farringdon

Chef Fergus Henderson champions nose-to-tail cooking, where a use is found for every part of the animal. Treats on the menu at his Clerkenwell restaurant include pig's head, ox tongue and chitterlings (pig intestines). The Spitalfields branch (p154) is also open for breakfast.

THE PLACE BELOW
Vegetarian £-££

☎ 7329 0789; www.theplacebelow.co.uk; St Mary-Le-Bow Church, Cheapside; 7.30am-3pm Mon-Fri; V; ⊖ Farringdon

Tucked into the crypt of St Mary-Le-Bow Church, The Place Below brings tasty vegetarian soups, pies, quiche and salads to the masses every weekday.

DRINK

Pubs and bars in the City tend to be mobbed on weekdays, quiet on Saturdays and closed on Sundays.

CAFÉ KICK *Bar*

☎ 7837 8077; www.cafekick.co.uk; Exmouth Market; noon-11pm Mon-Thu, noon-midnight Fri & Sat; ⊖ Farringdon

A little piece of the continent transplanted to Clerkenwell, Café Kick offers pavement tables, foosball, sausage-and-mustard sandwiches, and a stellar selection of imported bottled beers.

JERUSALEM TAVERN *Pub*

☎ 7490 4281; www.stpetersbrewery.co.uk/london; 55 Britton St; closed Sat & Sun; ⊖ Farringdon

In a world of dull chain pubs and chemical lagers, the Jerusalem Tavern is a brilliant blinding light. Owned by the tiny St Peters Brewery in rural East Anglia, this delightfully antiquated public house has been serving full-flavoured ales since 1720.

Ingesting news and brews at Ye Olde Mitre

VERTIGO 42 *Bar*

7877 7842; www.vertigo42.co.uk; 25 Old Broad St; noon-3pm & 5-11pm Mon-Fri; Bank

Every city worth its salt has a penthouse bar on top of a skyscraper. Vertigo 42 is London's contribution, perched on the 42nd floor of Tower 42, former home of the NatWest Bank. The views are magnificent, particularly at sunset, but the mood is corporate and reservations are essential for security reasons.

YE OLDE CHESHIRE CHEESE *Pub*

7353 6170; Wine Office Court, 145 Fleet St; 11am-11pm Mon-Fri, noon-11pm Sat, noon-3pm Sun; Blackfriars

A gloriously wonky building on Fleet St, this genial 17th-century pub was a popular refreshment stop for Charles Dickens, Mark Twain and Samuel Johnson, who used to waddle around the corner from Gough Sq.

YE OLDE MITRE *Pub*

7405 4751; 1 Ely Ct; closed Sat & Sun; Chancery Lane or Farringdon

People have been knocking back ales in this pint-sized pub since 1547. Come here for a healthy dose of nostalgia – just look for the alleyway between 8 and 9 Hatton Gardens.

PLAY

BARBICAN *Theatre*

☎ 7638 8891; www.barbican.org.uk; Silk St; ⊖ Moorgate or Barbican
The Barbican was one of those buildings that looked great as an architect's model, but never quite lived up to its potential as real bricks and mortar. Nevertheless, this concrete jungle of flats and piazzas is still worth a trip for its theatre, concert halls, cinema and art gallery. Shows tend to be avant-garde verging on the self-indulgent, much like the building.

FABRIC *Club*

☎ 7336 8898; www.fabriclondon.com; 77a Charterhouse St; 9.30pm-5am Fri, 10pm-7am Sat; ⊖ Farringdon
Fabric has been leading the scene for so long it feels like part of the architecture. Famous for its 'bodysonic' dance floor (it does what it says on the tin), this is London's top venue for techno, hip-hop, electro, house, drum & bass and break-beats.

SADLER'S WELLS *Dance*

☎ information 0844 412 4300, bookings 7863 8198; www.sadlers-wells.com; Rosebery Ave; ⊖ Angel
London's leading centre for ballet, dance and physical theatre, Sadler's Wells is a national treasure. Matthew Bourne regularly chooses Sadler's Wells for his edgy ballet performances (the all-male *Swan Lake* was first performed here in 1995).

VOLUPTÉ LOUNGE *Cabaret*

☎ 7831 1622; www.volupte-lounge.com; 9 Norwich St; 11.30am-1am Tue-Fri, 7.30pm-3am Sat; ⊖ Chancery Lane
Burlesque has gone back to its cheeky roots at this trendy cabaret lounge south of Chancery Lane. As many girls as guys show up for the peekaboo shows, which feature more parlour naughtiness than serious nudity.

WALKING TOUR

LONDON CHURCHES WALK

Only 24 of the 51 churches built by Sir Christopher Wren after the Great Fire of 1666 survived the bombing of WWII, but they are still the most eye-catching monuments on the City skyline.

Start your walk at **All-Hallows-by-the-Tower** (**1**), where Samuel Pepys watched the Great Fire sweep across the City. From here, stroll west along Byward St and Lower Thames St to **St Magnus-the-Martyr** (**2**), marking the original position of the medieval London Bridge. Next, walk north up Fish Hill past Wren's **Monument** (**3**; p62), and cross Gracechurch St for a peek at **St Clement's** (**4**), of nursery rhyme fame.

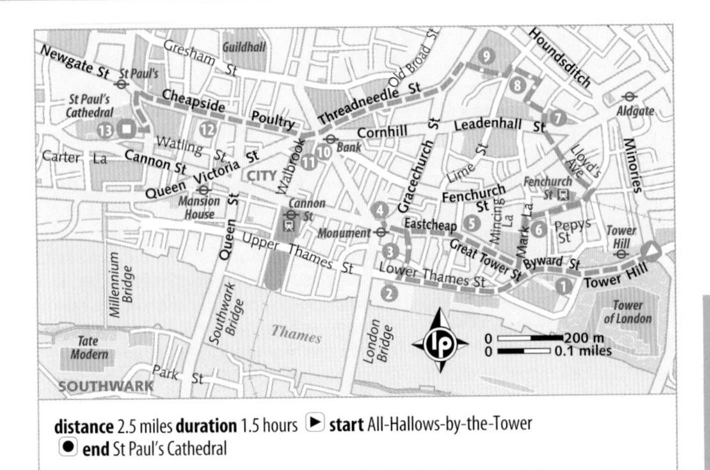

distance 2.5 miles **duration** 1.5 hours ▶ **start** All-Hallows-by-the-Tower
● **end** St Paul's Cathedral

Cut back along Eastcheap to **St Margaret Pattens** (**5**) then turn north up Mark Lane to **St Olave** (**6**), where Pepys was buried. Follow Crutched Friars and Lloyd's Ave to Fenchurch St, then duck down Fenchurch Buildings to Leadenhall St and Wren's tiny **St Katherine Cree** (**7**). Continue north up Creechurch Lane to Bury St, then cross behind the **Swiss Re Tower** (**8**; see boxed text, p65) to **St Helen's Bishopsgate** (**9**), with its wonderful medieval memorials.

Next, walk southwest along Bishopsgate and Threadneedle St to reach **Mansion House** (**10**; p62) and the perfectly proportioned interior of **St Stephen Walbrook** (**11**). Continue west along Poultry and Cheapside past the handsome Wren tower of **St Mary-Le-Bow** (**12**), then turn south on New Change to reach Wren's crowning glory, **St Paul's Cathedral** (**13**; p63).

For more on the City churches, visit www.london-city-churches .org.uk.

>SOUTH BANK

Every city has an area set aside by politicians for 'the arts'. Not all are successful, but London's South Bank has grown so far beyond its original remit as to be a work of art all by itself. Here you'll find the legendary Tate Modern, housed inside the old coal-fired power station that once painted the City black with soot. Follow the riverside promenade west along the Thames and you'll reach the British Film Institute, Southbank Centre and National Theatre; go east and you'll come to Shakespeare's Globe.

You don't have to be an art buff to enjoy the South Bank. Thousands flock to Westminster Bridge to ride the space-age Ferris wheel known as the London Eye, and close to London Bridge is the gastronomic tour de force that is Borough Market. Inland are some good pubs and restaurants and the Young and Old Vic theatres, home to young theatre talent and Kevin Spacey respectively. And don't overlook tiny Bermondsey, a miniature Hoxton on the south bank of the Thames.

SOUTH BANK

SEE

City Hall.......1 G2
County Hall.......2 B3
Design Museum.......3 H3
Hayward Gallery.......4 B2
Imperial War Museum.......5 C4
London Bicycle Tour Company.......6 C2
London Eye.......7 B3
Millennium Bridge.......8 D1
Old Operating Theatre Museum & Herb Garret.......9 F2
RIB London Voyages.......10 A3
Southbank Centre.......(see 35)
Tate Modern.......11 D2

SHOP

Borough Market.......12 F2
Cockfighter of Bermondsey.......13 G3
Konditor & Cook.......14 C2
Oxo Tower.......15 C1

EAT

Anchor & Hope.......16 C3
Baltic.......17 D3
Champor-Champor.......18 F3
Garrison.......19 G3
Roast.......(see 12)
Table.......20 D2
Tapas Brindisa.......21 E2
Tas Ev.......22 C2
Village East.......23 G3

DRINK

George Inn.......24 F2
King's Arms.......25 C2

PLAY

Barcode Vauxhall.......26 A6
BFI IMAX Cinema.......27 B2
British Film Institute.......28 B2
Ministry of Sound.......29 D4
National Theatre.......30 B2
Old Vic.......31 C3
Royal Vauxhall Tavern.......32 B6
Shakespeare's Globe.......33 E2
South Central.......34 B6
Southbank Centre.......35 B2
Young Vic.......36 C3

Please see over for map

SEE

The best options are covered here, but there are more tacky tourist attractions on Tooley St (F2).

CITY HALL

☎ 7983 4000; www.london.gov.uk; The Queen's Walk; admission free; 8am-8pm Mon-Fri, plus occasional weekends; ♿ ; ⊖ Tower Hill or London Bridge

Another of Norman Foster's future-organic creations, the official seat of the Mayor of London is a metallic sphere that resembles a motorcycle helmet, a sliced melon or a wonky egg – you decide. The top-floor viewing gallery is open to the public one weekend a month (check the website for dates).

COUNTY HALL

www.londoncountyhall.com; Westminster Bridge Rd; ♿ ; ⊖ Waterloo or Westminster

This imposing neo-baroque edifice on the riverside was built in 1908 to house the London County Council. Today it serves a purely commercial function as the home of the **London Aquarium** (www.londonaquarium.co.uk) and several other tacky and forgettable tourist attractions.

Spiralling into control – City Hall's interior staircase

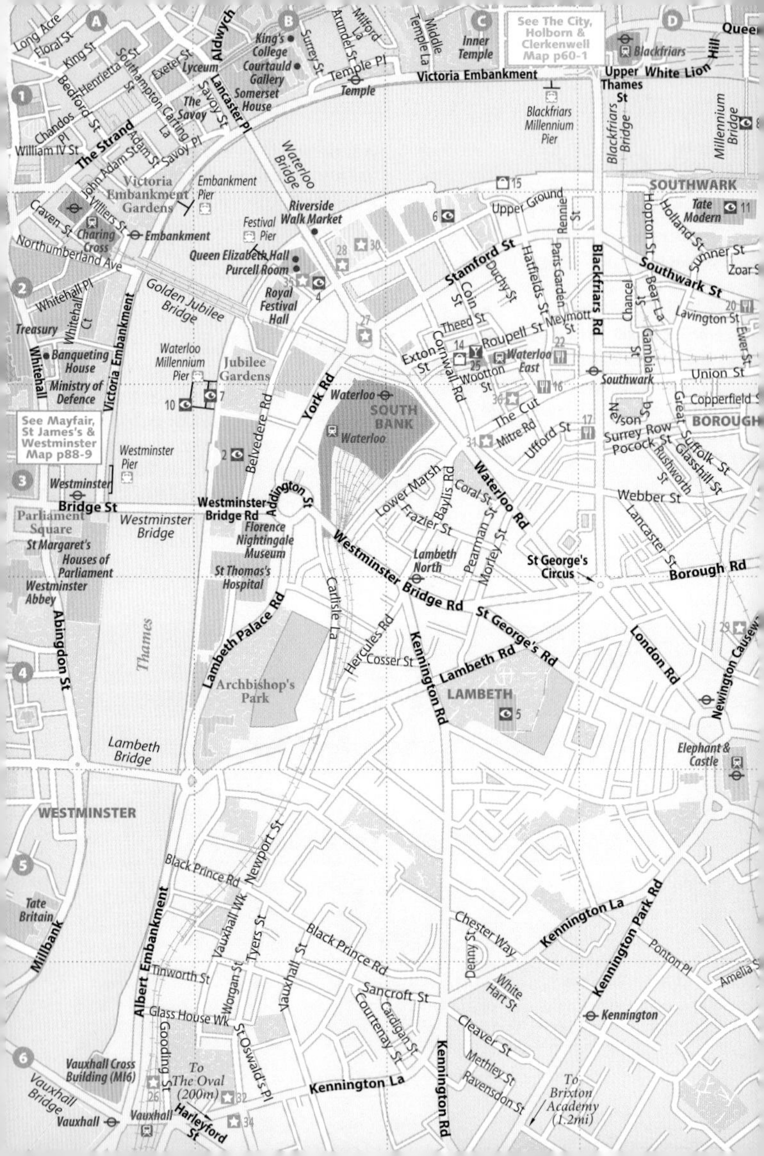

See The City, Holborn & Clerkenwell Map p60-1
See Mayfair, St James's & Westminster Map p88-9
Long Acre
Floral St
King St
Henrietta St
Southampton St
Exeter St
Lyceum
Aldwych
Bedford St
Chandos Pl
William IV St
The Strand
Adam St
Carting La
Savoy St
The Savoy
Savoy Pl
John Adam St
Lancaster Pl
King's College
Courtauld Gallery
Somerset House
Surrey St
Arundel St
Milford La
Temple Pl
Temple
Middle Temple La
Inner Temple
Victoria Embankment
Blackfriars
Upper Thames St
White Lion Hill
Queen
Blackfriars Millennium Pier
Blackfriars Bridge
Millennium Bridge
Waterloo Bridge
Victoria Embankment Gardens
Embankment Pier
Villiers St
Craven St
Charing Cross
Embankment
Northumberland Ave
Riverside Walk Market
Festival Pier
Queen Elizabeth Hall
Purcell Room
Royal Festival Hall
Golden Jubilee Bridge
Whitehall Pl
Whitehall Ct
Treasury
Whitehall
Banqueting House
Ministry of Defence
Victoria Embankment
Waterloo Millennium Pier
Jubilee Gardens
Westminster Pier
Westminster
Bridge St
Parliament Square
St Margaret's
Houses of Parliament
Westminster Abbey
Westminster Bridge
Westminster Bridge Rd
Addington St
Florence Nightingale Museum
St Thomas's Hospital
Belvedere Rd
York Rd
Waterloo
SOUTH BANK
Upper Ground
Rennie St
Stamford St
Coin St
Duchy St
Hatfields St
Paris Garden
Meymott St
Blackfriars Rd
Theed St
Roupell St
Exton St
Cornwall Rd
Waterloo East
Wootton St
The Cut
Mitre Rd
Ufford St
Southwark
SOUTHWARK
Tate Modern
Hopton St
Holland St
Sumner St
Zoar St
Southwark St
Chancel St
Bear La
Lavington St
Ewer St
Gambia St
Union St
Copperfield St
BOROUGH
Nelson Sq
Great Suffolk St
Surrey Row
Pocock St
Rushworth St
Glasshill St
Webber St
Lancaster St
Lower Marsh
Baylis Rd
Coral St
Frazier St
Waterloo Rd
Pearman St
Morley St
Lambeth North
St George's Circus
Borough Rd
Westminster Bridge Rd
St George's Rd
London Rd
Newington Causeway
Carlisle La
Lambeth Palace Rd
Hercules Rd
Cosser St
Kennington Rd
Lambeth Rd
LAMBETH
Archbishop's Park
Thames
Abingdon St
Lambeth Bridge
Elephant & Castle
WESTMINSTER
Black Prince Rd
Newport St
Tate Britain
Millbank
Albert Embankment
Vauxhall Wk
Tyers St
Black Prince Rd
Vauxhall St
Tinworth St
Worgan St
Glass House Wk
Gooding St
St Oswald's Pl
Sancroft St
Courtenay St
Cardigan St
Chester Way
Denny St
White Hart St
Kennington La
Kennington Park Rd
Ponton Pl
Amelia St
Kennington
Cleaver St
Methley St
Ravensdon St
Kennington Rd
Kennington La
Vauxhall Cross Building (MI6)
To The Oval (200m)
Vauxhall Bridge
Vauxhall
Harleyford St
To Brixton Academy (1.2mi)

Victoria St
Mansion House
Cloak La
CITY
Cannon St
King William St
Gracechurch St
Philpot La
Rood La
Fenchurch St
Fenchurch St
America Sq
Goodmans Yard
Tower Chamber St
Upper Thames St
Queen St
Cannon St
Monument
Eastcheap
Mincing La
Mark La
Seething La
Pepys St
Tower Hill
Minories
Tower Gateway DLR
Royal Mint St
Cousin La
Arthur St
Monument St
Gt Tower St
Trinity Sq
Shorter St
Cartwright St
Blue Anchor Yard
Angel Pa
Lower Thames St
Byward St
Tower Hill
Lwr Thames St
East Smithfield
Bankside Pier
Southwark Bridge
London Bridge
Old Billingsgate Market
Custom House
Tower Hill
Tower of London
Tower Bridge Approach
Thomas More St
Bankside
Southwark Tourist Information Centre
Thames
St Katharine's Dock
33
Park St
London Bridge City Pier
HMS Belfast
Tower Millennium Pier
Clink St
Park St
Stoney St
Montague Cl
Summer St
Southwark Cathedral
Southwark Crown Ct
Battle Br La
The Queen's Wk
WAPPING
Railway App
London Bridge
Tower Bridge
St Katharine's Pier
Great Guildford St
Thrale St
21
12
9
Weston St
Southwark Bridge Rd
St Thomas St
Tooley St
Redcross Way
Borough High St
24
Guy's Hospital
Horselydown La
Shad Thames
Ayres St
Shand St
Gainsford St
3
Newcomen St
18
Tooley St
Curlew St
Elizabeth St
Snowsfields
Crucifix La
Fair St
Mint St
Marshalsea Rd
Crosby Row
Porlock St
Guy St
Bermondsey St
White's Grounds
Shad Thames
Jacob St
Borough
Kipling St
13
Leathermarket St
19
Druid St
Mill St
George Row
Tanner St
Tanner St
Jamaica Rd
Gt Suffolk St
Borough
Tabard St
Swan St
Weston St
BERMONDSEY
23
Staple St
Long La
Riley Rd
Maltby St
Trinity St
Great Dover St
Wild's Rents
Decima St
Abbey St
Maltby St
Neckinger St
Tower Bridge Rd
Enid St
Law St
Harper Rd
Grange Rd
Spa Rd
Alscot Rd
New Kent Rd
Leroy St
Pages Walk
Rodney Pl
Balfour St
Willow Walk
Chatham St
Mandela Way
Hoygate St
Balaclava Rd
Alma Grove
Walworth Rd
Larcom St
Lyton Rd
Browning St
Old Kent Rd
Dunton Rd
East St
Humphrey St
Dawes St
Thurlow St
Portland St
East St
Cooper's Rd
Penrose St
Coburg Rd
Liverpool Grove
Roland Way
Merrow St
0
500 m
0
0.3 miles
E
F
G
H
1
2
3
4
5
6

TOP LONDON VIEWS

- > London Eye (below) – a big, big wheel with big, big views.
- > St Paul's Cathedral (p63) – a whole lot of steps but one hell of a view.
- > Tate Modern (p79) – city panoramas from the 4th-floor balconies.
- > Vertigo 42 (p69) – almighty views from the City's tallest skyscraper.
- > Galvin@Windows (see boxed text, p99) – views of the Queen's garden from the Hilton's top-floor eatery.
- > Parliament Hill (see p168) – the definitive view over London from the top of Hampstead Heath.

DESIGN MUSEUM

☎ 7403 6933; www.designmuseum.org; 28 Shad Thames; adult/under 12yr/concession £8.50/free/5-6.50; 10am-5.45pm, to 6.45pm Fri & Sat Jul-Aug, last entry 30min before closing; ♿ ; ⊖ Tower Hill or London Bridge

Although the displays are always slick and interesting, the best thing about London's Design Museum is the atmospheric approach beneath the overlapping gangways of Shad Thames. Fans of designer knickknacks will have to exercise discipline at the museum shop.

IMPERIAL WAR MUSEUM

☎ 7416 5000; www.iwm.org.uk; Lambeth Rd; admission free; 10am-6pm; ♿ ; ⊖ Lambeth North or Waterloo

A memorial to war, rather than a celebration of war, this excellent museum explores the realities of the conflicts that shaped the British Isles. Other parts of the museum include the Cabinet War Rooms (p90) and the HMS *Belfast*, moored permanently on the South Bank (see the website for details).

LONDON EYE

☎ 0870 500 0600; www.londoneye.com; Jubilee Gardens; adult/child/disabled/senior flight £15.50/7.75/12/12; 10am-8pm, to 9pm Jun & Sep, to 9.30pm Jul-Aug, closed 11-20 Jan; ♿ ; ⊖ Waterloo

The London Eye was only meant to be a temporary attraction, but the world's largest Ferris wheel is still rolling after eight years. The podlike carriages offer one of the best vantage points over central London, but there are long queues most days for the 30-minute 'flights'. Also see p10.

MILLENNIUM BRIDGE

Bankside; ♿ ; ⊖ Southwark, Mansion House or St Paul's

Constructed as a landmark for the 21st century, the sinuous Millennium Bridge provides a highly

Jennifer Tomlins

Medical student and assistant at Flour Power Bakery, Borough Market

Best buys in Borough Market There's some great seafood at Borough Market (p79) – the oysters are expensive but the mussels are cheap and they're really fresh. And you can buy specialist cheeses here that you won't find anywhere else. **Best London markets** Apart from Borough, I like Exmouth Market in Clerkenwell (p66) and the Sunday Up Market in Spitalfields (see boxed text, p150) for all the stalls selling little bits to eat as you walk around. **Best deli** Probably Carluccio's (see boxed text, p53), though there are lots of interesting Polish delis opening up around town. **Best pub on the South Bank** Just off Borough High St, the George Inn (p82) is a fabulous old pub. **Highlight of the South Bank** The Tate Modern (p79) because it appeals to such a range of people – there's serious art inside but it never feels intimidating.

scenic short cut from St Paul's to the Tate Modern – except during rush hour, when the views are obscured by thousands of commuters.

OLD OPERATING THEATRE MUSEUM & HERB GARRET

☎ 7188 2679; www.thegarret.org.uk; 9a St Thomas St; adult/child/concession/family £5.45/3/4.45/13.25; 10.30am-5pm; ⊖ London Bridge

The centrepiece of this quirky museum of surgery is the old wood-lined operating theatre for St Thomas's Hospital, where Victorian medics once carried out amputations without anaesthetics or antiseptics! Upstairs in the garret is an atmospheric re-creation of a traditional apothecary's shop (no access for the mobility impaired).

SOUTHBANK CENTRE

☎ 0871 663 2500; www.southbankcentre.co.uk; Belvedere Rd; admission varies; exhibitions at the Hayward 10am-6pm, to 10pm Fri; ; ⊖ Waterloo

A taste of your old medicine – the Old Operating Theatre Museum and Herb Garret

This concrete complex of cultural institutions includes the **Hayward Gallery**, whose lavish art shows often spill onto the South Bank promenade. For cultural performances, see p85.

TATE MODERN

☎ 7887 8888; www.tate.org.uk; Bankside; admission free, £3 donation requested, prices for special exhibitions vary; 10am-6pm Sun-Thu, 10am-10pm Fri & Sat; ; ⊖ Southwark

A former power station seems a fitting location for a powerhouse of modern art. The world's most successful contemporary art gallery is crammed with works from visionaries like Picasso, Pollock, Miro, Magritte, Francis Bacon, Giacometti, Mondrian and Roy Liechtenstein. The Tate boat service powers to Tate Britain (p126) every 40 minutes (£4.30, Travelcard holders £2.85). Also see p13.

SHOP

As well as gastro-tastic Borough Market, there are more shops along the South Bank walk.

BOROUGH MARKET *Market*

☎ 7407 1002; www.boroughmarket.org.uk; Borough High St; 11am-5pm Thu, noon-6pm Fri, 9am-4pm Sat; ⊖ London Bridge

London largely has Borough Market to thank for the surging popularity of deli cafés and artisan producers. Beneath the railway arches are dozens of stalls selling rare cheeses, marinated olives, home-baked bread and pastries, and organic fruit, veg and meat. Try before you buy – grab lunch at one of the many stands selling stuffed pittas and posh burgers.

MORE TATE FOR YOUR MONEY?

Not content with occupying one of London's most eye-catching buildings, Tate Modern (left) is constructing a new extension at the back to display more of its vast collection of modern art. The planned building looks a bit like a crushed Tetra Pak, with long glass-fronted galleries that make the best of the views along the riverbank. Construction is set to begin in 2009 – visit www.tate.org.uk/modern/transformingtm for more information.

KONDITOR & COOK *Food*

☎ 7261 0456; www.konditorandcook.com; 22 Cornwall Rd; 7.30am-6.30pm Mon-Fri, 8.30am-3pm Sat; ⊖ London Bridge

Serving the kind of biscuits and cakes that were used to build the *Hansel and Gretel* house, Konditor & Cook has transformed baking into an artform. There's another branch by Borough Market (left).

GETTING AROUND THE SOUTH BANK

The Southbank's revitalised riverside promenade forms part of the **Thames Path** (www.nationaltrail.co.uk/Thamespath), which continues all the way to the source of the river in the West Country. One popular walk is the leisurely stroll from London Bridge along the river to the London Eye – see www.southbanklondon.com/walkthisway for more walking itineraries.

If walking seems too much like hard work, you can rent good quality bikes from **London Bicycle Tour Company** (☎ 7928 6838; www.londonbicycle.com; 1a Gabriel's Wharf; ⊖ Blackfriars or Southwark), just west of the Oxo Tower. Prices start from £18 per day, with big discounts for longer rentals.

Alternatively, take to the river. As well as the scheduled riverboat services (see boxed text, p222), you can charge down the river by rigid inflatable boat (RIB) on a tour with **RIB London Voyages** (☎ 7928 8933; www.londonribvoyages.com; Waterloo Millennium Pier, Westminster Bridge Rd; ⊖ Waterloo); prices start at £32.50/19.50 per adult/child.

OXO TOWER *Accessories*

☎ 7021 1600; www.coinstreet.org; Barge House St; ⊖ Southwark, London Bridge or Blackfriars

The Oxo Tower was originally built as a power station and then later used as a cold store for the beef that was boiled up to make Oxo stock cubes. Today it contains an upscale shopping arcade, with two floors of bijou craft and fashion shops and an exclusive restaurant looking out from the 8th floor (predictably, you pay a premium for the view).

EAT

As well as the various eateries along the river, there are more good eats around the Cut (C3) and Bermondsey (see boxed text, p84).

ANCHOR & HOPE

Gastropub £

☎ 7928 9898; 36 The Cut; 6-10.30pm Mon, 11am-10.30pm Tue-Sat, 12.30-5pm Sun; ⊖ Southwark

Despite the predictable brown-painted interior and High St vibe, this gastropub serves delectable modern British cooking – terrines, Old Spot pork, game birds and traditional roasts. Reservations are only taken on Sunday.

BALTIC

Eastern European ££-£££

☎ 7928 1111; www.balticrestaurant.co.uk; 74 Blackfriars Rd; noon-3.30pm & 6-11.15pm; ⊖ Southwark

Polish flavours are becoming more familiar with the influx of Eastern European workers, but Baltic still offers some surprises – take the *chlodnik* iced cherry soup. It's a swish-looking eatery and the menu features dishes from across the Baltic region.

ROAST *British* *££-£££*

☎ 7940 1300; www.roast-restaurant.com; Floral Hall, Borough Market, Borough High St; noon-2.30pm daily, 5.30-10.30pm Mon-Sat, breakfast from 7am Mon-Thu, 9.30am Fri, 8am Sat; ⊖ London Bridge

Reached via an elevator in the middle of Borough Market, this busy modern British eatery sources the meat for its popular roasts from the artisan producers downstairs. The arched Victorian windows offer grand views of St Paul's and you're guaranteed a hearty feed.

TABLE *British* *£*

☎ 7401 2760; www.thetablecafe.com; 83 Southwark St; 7.30am-5.30pm Mon-Wed, 7.30am-10.30pm Thu & Fri, 9am-4pm Sat; ⊖ Southwark or London Bridge

Table specialises in organic, free-range and free-trade food, served at bench tables the Wagamama (p109) way. It's below an architect's practice and the design ethos shows in the décor.

Eastern toast – Baltic

TAPAS BRINDISA

Spanish £££

☎ 7357 8880; www.brindisa.com; 18-20 Southwark St; 11am-11pm Mon-Thu, 9-11am, noon-4pm & 5.30-11pm Fri & Sat; ⊖ London Bridge

Brindisa attracts a well-heeled, foodie crowd who know good tapas when they taste it. No bookings are accepted, so come early if you want to bag an outside table on market days.

TAS EV *Turkish* ££

☎ 7620 6191; www.tasrestaurant.com; The Arches, 97-99 Isabella St; noon-11.30pm, to 10.30pm Sun; ⊖ Southwark

The Tas Turkish restaurant chain has branches all over the South Bank, but Tas Ev is far and away the best. Set under the railway arches in a Mediterranean pot plant–filled courtyard, it's part deli, part bar and part restaurant.

DRINK

None of the riverbank pubs are particularly exciting – the following places are more inviting.

GEORGE INN *Pub*

☎ 7407 2056; Talbot Yard, 77 Borough High St; ⊖ London Bridge

Owned by the National Trust, London's last surviving galleried coaching inn swims with history. The building is delightfully wonky, but ye gods it gets busy, particularly in summer, when the courtyard is often full to bursting point.

KING'S ARMS *Pub*

☎ 7207 0784; 25 Roupell St; ⊖ Waterloo

No, you haven't wandered onto the set of a wartime British drama – real people live in the

GAY VAUXHALL

Less famous than Soho, but more edgy, the area south of Vauxhall Bridge is the heart of the gay scene in south London. Top venues include the following (all accessible by Vauxhall tube or Vauxhall station):

Barcode Vauxhall (☎ 7582 4180; www.barcode.co.uk; Arch 69, Albert Embankment) Partner to the Barcode in Soho, with a good-looking dancefloor-oriented crowd.

Royal Vauxhall Tavern (☎ 7820 1222; www.theroyalvauxhalltavern.co.uk; 372 Kennington Lane) Inclusive pub and bar, and venue for the famous kitsch cabaret Duckie (www.duckie.co.uk).

South Central (☎ 7793 0903; www.southcentrallondon.co.uk; 349 Kennington Lane) Hardcore venue for guys, hosting the cruise-oriented Eagle London (www.eaglelondon.com) sessions.

immaculate terraces along Roupell St, and this pub is their delightful local boozer.

PLAY

BRITISH FILM INSTITUTE

Cinema

BFI; ☎ 7928 3232; www.bfi.org.uk; South Bank; 11am-11pm, Mediatheque 11am-8pm Tue-Sun; ⊖ Waterloo or Embankment

The BFI maintains the national archives of films and moving images, many of which can be viewed for free in the 'Mediatheque' library. As well as red-carpet events for new releases, there's a regular programme of golden oldies and giant-sized screenings at the **Imax** screen (in the middle of the Waterloo Rd roundabout). If you can, visit during the London Film Festival (p31) in October.

BRIXTON ACADEMY

Live Music

☎ 7771 3000; www.brixton-academy.co.uk; 211 Stockwell Rd; ⊖ Brixton

A hike south of the South Bank, the Academy is one of London's classic music venues. Set in the lively Afro-Caribbean sprawl of Brixton, this huge, grungy music-hall has hosted everyone from Motorhead and the Pogues to Billy Ocean.

IT'S ALL HAPPENING AT THE MOVIES

The British Film Institute (left) screens a fascinating range of vintage movies, and you'll find more import, art house and alternative films at the following theatres:

> Curzon Soho (p55)
> Prince Charles Cinema (p57)
> Curzon Mayfair (p101)
> Chelsea Cinema (p131)
> Screen on Baker St (p119)
> Screen on the Green (p165)
> Electric Cinema (p142)

MINISTRY OF SOUND *Club*

☎ 0870 060 0100; www.ministryofsound.com; 103 Gaunt St; 10.30pm-6am Fri, 11pm-7am Sat; ⊖ Elephant & Castle

More than just a club, the Ministry of Sound is a global phenomenon. This booming nightclub is where it all kicked off 17 years ago – high-profile DJs like Sasha, Pete Tong and Paul Oakenfold still lift the roof off every weekend.

NATIONAL THEATRE *Theatre*

☎ 7452 3000; www.nationaltheatre.org.uk; South Bank; ⊖ Waterloo

Another massive concrete block on the South Bank, the National Theatre is going from strength to strength under the directorship of Nicholas Hytner. The three theatre spaces – the Olivier,

WORTH A DETOUR – BERMONDSEY

Once run-down and forlorn, Bermondsey is rapidly transforming into a mini Hoxton south of the river. A popular live-in locale for architects and designers, it exudes a villagey vibe thanks to its tiny park and the cute shops and eateries on Bermondsey St, a short stroll southeast from London Bridge tube. The following are our highlights:

Champor-Champor (☎ 7403 4600; www.champor-champor.com; 62-64 Weston St; noon-2pm Thu-Fri, 6.15-10.15pm Mon-Sat) Unique is the best way to describe this fabulously funky place; the upmarket Malaysian food more than justifies the trip out here.

Cockfighter of Bermondsey (☎ 7357 6482; www.cockandmagpie.com; 96 Bermondsey St; 11am-7pm Tue-Fri, 10am-6pm Sat) T-shirts with attitude are found in this small boutique popular with DJs and pop stars.

Garrison (☎ 7089 9355; www.thegarrison.co.uk; 99-101 Bermondsey St; 12.30-3.30pm & 6.30-10pm Mon-Sat, to 9.30pm Sun, breakfast 8-11.30am, from 9am Sat & Sun) Midway between gastropub and beach hut, the trendified pub is a solid local lunch stop.

Village East (☎ 7357 6082; www.villageeast.co.uk; 171-3 Bermondsey St; noon-11pm Mon-Thu, noon-1.30am Fri, 11am-1.30am Sat, 11am-11pm Sun) Like a New York brasserie teleported into Bermondsey, Village East is moody and futuristic, with good steaks and Asian fusion dishes on the menu.

Lyttleton and Cottesloe – stage different work, so there's always a selection of performances to choose from.

OLD VIC *Theatre*

☎ 0870 060 6628; www.oldvictheatre.com; The Cut; ⊖ Waterloo

Many were hoping for more from Kevin Spacey's tenure as the Old Vic's artistic director, but recent performances seem to have restored the theatre's good reputation. Spacey appears in a couple of shows every year, and Sir Ian McKellen frequently crops up in the Christmas pantomime.

SHAKESPEARE'S GLOBE *Theatre*

☎ 7902 1500; www.shakespearesglobe.org; 21 New Globe Walk; exhibition & tour adult/child/concession/family £10.50/6.50/8.50/28; 9am-5pm Apr-Oct, from 10am rest of the year, tours every 30min until noon; ⊖ London Bridge

The original theatre where Shakespeare staged his famous plays burned down in 1613, but American director Sam Wanamaker decided to build a perfect replica in 1997. Although touristy, shows here conjure up a powerful sense of the way Shakespeare's work was originally meant to be performed. If you don't have time to catch a performance, you can still enjoy the permanent exhibition and tour.

SOUTHBANK CENTRE *Live Music*

☎ 0871 663 2500; www.southbankcentre.co.uk; Belvedere Rd; ⊖ Waterloo
Proof that you can't judge a book by its cover, the concrete blocks of the Royal Festival Hall and the smaller Queen Elizabeth Hall and Purcell Room are some of Britain's finest spaces for classical music, dance, theatre and monologues. Check the website for upcoming performances.

THE OVAL *Sport*

☎ 0871 246 1100; www.surreycricket.com; Kennington Oval; ⊖ Oval
Surrey Cricket Club, commonly known as the Oval, is London's other big cricket ground and the setting for the last test of the summer (August or September). Tickets are available through the website. For more, see boxed text, p115.

YOUNG VIC *Theatre*

☎ 7922 2922; www.youngvic.org; 66 The Cut; ⊖ Waterloo
The stylish Young Vic more than lives up to its name, providing a forum for new acting and directing talent. Plays here are edgy affairs, winning plaudit after plaudit from the critics.

>MAYFAIR, ST JAMES'S & WESTMINSTER

The districts of Mayfair, St James's and Westminster form the political heart of London, and the level of pomp and circumstance in these streets has to be seen to be believed – state occasions are marked by convoys of gilded carriages, elaborate parades, and in the case of the opening of Parliament, by a man in a black coat banging on the front door with a jewelled sceptre.

Westminster was founded in the 11th century, when Edward the Confessor moved the royal court out of the City, but after the public beheading of Charles I, the royals wisely decided to create a new royal enclave at St James's. Millions of tourists still flock here to marvel at Buckingham Palace and the neo-Gothic Houses of Parliament.

Further north is Mayfair, a playground for the obscenely wealthy. There's a golden rule for shopping in Mayfair – if you have to ask how much it costs, you probably can't afford it.

MAYFAIR, ST JAMES'S & WESTMINSTER

SEE

Apsley House....1 A4
Buckingham Palace....2 C5
Cabinet War Rooms....3 F5
Churchill Museum....(see 3)
Horse Guards Parade....4 F4
Houses of Parliament....5 G6
Institute of Contemporary Arts....6 F3
No 10 Downing Street....7 G4
Queen's Gallery....8 C6
Royal Academy of Arts....9 D2
Royal Mews....10 B6
St James's Park....11 E5
Wellington Arch....12 A5
Westminister Abbey....13 G6
White Cube Gallery....14 D3

SHOP

Abercrombie & Fitch....15 D2
Bonhams....16 B1
Burberry....17 C1
Burlington Arcade....18 D2
Christie's....19 D3
Dover Street Market....20 C2
Fortnum & Mason....21 D3
Henry Poole & Co....22 D2
Kilgour....23 D2
Lillywhites....24 E2
Matthew Williamson....(see 31)
Mulberry....25 C1
Ozwald Boateng....26 D2
Paul Smith Sale Shop....27 B1
Piccadilly Arcade....28 D3
Royal Arcade....29 C2
Sotheby's....30 C1
Stella McCartney....31 B2
Waterstone's....32 D2

EAT

Cinnamon Club....33 F6
Galvin@Windows....34 A4
Gordon Ramsay at Claridges....35 B1
Inn the Park....36 F4
Japan Centre....37 E2
Maze....38 A1
Momo....39 D1
Nahm....40 A5
Nobu....41 A4
Sketch....42 C1
Wolseley....43 C3

DRINK

Nobu Berkeley St....44 C3
Ye Grapes....45 B3

PLAY

Curzon Mayfair....46 B3
Pigalle Club....47 E2

Please see over for map

SEE

APSLEY HOUSE & WELLINGTON ARCH

7499 5676; www.english-heritage.org.uk; Hyde Park Corner; adult/child/concession£5.50/2.80/4.40; 11am-5pm Wed-Sun, to 4pm Nov-Mar; ; Hyde Park Corner

The whole of Hyde Park Corner was conceived as a monument to the defeat of Napoleon by the Duke of Wellington. His palatial home at Aspley House is now a museum, with a fine collection of Napoleonic memorabilia, including a giant nude sculpture of Napoleon and Wellington's original Wellington boots. For a small additional fee, you can climb the nearby Wellington Arch.

BUCKINGHAM PALACE

7766 7300; www.royalcollection.org.uk; Buckingham Palace Rd; adult/under 5yr/5-17yr/concession £15.50/free/8.75/14; 9.45am-6pm late Jul-Sep; ; St James's Park, Victoria or Green Park

The official residence of Her Royal Highness Queen Elizabeth II – Lilibet to those who know her – is a stunning piece of Georgian architecture, crammed with the kind of gold- and gem-encrusted chintz that royals like to surround themselves with. The palace only opens to the public in summer – there's a timed entry system and tickets should be bought from the kiosk in Green Park. At other times you can look from outside. If the royal flag is flying, she's home.

CHANGING OF THE GUARD

If you want to see grown men shouting at each other with what look like giant black Q-tips on their heads, head to Buckingham Palace. Every two days (daily from May to July) at 11.30am, the guards who protect the Queen come off duty to be replaced by a new roster of soldiers from the Household Regiment, all dressed in flamboyant ceremonial costume. Tourists gather in their thousands to watch the spectacle – the reaction of the audience is almost as interesting as the pomp and ceremony in the courtyard.

A second ritual, the changing of the Horse Guard, takes place daily at 11am (10am on Sunday) at Horse Guards Arch, which is also the setting for the Trooping of the Colour parade for the Queen's official birthday in June (p29). At other times people come to Horse Guards to pull faces at the guardsmen, who are under strict orders not to respond, no matter how annoying the provocation. As an interesting side note, Horse Guards Parade will also be the setting for the beach volleyball competition during the London 2012 Olympics, which should give the guards something to smile about.

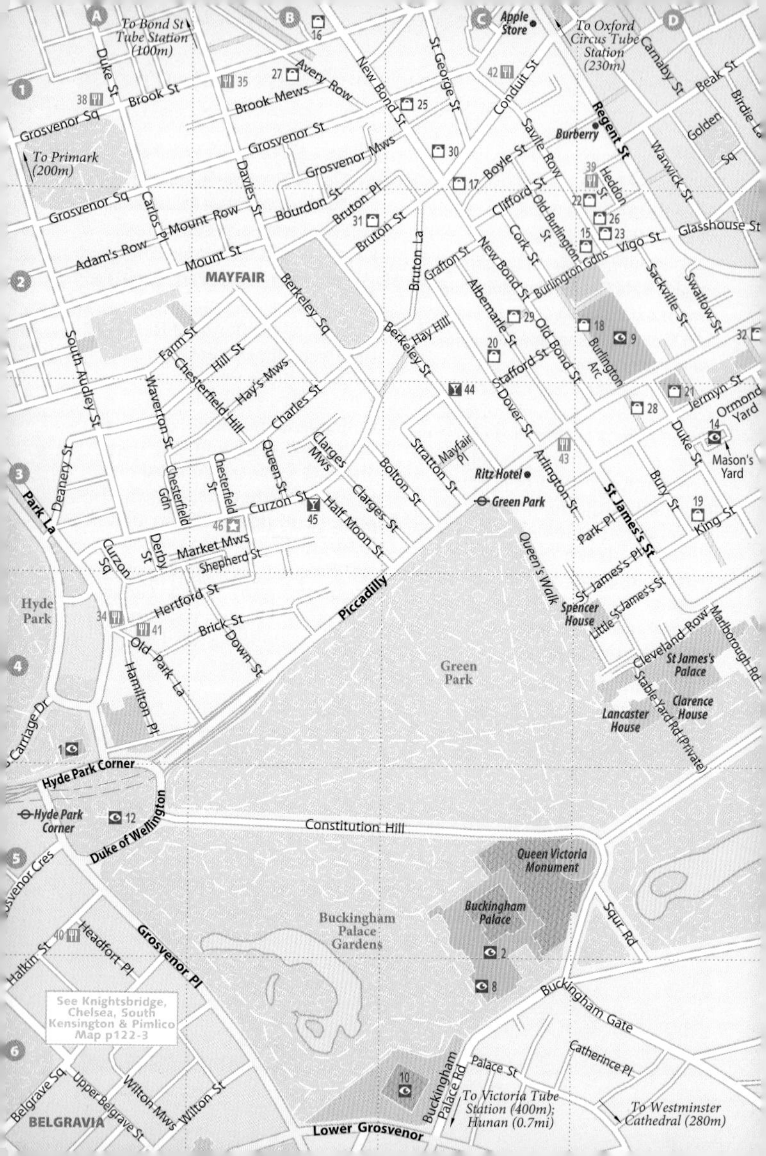

A
B
C
D
1
2
3
4
5
6
To Bond St Tube Station (100m)
To Oxford Circus Tube Station (230m)
Apple Store
To Primark (200m)
MAYFAIR
Burberry
Regent St
Ritz Hotel
Green Park
Hyde Park
Green Park
Spencer House
St James's Palace
Clarence House
Lancaster House
Hyde Park Corner
Constitution Hill
Queen Victoria Monument
Buckingham Palace
Buckingham Palace Gardens
Piccadilly
Mason's Yard
See Knightsbridge, Chelsea, South Kensington & Pimlico Map p122-3
BELGRAVIA
To Victoria Tube Station (400m); Hunan (0.7mi)
To Westminster Cathedral (280m)
Lower Grosvenor
Buckingham Gate
Buckingham Palace Rd

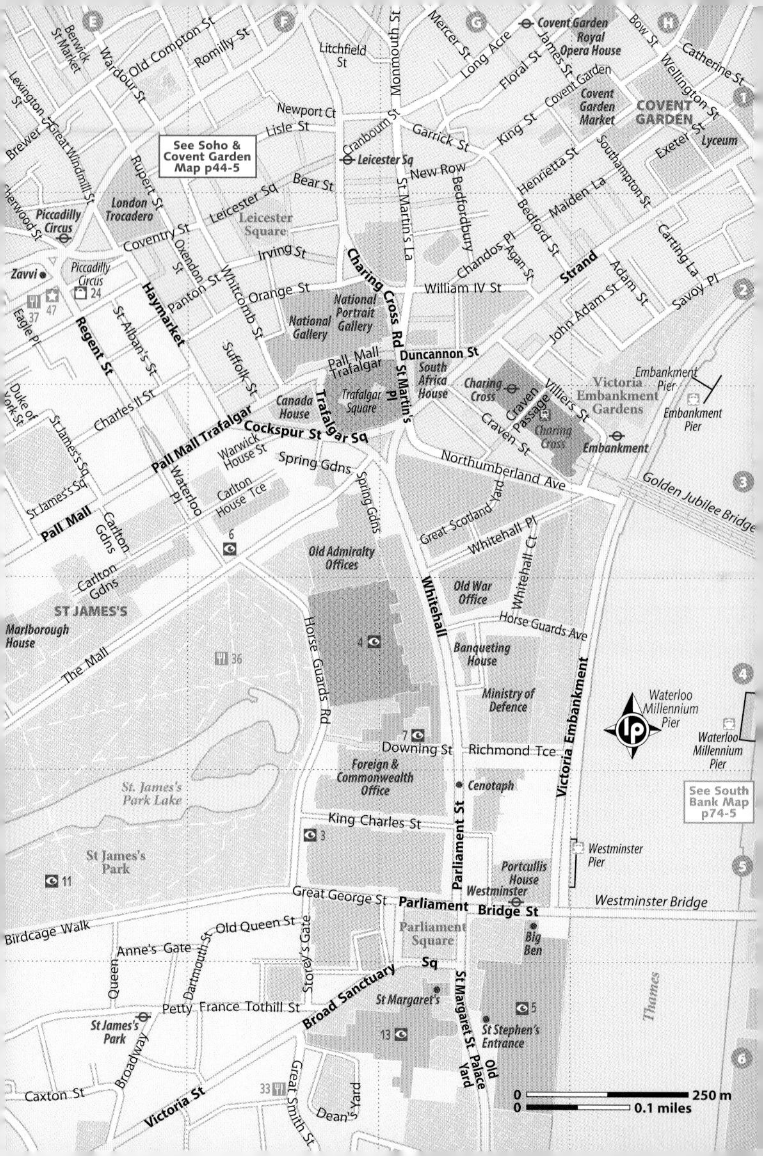

See Soho & Covent Garden Map p44-5
See South Bank Map p74-5
Leicester Square
Piccadilly Circus
Covent Garden
Charing Cross
Embankment
Westminster
St James's Park
COVENT GARDEN
ST JAMES'S
London Trocadero
National Gallery
National Portrait Gallery
Trafalgar Square
Canada House
South Africa House
Old Admiralty Offices
Old War Office
Banqueting House
Ministry of Defence
Foreign & Commonwealth Office
Cenotaph
Portcullis House
Big Ben
St Margaret's
St Stephen's Entrance
Parliament Square
Marlborough House
St. James's Park Lake
St James's Park
Victoria Embankment Gardens
Royal Opera House
Covent Garden Market
Lyceum
Zavvi
Embankment Pier
Waterloo Millennium Pier
Westminster Pier
Golden Jubilee Bridge
Westminster Bridge
Thames
Strand
Whitehall
Victoria Embankment
Pall Mall
The Mall
Haymarket
Regent St
Birdcage Walk
Victoria St
Broad Sanctuary
0 250 m
0 0.1 miles

The royal residence – Buckingham Palace (p87)

CABINET WAR ROOMS & CHURCHILL MUSEUM

7930 6961; cwr.iwm.org.uk; Clive Steps, King Charles St; adult/under 16yr/concession £12/free/9.50; 9.30am-6pm, last entry 5pm; ; Charing Cross

For much of WWII, Britain's greatest prime minister lived like a rabbit in a warren of tunnels under Whitehall, coordinating the Allied resistance on an old-fashioned Bakelite telephone. You can still see the austere Cabinet War Rooms much as Churchill left them, complete with maps of Allied advances on the walls. The attached Churchill Museum features spine-tingling recordings of the great orator's speeches.

HOUSES OF PARLIAMENT

tours 0870 906 3773; www.parliament.uk; St Stephen's Entrance, St Margaret St; admission to debates free, tours adult/concession £12/8; parliamentary sessions 2.30-10.30pm Mon & Tue, 11.30am-7pm Wed, 11.30am-6.30pm Thu, 9.30am-3pm Fri; ; Westminster

Ever since Guy Fawkes tried to blow the place up in 1605, security at the official seat of the British government has been tight. Unless attending public debates, visitors are only allowed to enter parliament on an organised 75-minute tour during summer when the MPs are safely away on holiday. Even if you can't get inside, come here to marvel at Barry's stunning building and its famous tower, topped by the 'Big Ben' bell. Check the website for tour times and bag restrictions. See also p14.

INSTITUTE OF CONTEMPORARY ARTS

ICA; ☎ 7930 3647; www.ica.org.uk; The Mall; day membership during exhibitions adult/concession Mon-Fri £2/1.50, Sat & Sun £3/2; noon-11pm Mon, noon-1am Tue-Sat, noon-10.30pm Sun; ; Charing Cross or Piccadilly Circus

The ICA was wracked by controversy recently when its chairman dismissed conceptual art as 'pretentious, self-indulgent, craftless tat'. Don't worry – the shows, exhibitions and films still lean towards the modern end of the spectrum. The trendy café-bar is open to anyone who pays the day membership fee.

NO 10 DOWNING STREET

www.number10.gov.uk; 10 Downing St; Westminster

It's rather charming that the official seat of the British prime minister is a bog-standard Georgian townhouse in Whitehall. Unless you have permission to file a petition, the closest you'll get to the famous black door is the gate on Whitehall. See also p14.

ROYAL ACADEMY OF ARTS

☎ 7300 8000; www.royalacademy.org.uk; Burlington House, Piccadilly; admission varies; 10am-6pm, to 10pm Fri; ; Green Park

Over its 240-year history, Britain's first art school has counted Reynolds, Turner, Constable and Gainsborough among its members. Today the Academy's grand Portland Stone headquarters host high-profile temporary art exhibitions and a small permanent show in the John Madejski Rooms. The Summer Exhibition (June to August) displays art by the public, both dreadful and sublime.

ST JAMES'S PARK

☎ 7930 1793; www.royalparks.gov.uk; The Mall; ; St James's Park

Despite its proximity to Buckingham Palace and Downing St, St James's Park remains a place for ordinary people, who come here to feed the ducks and tame squirrels, gawp at the pelicans and relax on the rentable deck chairs. In the middle is the Inn the Park café (see p98). See also p15.

WORTH A TRIP – KEW GARDENS & HAMPTON COURT PALACE

The Thames provides access to some interesting attractions outside the city centre, including Kew Gardens and Hampton Court Palace.

A World Heritage Site, **Kew Gardens** (☎ 8332 5655; www.kew.org; Kew Rd; adult/under 16yr/concession £13/free/12; 9.30am-6.30pm Mon-Fri, to 7.30pm Sat & Sun, early closing Aug-Mar; ; Kew Gardens) is a delightful Victorian botanical garden. Must-sees include the exquisite glass Palm House and Waterlilly House and the 10-storey Chinese pagoda. A hop-on hop-off shuttle runs to the main sights (adult/child £4/1).

Hampton Court Palace (☎ 0844 482 7777; www.hrp.org.uk; East Molesy; all-inclusive ticket adult/under 5yr/5-15yr/concession/family £13.30/free/6.65/11.30/37; 10am-4.30pm, last entry 3.30pm; ; Hampton Ct from Waterloo, or Wimbledon), the former palace of Henry VIII, is said to be stalked by the ghost of one of his six wives. Perhaps you'll see her stalking through the regal state apartments or searching for her missing head in the famous palace maze.

From Easter to September, **Thames River Boats** (☎ 7930 2062; www.wpsa.co.uk) runs boats from Westminster Pier (Map p88-9, H5) to Kew (one way adult/child £10.50/5.25, 1½ hours) and Hampton Court (£13.50/6.75, four hours). See the website for times. Alternatively, you can reach Kew by tube (Kew Gardens) and Hampton Court by train from Waterloo (Map p74-5, B3).

THE QUEEN'S GALLERY & ROYAL MEWS

☎ 7766 7301; www.royalcollection.org.uk; Buckingham Palace Rd; adult/under 5yr/5-17yr/concession £8.50/free/4.25/7.50; 10am-5.30pm, closed late Sep–mid-Oct; ; St James's Park, Victoria or Green Park

If you can't get into the palace, a good second-best is this gallery full of paintings and treasures from the Royal Collection. Around 450 works are on display at any one time, from a total of more than one million. At the adjacent Royal Mews (see the website for details), you can check out the outrageous carriages used by the royals for state functions.

WESTMINSTER ABBEY

☎ 7222 5152; www.westminster-abbey.org; Dean's Yard; adult/child/concession £12/free/9; 9.30am-4.30pm Mon-Fri, to 7pm Wed, 9.30am-2.30pm Sat, last entry 1hr before closing; ; Westminster

The ceremonial heart of the Anglican Communion, Westminster Abbey is Britain's finest medieval church – architecture buffs will note strong similarities with the great Gothic cathedrals of Paris and Reims. The awesome interior is used for the coronations and funerals of British monarchs, many of whom are buried in the abbey, along with national heroes like Charles Darwin and Sir Isaac Newton. Also see p14.

WESTMINSTER CATHEDRAL

☎ 7798 9055; www.westminstercathedral.org.uk; Victoria St; cathedral admission free, tower adult/concession £5/2.50; cathedral 7am-6pm, tower 9.30am-12.30pm & 1-5pm; ; ⊖ Victoria

Looking almost Byzantine, or dare we say it, Turkish, the headquarters of the Roman Catholic Church in Britain is seeing something of a revival with the influx of Eastern European Catholics into the UK. The main reason to visit is to ride the lift to the top of the 83m bell tower.

WHITE CUBE GALLERY

☎ 7930 5373; www.whitecube.com; 25-26 Mason's Yard; admission varies; 10am-6pm Tue-Sat; ⊖ Green Park or Piccadilly Circus

Along with adman Charles Saatchi, White Cube director Jay Jopling was one of the pioneers of Brit Art, so presumably we have him to thank for all the pickled sharks and 'my life as art' installations. There's a second White Cube in Hoxton (p147).

SHOP

ABERCROMBIE & FITCH

Clothing

☎ 0844 412 5750; www.abercrombie.com; 7 Burlington Gardens; 10am-7pm Mon-Sat, noon-6pm Sun; ⊖ Piccadilly Circus

It's hard to know if Abercrombie & Fitch is a clothing store or a disco – music blasts out at 90 decibels, and staff are handpicked for their chiselled good looks. Prices are pretty steep for what is essentially casual jeanswear but it's certainly a novel retail experience.

BONHAMS *Auctions*

☎ 7447 7447; www.bonhams.com; 101 New Bond St; auction times vary; ⊖ Bond St

One of London's most famous auction houses, Bonhams sees some real treasures going under the hammer, from paintings by old masters to Scottish broadswords. There's a branch in Knightsbridge.

BURBERRY *Clothing*

☎ 7968 0000; www.burberry.com; 21-23 New Bond St; 10am-7pm Mon-Sat, noon-6pm Sun; ⊖ Bond St or Green Park

This iconic British fashion label is famous for the 'Burberry check', created as a lining for trench coats

MAYFAIR ARCADES

The fashion-conscious Georgians came up with the idea of covered shopping streets, and several swanky 19th-century shopping arcades survive to this day. **Burlington Arcade** (www.burlington-arcade.co.uk; 51 Piccadilly) is still patrolled by its own police force, the Burlington Beadles, who monitor shoppers to prevent unseemly activities, such as running, whistling or chewing gum. Nearby are the similarly regal **Piccadilly Arcade** (www.piccadilly-arcade.com; Piccadilly) and **Royal Arcade** (28 Old Bond St). All are accessible from the nearest tube station, Piccadilly Circus.

after WWI. The sleek new Prorsum range aims to get away from the tartan look, now associated with unsophisticated social climbers. There are branches on Regent St and in Knightsbridge.

CHRISTIE'S *Auctions*

7839 9060; www.christies.com; 8 King St; viewing 9am-5pm Mon-Fri, noon-5pm Sat & Sun; South Kensington

If you fancy an original Monet or Picasso to hang on the wall, just bring a cashier's check for £40 million to Christie's, Britain's most famous auction house. There's a second showroom in Knightsbridge.

DOVER STREET MARKET *Clothing*

7518 0680; www.doverstreetmarket.com; 17-18 Dover St; 11am-6pm Mon-Sat, to 7pm Thu; Green Park

Showcasing the colourful creations of Tokyo fashion-darlings Comme des Garçons, among other labels, Dover Street Market is the place to come for that shirt you only wear on special occasions. There are four floors of clothing for men and women, all artfully displayed.

FORTNUM & MASON *Department Store*

7734 8040; www.fortnumandmason.co.uk; 181 Piccadilly; 10am-8pm Mon-Sat, 11.30am-6pm Sun; Piccadilly Circus or Green Park

This legendary store started business in 1707, recycling half-burned candles from the royal household and selling them on at a profit. Today Fortnum's is London's most elegant department store – it's worth a trip just to see the window displays.

LILLYWHITES *Sporting Goods*

0870 333 9600; 24-36 Regent St; 10am-9pm Mon-Wed, 10am-9.30pm Thu-Sat, noon-6.30pm Sun; ; Piccadilly Circus

London's largest sports shop offers five floors of tennis racquets, trainers and track suits. Despite the crowds and erratic service, it's worth a visit for the regular 70% off sales.

MATTHEW WILLIAMSON *Clothing*

☎ 7629 6200; www.matthewwilliamson.com; 28 Bruton St; 10am-6pm Mon-Sat; ⊖ Green Park

The upscale bohemian creations of Matthew Williamson show clear the influence of the time he spent in India. Starlets like Sienna Miller and Keira Knightley are regularly seen stepping out in his designs.

MULBERRY *Accessories*

☎ 7491 3900; www.mulberry.com; 41-42 New Bond St; 10am-6pm Mon-Sat, to 7pm Thu; ⊖ Bond St

Girls who come to Mulberry know what they want – stylish leather shoes and clutch bags in patent leather. This Brit label is rapidly expanding, so catch it while it still feels exclusive. You'll find branches all over London.

Look and yearn – window shopping in Burlington Arcade

PAUL SMITH SALE SHOP *Clothing*

☎ 7493 1287; 23 Avery Row W1; 10am-6.30pm Mon-Sat, to 7pm Thu, 1-5.30pm Sun; ⊖ Bond St

Discounted streetwise fashions from the classic British designer (see also p50).

PRIMARK *Clothing*

☎ 7495 0420; www.primark.co.uk; 499-517 Oxford St; 9am-9pm Mon-Fri, 9am-8pm Sat, noon-6pm Sun; ⊖ Baker St

Despite some recent bad press about its manufacturing methods, the flagship store of Primark is still crammed to the rafters with women hunting for bargain fashions that look like haute couture. They don't call it 'Primani' for nothing.

SOTHEBY'S *Auctions*

☎ 7293 5000; www.sothebys.com; 34-35 New Bond St; auction times vary; ⊖ Bond St

The oldest auction house in Britain, Sotheby's has sold some unique treasures in its time, including Picasso's *Garçon á la pipe* and one of the seven printed copies of JK Rowling's *The Tales of Beedle the Bard,* purchased by the owners of Amazon.com for a staggering £3.8 million.

STELLA MCCARTNEY *Clothing*

☎ 7518 3100; www.stellamccartney.com; 30 Bruton St; 10am-6pm Mon-Sat, to 7pm Thu; ⊖ Green Park

Many suspect that Stella McCartney got a helping hand

TAILOR-MADE

Tailors started setting up shop along Savile Row in the 1770s to cater to flamboyant Georgian gents like Beau Brummell, the playboy dandy credited with introducing the men's suit into the fashion repertoire. There are fewer shops today, but the remaining tailors are masters of their art. If you can stretch to £1500 or more for a suit, consider the following (all are accessible from Piccadilly Circus tube):

Henry Poole & Co (☎ 7734 5985; www.henrypoole.com; 15 Savile Row) The tailoring house that invented the tuxedo, still going strong after 200 years.

Kilgour (☎ 7734 6905; www.kilgour.eu; 8 Savile Row) Traditional without being conservative, Kilgour sells ready-to-wear as well as tailor-made suits from £1500.

Ozwald Boateng (☎ 7437 0620; www.ozwaldboateng.co.uk; 30 Savile Row) Bringing the flamboyance back to Savile Row, Boateng uses striking colours and rich fabrics in his £3000-plus bespoke suits.

MAYFAIR SHOPPING STREETS

Top spots to splash out on the platinum credit card:

- Bruton St (C2) – flamboyant designer creations from the likes of Stella McCartney.
- Old Bond St (C2) – De Beers' diamonds and fashions from designers like Alexander McQueen.
- New Bond St (C2) – painfully cool brands like Moschino, Ermenegildo Zegna and Jimmy Choo.
- Savile Row (D2) – bespoke suits from the kind of tailors who still say 'suits you, sir'.

with her career from her Beatle dad, but her fashion creations are undeniably popular with the in-crowd. Trademark McCartney products include tailored trouser suits, sweater dresses and flamboyant fragrances.

WATERSTONE'S *Books*

7851 2400; www.waterstones.com; 203-206 Piccadilly; 10am-10pm Mon-Sat, noon-6pm Sun; Green Park

The largest bookshop in Europe, with five floors of reading material and a top-floor café where you can pull up a chair and enjoy sweeping views across London's rooftops.

EAT

CINNAMON CLUB

Indian £££

7222 2555; www.cinnamonclub.com; Old Westminster Library, 30-32 Great Smith St; 7.30am-9.30am Mon-Fri, noon-2.45pm & 6-10.45pm Mon-Sat; St James's Park

A posh Indian restaurant that opened long before the current fad for posh Indian restaurants, with a refined location in the former Westminster Library. Despite the nostalgic styling, the food is refreshingly modern and inventive.

GORDON RAMSAY AT CLARIDGES

Modern British £££

7499 0099; www.gordonramsay.com; 53 Brook St; noon-2.45pm & 5.45-11pm Mon-Fri, noon-3pm & 6-11pm Sat, 6-10.30pm Sun; Bond St

The nobbly chinned superchef brings his magic to Mayfair's most exclusive hotel. Ramsay aficionados book months ahead to secure a seat in the gorgeous, flower-filled Art Deco dining room and sample such delights as roasted John Dory and lobster and salmon ravioli.

GORDON RAMSAY – AN F-ING GREAT BRITISH SUCCESS STORY

The ultimate TV chef, former footballer Gordon Ramsay has built up a veritable dining empire, with more than a dozen restaurants in London, Europe, the USA, Japan and the Middle East. The recipient of 12 Michelin stars, Ramsay is probably the most recognised chef on the planet thanks to his internationally syndicated TV shows *Hell's Kitchen* and the *F-Word*. Nevertheless, the expletive-loving culinarian is not universally loved by those who work under him. Pétrus head chef Marcus Wareing, who trained under Ramsay, recently announced that he would rather 'kill himself' than be overshadowed by 'that sad bastard'. All's fair in cooking and war, apparently…

Top Ramsay eateries in London:

> Boxwood Café (see boxed text, p130)
> Gordon Ramsay (p129)
> Gordon Ramsay at Claridges (p97)
> Maze (below)

INN THE PARK *British* £££

☎ 7451 9999; www.innthepark.co.uk; St James's Park; 8am-8.30pm Mon-Fri, 9am-8.30pm Sat & Sun; ⊖ St James's Park

This posh park café has delusions of grandeur, but we love the setting in St James's Park. Oliver Peyton, who also runs the restaurants at the National Gallery and Wallace Collection, uses ingredients from small producers to create a familiar but pleasing modern British menu.

JAPAN CENTRE *Japanese* £-££

☎ 7255 8270; www.japancentre.com; 212-213 Piccadilly; teashop 10am-7pm Mon-Sat, 11am-7pm Sun, restaurant noon-10pm Mon-Sat, noon-9pm Sun; ⊖ St James's Park

Part supermarket, part bookshop, part café, part restaurant and part sushi bar, Japan Centre offers the full Tokyo experience. Come for fast sushi, slurpy ramen noodles and cute Japanese homewares.

MAZE *International* £££

☎ 7107 0000; www.gordonramsay.com; 10-13 Grosvenor Sq; noon-2.30pm & 6-10.30pm; ⊖ Bond St

If you want to sample the Gordon Ramsay magic in small doses, try this light-filled restaurant on Grosvenor Sq. The Michelin star–winning menu fuses French haute cuisine and subtle Asian flavours – tasting is the idea here, so order several plates to make a meal.

MOMO *North African* £££

☎ 7434 4040; www.momoresto.com; 46 Grosvenor Pl; noon-2.30pm & 6.30-11.30pm Mon-Sat, 6.30-11.30pm Sun; ⊖ Oxford Circus

Moroccan but more fun is the motto at Momo (don't think we can fit any more 'm' words into that sentence). People come as much for the bazaar-style décor and party mood as the food, though we can't fault its couscous, tagines and roasts.

NAHM *Thai* £££

☎ 7333 1234; halkin.como.bz; The Halkin, Halkin St; noon-2.30pm & 7-11pm Mon-Fri, 7-11pm Sat, 7-10pm Sun; ⊖ Hyde Park Corner

If you want top-notch Thai food, visiting the former chef of the king of Thailand is probably a good start. David Thomson has bought the tricks he learnt at the royal palace to this posh dining room at the Halkin hotel.

NOBU *Japanese* £££

☎ 7447 4747; Metropolitan Hotel, 19 Old Park Lane; noon-2.15pm & 6-10.15pm, to 11pm Sat, to 9.30pm Sun; ⊖ Hyde Park Corner

The original Nobu is facing stiff competition from sister-restaurant Nobu Berkeley St (p100), but you'll still see a few famous faces here if celebrity-spotting tickles your teriyaki. A better reason to come is the innovative Japanese food – for pudding, try the chocolate 'Bento boxes' with green tea icecream.

SKETCH *International* £££

☎ 7659 4500; www.sketch.uk.com; 9 Conduit St; Gallery 7-11pm Mon-Sat, Glade noon-3pm Mon-Sat, Lecture Room noon-3pm & 7-10.30pm Tue-Fri, 7-10.30pm Sat; ⊖ Oxford Circus

Not quite as cool or essential as a few years ago, Sketch still scores points for total lack of restraint. The downstairs Gallery and Glade rooms serve classy Modern European food while the upstairs Lecture Room has haute cuisine at very haute prices. Be sure to check out the egg-shaped loos.

A PLATE WITH A VIEW

For a dinner with a view to die for, book a table at **Galvin@Windows** (☎ 7208 4021; www.galvinatwindows.com; London Hilton On Park Lane, 22 Park Lane; 7.30-10am Tue-Sun, noon-3pm Sun-Fri, 5.30-11pm Mon-Sat; ⊖ Hyde Park Corner), perched on a raised platform on the 28th floor of the Hilton on Park Lane. The restaurant offers swoon-inducing views over Belgravia, including a peek into the Queen's back garden, and an excellent menu of modern French cuisine.

Cuisine with class at Sketch restaurant (p99)

WOLSELEY

Modern European £££

☎ 7499 6996; www.thewolseley.com; 160 Piccadilly; 7am-midnight Mon-Fri, 8am-midnight Sat, 8am-11pm Sun; ⊖ Green Park

Wolseley harks back to a more genteel time. Following the tradition of the grand cafés of Europe, this elegant Art Deco dining room serves fine teas, sandwiches, scones, cakes and *viennoiserie* to members of the local gentry and glitterati.

DRINK

NOBU BERKELEY ST *Bar*

☎ 7290 9222; www.noburestaurants.com; 15 Berkeley St; noon-1am Mon-Wed, noon-2am Thu-Fri, 6pm-2am Sat, 6-11pm Sun; ⊖ Green Park

This glimmering bar full of metal, wood and twinkling tree sculptures feels like the kind of place James Bond would take Moneypenny for a drink after work – ie smooth, classy and expensive. The restaurant serves the same imaginative Japanese cuisine as Nobu (p99) at Hyde Park Corner.

YE GRAPES *Pub*

☎ 7499 1563; 16 Shepherd Market; ⊖ Green Park

There's a villagey vibe to Shepherd Market, lending a local feel to this handsome Victorian corner pub. The interior is full of stuffed animals, which presumably appeals to the well-heeled Mayfair clientele, who mob the place after work on weekdays.

PLAY

CURZON MAYFAIR *Cinema*

☎ information 7495 0501, bookings 0871 703 3989; www.curzoncinemas.com; 38 Curzon St; ⊖ Green Park or Hyde Park Corner

The original Curzon cinema, this arty movie house screens some excellent films from small, independent film makers. The 'futuristic' 1970s décor makes the theatre look a bit like the bridge of a Klingon battle cruiser.

PIGALLE CLUB *Cabaret*

☎ information 7644 1420, bookings 0845 345 6053; www.thepigalle.co.uk; 215 Piccadilly; 🕓 from 7pm; ⊖ Piccadilly Circus

Put on your best Dean Martin swagger and relive the glory days of the 1940s at this wonderfully nostalgic supper club. The cabaret repertoire runs the gamut from crooners and big bands to the genteel burlesque of Immodesty Blaize. Also see p22.

>BLOOMSBURY & FITZROVIA

Drug-taking. Extra-marital affairs. Eastern mysticism. Bisexuality. If the conservative residents of Bloomsbury had known what was going on at the parties hosted by Virginia Woolf and the other members of the Bloomsbury Group, there would have been outrage. But Bloomsbury has always attracted free thinkers – Darwin lived on Gower St and George Bernard Shaw conceived his withering critiques of the class system in the Reading Room at the British Museum.

Similar creative rumblings were taking place in Fitzrovia on the far side of Tottenham Court Rd. Authors like George Orwell, Quentin Crisp and Dylan Thomas traded stories and pints in pubs like the Fitzroy Tavern, while a few blocks away, Aleister Crowley dabbled in the dark arts. The creative spark lives on in the precincts of the University of London and the works of modern-day residents like Ian McEwan and Ricky Gervais.

For visitors, the undisputed main attraction is the British Museum, but it's also worth exploring the well-stocked computer and furniture shops on Tottenham Court Rd and the smart restaurants in the backstreets of Fitzrovia.

BLOOMSBURY & FITZROVIA

SEE

British Museum.............1 C5
Cartoon Museum...........2 C6
New London Architecture...............3 B5
Petrie Museum..............4 B4
Pollock's Toy Museum....5 B5
Wellcome Collection......6 B4

SHOP

Comicana...................(see 10)
Gosh!.............................7 C6
Habitat..........................8 B5
Heal's.............................9 B5
James Smith & Sons Umbrellas.........10 C6
Paperchase..................11 B5
Topshop.......................12 A6

EAT

Busaba Eathai 13 B5
Fino..............................14 B5
Giraffe ....(see 27)
Hakkasan......................15 B6
Rasa Samudra 16 B6
Salt Yard.......................17 B5
Wagamama..................18 C6

DRINK

Annexe 3 ....19 A6
Bradley's Spanish Bar....20 B6
Fitzroy Tavern...............21 B5
Lamb 22 D4
Princess Louise.............23 D6

PLAY

100 Club 24 B6
All Star Lanes 25 D5
Bloomsbury Bowling ...26 C4
Renoir Cinema..............27 D4

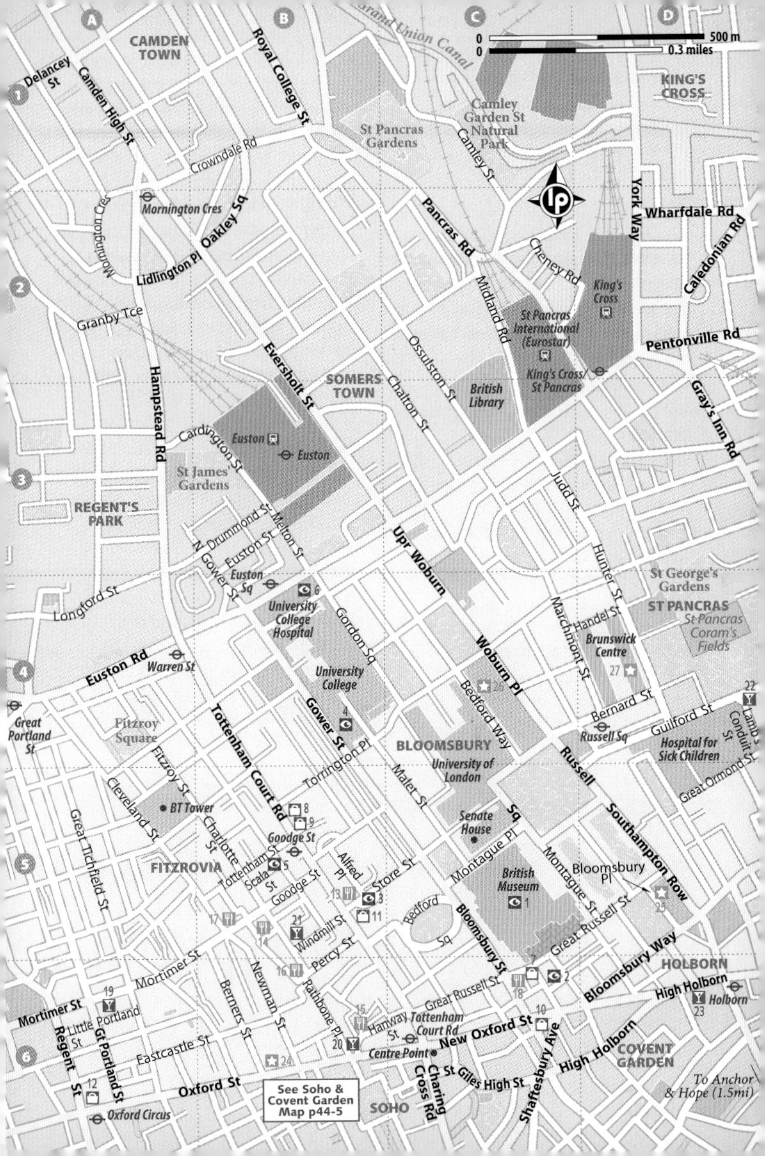

A
B
C
D
1
2
3
4
5
6
0
500 m
0
0.3 miles
Grand Union Canal
CAMDEN TOWN
KING'S CROSS
Delancey St
Camden High St
Royal College St
St Pancras Gardens
Camley Garden St Natural Park
Camley St
Crowndale Rd
Mornington Cres
Mornington Cres
Oakley Sq
Lidlington Pl
Pancras Rd
Cheney Rd
York Way
Wharfdale Rd
Caledonian Rd
Midland Rd
King's Cross
St Pancras International (Eurostar)
Pentonville Rd
King's Cross/St Pancras
Granby Tce
Eversholt St
SOMERS TOWN
Chalton St
Ossulston St
British Library
Gray's Inn Rd
Hampstead Rd
Cardington St
Euston
Euston
St James' Gardens
REGENT'S PARK
Judd St
Drummond St
Melton St
Euston St
N Gower St
Euston Sq
Upr Woburn
Hunter St
St George's Gardens
6
University College Hospital
Longford St
Gordon Sq
Marchmont St
Handel St
ST PANCRAS
St Pancras Coram's Fields
Brunswick Centre
Euston Rd
Warren St
University College
Woburn Pl
26
27
Bedford Way
Bernard St
22
Great Portland St
Fitzroy Square
Tottenham Court Rd
Gower St
4
Torrington Pl
BLOOMSBURY
Russell Sq
Guilford St
Lamb's Conduit St
Hospital for Sick Children
Russell
University of London
Fitzroy St
Cleveland St
BT Tower
Malet St
Great Ormond St
8
9
Goodge St
Senate House
Sq
Southampton Row
Great Titchfield St
Charlotte St
Tottenham St
5
FITZROVIA
Scala St
Alfred Pl
Store St
Montague Pl
British Museum
1
Montague St
Bloomsbury Pl
Goodge St
13
3
Bedford Sq
Bloomsbury St
25
17
21
11
Windmill St
Great Russell St
14
Percy St
Bloomsbury Way
HOLBORN
Mortimer St
16
7
2
18
Great Russell St
High Holborn
Holborn
19
Berners St
Newman St
Rathbone Pl
15
Hanway St
Tottenham Court Rd
10
23
Mortimer St
Little Portland St
Gt Portland St
Regent St
Eastcastle St
20
Centre Point
New Oxford St
Shaftesbury Ave
High Holborn
COVENT GARDEN
24
Charing Cross Rd
St Giles High St
12
Oxford St
See Soho & Covent Garden Map p44-5
SOHO
To Anchor & Hope (1.5mi)
Oxford Circus

SEE

BRITISH MUSEUM

☎ 7323 8000, tours 7323 8181; www.britishmuseum.org; Great Russell St; admission free, £3 donation suggested, fees apply for special exhibitions & some tours; galleries 10am-5.30pm, to 8.30pm Thu & Fri, Great Court 9am-6pm Sun-Wed, 9am-11pm Thu-Sat; ; Tottenham Court Rd or Russell Sq

Sir Hans Sloane bequeathed his humble 'cabinet of curiosities' to the nation in 1753. Since then adventurers and archaeologists have expanded the collection to include some of the greatest antiquities on earth. Inside you can see such priceless treasures as the Rosetta Stone and the Elgin Marbles (see p11 for more). Special tours cover the highlights.

NEW LONDON ARCHITECTURE

☎ 7636 4044; The Building Centre, 26 Store St; admission free; 9am-6pm Mon-Fri, 10am-5pm Sun; Goodge St

For updates on the latest architectural monsters and marvels appearing around London, check out the maps, models and exhibitions at New London Architecture.

POLLOCK'S TOY MUSEUM

☎ 7639 3452; www.pollockstoymuseum.com; 1 Scala St; adult/child/concession £5/2/3; ; 10am-5pm Mon-Sat; Goodge St

This quirky children's museum is crammed with old-fashioned board games, wind-up toys, dolls, science toys, cardboard theatres and traditional folk toys. Go on, indulge your inner child.

SECRET MUSEUMS

Everybody knows about the British Museum, with its astounding collection of Egyptian treasures, but fewer visitors have heard of the tiny **Petrie Museum** (☎ 7679 2884; www.petrie.ucl.ac.uk; University College, Malet Pl; admission free; 1-5pm Tue-Fri, 10am-1pm Sat; Goodge St) with its cabinets of Egyptian statues, ceramics, stone inscriptions and mummies.

Nearby is the **Wellcome Collection** (☎ 7611 2222; www.wellcomecollection.org; 183 Euston Rd; admission free; 10am-6pm Mon-Sat, to 10pm Thu, 10am-6pm Sun; ; Euston Sq), a bizarre assortment of surgical and ethnological objects collected by the Victorian philanthropist Sir Henry Wellcome, alongside some futuristic displays on modern medicine.

Ben Roberts

Curator of European Bronze Age History, British Museum

Highlights of the museum There's so much history attached to the Rosetta Stone and the Elgin Marbles, but my personal favourite is the Mold gold cape from North Wales – nobody knew what it was until they put all these little pieces of gold together. **Other great London museums** The Natural History Museum (p124) is an absolute must – the Darwin Centre makes it even better than it was before – and the V&A (p126) if you're interested in anything to do with art and design. And Sir John Soane's Museum (p62) is wonderfully bonkers. **Best cultural experience** Take a boat trip from Westminster (see p86) to Greenwich (see p176) – you get a totally different view of London. And don't overlook the City churches (p70); they have a sense of silence and peace you won't find anywhere else.

onations

A lion presides over the Great Court at the British Museum (p104)

SHOP

Overcrowded, underwhelming Oxford St is the main shopping street; there are branches of music superstores HMV and Zavvi. The south end of Tottenham Court Rd is lined with shops selling computers, audio equipment and electronics, often at negotiable prices.

HABITAT

Furniture & Homewares

☎ 0844 499 1122; www.habitat.net; 196-199 Tottenham Court Rd; 10am-6pm Mon-Wed, to 8pm Thu, to 6.30pm Fri, 9.30am-6.30pm Sat, noon-6pm Sun; ⊖ Goodge St

More accessible than Heal's, Habitat takes the European modernist style of home furnishing and markets it to the masses. There are branches on Regent St and the King's Rd in Chelsea.

HEAL'S

Furniture & Homewares

☎ 7636 1666; www.heals.co.uk; 196 Tottenham Court Rd; 10am-6pm Mon-Wed, to 8pm Thu, to 6.30pm Fri, 9.30am-6.30pm Sat, noon-6pm Sun; ⊖ Goodge St

Heal's has been kitting out the houses of wealthy Londoners for generations. Come here for sleek designer furniture and high-quality homewares in the latest 'maximalist' style. There's a branch on the King's Rd in Chelsea.

JAMES SMITH & SONS UMBRELLAS *Umbrellas*

☎ 7836 4731; www.james-smith.co.uk; 53 New Oxford St; 9.30am-5.25pm Mon-Fri, 10am-5.25pm Sat; ⊖ Tottenham Court Rd

This venerable Victorian umbrella and walking-stick shop has hardly changed in centuries, though the sword sticks advertised in the window were sensibly banned in 1988.

PAPERCHASE *Stationary*

☎ 7636 1666; www.paperchase.co.uk; 196 Tottenham Court Rd; 9.30am-7pm Mon-Wed, Fri & Sat, 9.30am-8pm Thu, noon-6pm Sun; ⊖ Goodge St

Funky pencils, colourful notepads, novelty erasers and fabulous wrapping paper make this one of the best places in town to shop for kids. There are branches in Notting Hill and Chelsea.

TOPSHOP *Clothing*

☎ 7636 7700; www.topshop.co.uk; 36-38 Great Castle St; 9am-8pm Mon-Sat, to 9pm Thu, noon-6pm Sun; ⊖ Oxford Circus

This flagship fashion emporium offers high-street fashion at its most accessible, with lots of spangly tops, low-rise jeans and new lines 'designed' (with a little assistance) by celebs like Kate Moss.

COMIC INTERLUDE

Between Bloomsbury and Soho is a curious twilight world where middle-aged men revert to childhood and fantasies about thigh-high boots and capes are positively encouraged. This is the comic capital of London, the setting for an entire counterculture based on the fantasy worlds of comic-strip heroes like Neil Gaiman and Alan Moore.

Classic British comics and political cartoons are showcased at the interesting **Cartoon Museum** (☎ 7580 8155; www.cartoonmuseum.com; 35 Little Russell St; adult/child £4/free; 10.30am-5.30pm Tue-Sat, noon-5.30pm Sun; ⊖ Tottenham Court Rd). For back issues and collectables, head to **Forbidden Planet** (p48) in Soho or the following comic stores dotted around Tottenham Court Rd tube:

Comicana (☎ 7836 5630; www.comicana.com; 237 Shaftesbury Ave)

Gosh! (☎ 7636 1011; www.goshlondon.com; 39 Great Russell St)

EAT

As well as the following restaurants, there are more good eateries around Goodge St (B5) and Charlotte St (B5).

BUSABA EATHAI

Thai £-££

☎ 7299 7900; 22 Store St; noon-11pm Mon-Thu, noon-11.30pm Fri & Sat, noon-10.30pm Sun; ⊖ Goodge St

The Alan Yau take on Thai food. Diners sit down together at long communal tables, ordering from a stripped-down Thai menu, but the sultry wood-lined interior adds a touch of class. There's a branch in Marylebone (p117).

FINO *Spanish* £££

☎ 7813 8010; www.finorestaurant.com; 33 Charlotte St; noon-2.30pm & 6-10.30pm Mon-Sat; ⊖ Goodge St

Fino is tapas perfection – the small dishes here are created using the best Spanish ingredients. It's a favourite of West End media types; lesser mortals may baulk at the prices. The entrance is on Rathbone St.

GIRAFFE

International £-££

☎ 7812 1336; www.giraffe.net; 19-21 Brunswick Centre, Brunswick Sq; 8am-11pm Mon-Fri, 9am-10.30pm Sat & Sun; ; ⊖ Russell Sq

This accessible chain restaurant inside the Brunswick Centre (see boxed text, p111) specialises in fun, colourful food made with fresh ingredients. Kids won't even realise they're being healthy.

HAKKASAN

Chinese £££

☎ 7907 1888; www.hakkasan.com; 8 Hanway Pl; noon-3pm & 6pm-midnight Mon-Fri, noon-5pm & 6pm midnight Sat, noon-5pm & 6-11pm Sun; ⊖ Tottenham Court Rd

Hakkasan is everything you would expect of a cult eatery. The food is exquisite, the clientele exclusive, the décor inspiring and the location obscure. Owned by Alan Yau, this was the first Chinese restaurant to win a Michelin star, and the food easily warrants the high prices. Reserve well ahead.

RASA SAMUDRA

South Indian ££

☎ 7637 0222; www.rasarestaurants.com; 5 Charlotte St; noon-3pm & 6-11pm Mon-Sat, 6-11pm Sun; ⊖ Goodge St or Tottenham Court Rd

Behind Rasa's shocking-pink façade you can sample *kappayum meenum* (kingfish in spicy cassava gravy) and other vegetarian and seafood delights from Kerala.

SALT YARD *Spanish* £££

☎ 7637 0657; www.saltyard.co.uk; 54 Goodge St; noon-11pm Mon-Fri, 5-11pm Sat; ⊖ Goodge St

Describing itself as a 'charcuterie bar', Salt Yard specialises in hybrid tapas, with influences from Spain and Italy. The menu runs to delicious tuna carpaccio, confit pork-belly with cannellini beans and some tasty but expensive cuts of ham.

WAGAMAMA

Pan-Asian £-££

☎ 7323 9223; www.wagamama.com; 4 Streatham St; noon-11pm Mon-Sat, noon-10pm Sun; ⊖ Tottenham Court Rd

The chain that made Alan Yau famous started here in this anonymous basement just off Gower St. The menu focuses on Japanese soba, ramen and udon noodles, served fried or in big bowls of curry or soup. There are Wagamamas all over London – just look out for the red star on a black background.

DRINK

ANNEXE 3 *Bar*

☎ 7631 0700; www.loungelover.com; 6 Little Portland St; 5pm-midnight Mon-Fri, 6pm-midnight Sat; ⊖ Oxford Circus

This West End bar is an explosion of camp, like a dance party in a 1960's chandelier warehouse. Every available surface is covered in chintz and the cocktail menu is probably the most imaginative this side of Malibu.

BRADLEY'S SPANISH BAR

Bar

☎ 7636 0359; 42-44 Hanway St; ⊖ Tottenham Court Rd

Not much bigger than a tapas plate, this tiny hostelry looks like an English pub that has collided with a truckload of Spanish holiday souvenirs. It's chintzy, unrefined and undeniably cosy.

FITZROY TAVERN *Pub*

☎ 7580 3714; 16 Charlotte St; closed Sun; ⊖ Goodge St

In the years before and after WWII, the Fitzroy was the hangout of literary giants like George Orwell and Dylan Thomas. Today it's a typical downtown boozer, and part of the popular Sam Smith's chain, which means plenty of ales and specialist beers at bargain prices.

LAMB *Pub*

☎ 7405 0713; 94 Lamb's Conduit St; closed Sun; ⊖ Russell Sq

Hidden away among deli cafés on Lamb's Conduit St, this atmospheric pub retains most of its original Victorian décor. Ignore the sign – the pub and lane were named after the philanthropist William Lamb.

Quenching a thirst in the pretty Princess Louise

PRINCESS LOUISE *Pub*

7405 8816; 208-9 High Holborn; closed Sun; Holborn

One of the prettiest pubs in the West End, this heritage-listed pub is a flashback to the golden age of British public houses. The atmosphere is typical for a downtown pub, but the décor warrants a look in.

PLAY

100 CLUB *Live Music*

7636 0933; www.the100club.co.uk; 100 Oxford St; event times vary; Leicester Sq

One of London's oldest venues, the 100 club has hosted everyone from the Sex Pistols and the White Stripes to Muddy Waters and Louis Armstrong. The swing and boogie-woogie nights held on Monday always pull in a crowd – see the boxed text, p163.

ALL STAR LANES *Bowling*

7025 2676; www.allstarlanes.co.uk; Victoria House, Bloomsbury Pl; 5-11.30pm Mon-Wed, 5pm-midnight Thu, noon-2am Fri & Sat, noon-11pm Sun; Holborn or Russell Sq

Riding the crest of the 1990's *Big Lebowski* bowling revival, All Star Lanes faithfully reproduces the

A NEW LIFE FOR THE BRUNSWICK

Set incongruously on the edge of genteel Brunswick Sq, the **Brunswick Centre** (☎ 7833 6066; www.brunswick.co.uk; Brunswick Sq; ⏲ 9am-7pm Mon-Sat, 11am-5pm Sun; ⊖ Russell Sq) was one of London's original brutal modernist buildings. Despite lofty ambitions, this huge concrete edifice was used as council housing until a massive £22 million revamp in 2002. Now repainted in gleaming white, the revitalised Brunswick Centre has been filled with upmarket shops and chain restaurants, including a branch of kid-friendly **Giraffe** (p108). Also here is the Curzon group's **Renoir Cinema** (☎ 0871 703 3991; www.curzoncinemas.com), screening art house films for the arty residents of the modernist flats upstairs.

style and glamour of a vintage American bowling alley. There's a branch in the Whiteleys Centre in Bayswater.

BLOOMSBURY BOWLING

Bowling

☎ 7183 1979; www.bloomsburybowling.com; Tavistock Hotel, Bedford Way; ⏲ noon-2am Mon-Thu, noon-3am Fri & Sat, 1pm-midnight Sun; ⊖ Russell Sq

The former car park of the Tavistock Hotel has been turned into another 1950's-style ten-pin bowling alley. It doesn't have quite the same retro-chic as All Star Lanes (opposite), but it's easier to get a lane here.

>MARYLEBONE & REGENT'S PARK

Most people assume that Marylebone (mar-lee-bone) comes from the French, Marie La Bonne, when the name is actually derived from the medieval word *bourne* meaning stream, a reference to the stream that used to trickle along Marylebone Lane. That's the history lesson over – today Marylebone is the calmer, less showy cousin of Mayfair, and one of London's most desirable residences.

The fact that people actually live here works in Marylebone's favour, as it means the streets are dotted with restaurants, cafés, pubs and genuinely useful shops. Most of the action takes place around Marylebone High St, which runs north from Oxford St towards peaceful Regent's Park and London Zoo. Before you reach the park, you'll have to cross Marylebone Rd, best known as the location of the tacky Madame Tussauds. Further west is Baker St, where Sherlock Holmes would have lived if he actually existed.

Highlights of the area include the famous department stores along Oxford St, the cool breezes that blow over Regent's Park and the wonderful Wallace Collection in one of London's finest stately homes. Lowlights include the shops selling reproduction Sherlock Holmes pipes and deerstalkers along Baker St.

MARYLEBONE & REGENT'S PARK

SEE

London Zoo 1 C1
Lord's Cricket Ground 2 A2
Madame Tussauds 3 C4
Regent's Park 4 C2
Wallace Collection 5 C5

SHOP

Alfie's Antiques Market 6 A4
Cabbages & Frocks Market 7 C4
Daunt Books 8 C4
John Lewis 9 D5
Selfridges 10 C5
Tracey Neuls 11 C5

EAT

Busaba Eathai 12 C5
Chaophraya 13 D5
Eat & Two Veg 14 C4
Golden Hind 15 C5
La Fromagerie 16 C4
Locanda Locatelli 17 C5
Maroush 18 B5
Providores & Tapa Room 19 C5
Texture 20 C5

DRINK

deVigne .... (see 11)
Volunteer 21 B4

PLAY

Screen on Baker Street 22 C4

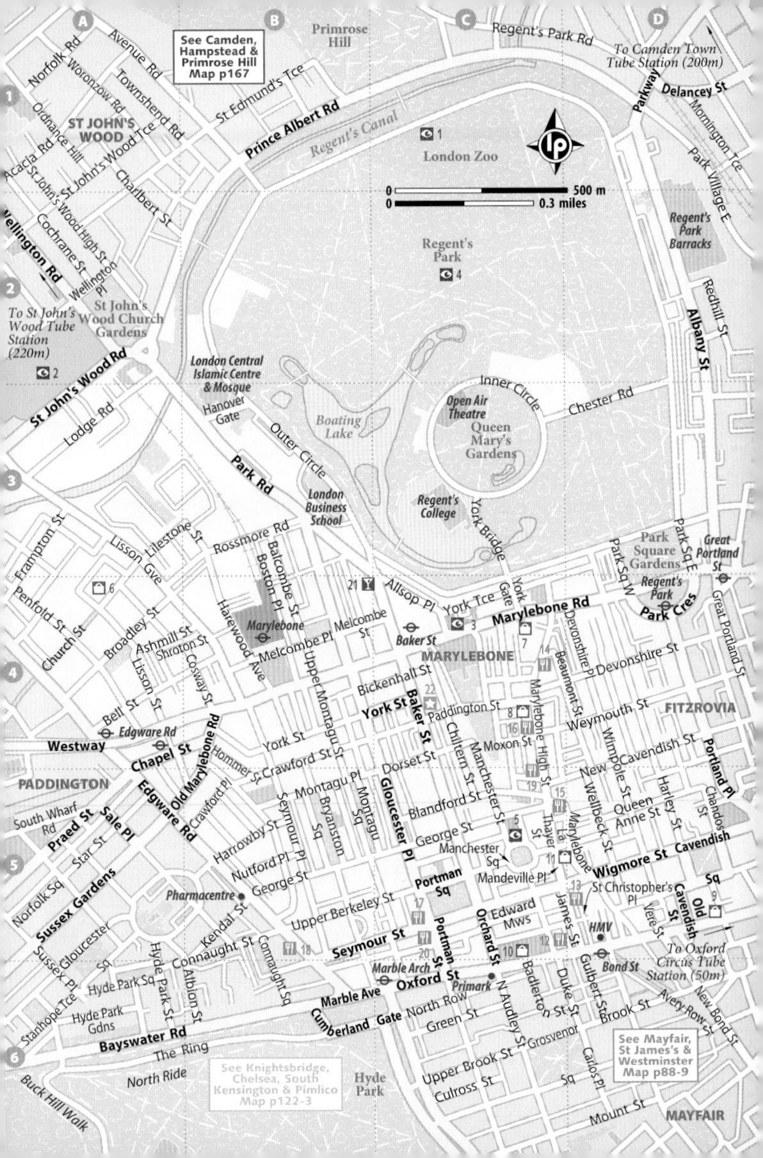

See Camden, Hampstead & Primrose Hill Map p167
Primrose Hill
Regent's Park Rd
To Camden Town Tube Station (200m)
ST JOHN'S WOOD
Prince Albert Rd
Regent's Canal
London Zoo
500 m
0.3 miles
Regent's Park
Regent's Park Barracks
To St John's Wood Tube Station (220m)
St John's Wood Church Gardens
London Central Islamic Centre & Mosque
Hanover Gate
Boating Lake
Inner Circle
Open Air Theatre
Queen Mary's Gardens
Chester Rd
Outer Circle
Park Rd
London Business School
Regent's College
York Bridge
Park Square Gardens
Great Portland St
Regent's Park
Park Cres
Marylebone
Baker St
MARYLEBONE
Marylebone Rd
FITZROVIA
Edgware Rd
Westway
PADDINGTON
Pharmacentre
Portman Sq
Manchester Sq
Wigmore St
HMV
Bond St
To Oxford Circus Tube Station (50m)
Marble Arch
Oxford St
Primark
Bayswater Rd
See Knightsbridge, Chelsea, South Kensington & Pimlico Map p122-3
Hyde Park
See Mayfair, St James's & Westminster Map p88-9
MAYFAIR

SEE

LONDON ZOO

7722 3333; www.zsl.org/zsl-london-zoo; Regent's Park; adult/child/concession £15.40/11.90/13.90; 10am-6pm, last admission 5pm; ; Baker St or Camden Town

Londoners have fallen in and out of love with their famous zoo over the years, but the Royal Zoological Society seems to be winning the public over with more naturalistic enclosures, including the new lowland gorilla pen and the walk-through 'Meet the Monkeys' cage.

LORD'S CRICKET GROUND

tickets 7432 1000, tours 7616 8595; www.lords.org; St John's Wood Rd; tours adult/child/concession/family £12/6/7/31; tours 10am, noon & 2pm Apr-Sep, noon & 2pm Oct-Mar; ; St John's Wood

The Marylebone Cricket Club, aka Lords, is the official guardian of the Rules of Cricket and the Spirit of Cricket, which tells you a lot about the game. Fans can enjoy a tour of the club buildings when there are no games in play, including a quick peek at the urn containing the Ashes.

MADAME TUSSAUDS

0870 999 0046; www.madametussauds.com; Marylebone Rd; adult/under 16yr/family incl planetarium £25/21/85, discounts online; 9.30am-5.30pm; ; Baker St

On one level, Madame Tussauds is impressive – it's been pulling in punters using wax models of celebrities ever since 1884. On the other hand, this is the most tacky, commercial and expensive tourist attraction in London. You know

Spot the superficial celebs at Madame Tussauds

LEATHER & WILLOW

What could be more sublime on a sunny summer afternoon than watching men in white trousers play a game that can run for six days and still end in a draw? Cricket is one of those sports you either get or you don't. The rules are complicated, the action takes place at a distance and hours can pass by without anything exciting happening, but those who love cricket, love it with a passion. If you hope to catch a game at Lords (see opposite) or the Oval (p85), you'll need to book well in advance.

exactly what to expect – lots of wax celebrities and giggling, posing school girls.

REGENT'S PARK

☎ 7486 7905; www.royalparks.gov.uk; ; ⊖ Baker St or Regent's Park

The most formal of London's Royal Parks, the Regent's Park was created around 1820 by architect John Nash. Backing onto the Regent's Canal, this peaceful spot is best known as the location for London Zoo (opposite), but it's also worth visiting for the formal gardens and boating lake.

WALLACE COLLECTION

☎ 7563 9515; www.wallacecollection.org; Hertford House, Manchester Sq; admission free; 10am-5pm; ; ⊖ Bond St

A national treasure, this stunning stately home is piled floor to ceiling with rare antiquities. The painting collection includes Frans Hals' *Laughing Cavalier* and works by brush-wielders such as Rembrandt, Titian, Rubens, Van Dyck, Reynolds and Gainsborough.

SHOP

Look out for more bijou boutiques along Marylebone High St (C4).

ALFIE'S ANTIQUES MARKET

Antiques

☎ 7723 6066; www.alfiesantiques.com; 13-25 Church St; 10am-6pm Tue-Sat; ⊖ Edgware Rd or Marylebone

For fine antiques at premium prices, explore this higgledy-piggledy tangle of pocket-sized shops selling vintage trinkets, furniture, fashion, ceramics and homewares.

DAUNT BOOKS *Books*

☎ 7224 2295; www.dauntbooks.co.uk; 83-84 Marylebone High St; 9am-7.30pm Mon-Sat, 11am-6pm Sun; ⊖ Baker St

Daunt books is an old-fashioned bookshop with a gorgeous,

Shoppers worship at Selfridges – one of London's high churches of consumerism

light-filled Edwardian book-hall at the back. There's a branch in Hampstead.

JOHN LEWIS

Department Store

☎ 7629 7711; www.johnlewis.com; 278-306 Oxford St; 9.30am-8pm Mon-Fri, to 9pm Thu, 9.30am-7pm Sat, 11.30am-6pm Sun; ⊖ Bond St

Less grand than Selfridges but more accessible, John Lewis is a department store for the masses. It's good for fashions, homewares and luggage.

SELFRIDGES

Department Store

☎ 7629 1234; www.selfridges.com; 400 Oxford St; 9.30am-9pm Mon-Sat, to 10pm Thu, 11.30am-6.15pm Sun; ⊖ Bond St

A cathedral to shopping, this grand Art Deco department store has floor after floor of fashions, homewares, cosmetics, fragrances and foodstuffs, hidden behind some of the most inventive window displays in London. Make time for the confectionary hall on the ground floor.

TOP DEPARTMENT STORES

You can find everything imaginable in London's department stores from beluga caviar to slingback heels. Following is a list of the best:

- Fortnum & Mason (p94)
- Harrods (p127)
- Harvey Nichols (p127)
- John Lewis (left)
- Liberty (p50)
- Peter Jones (p128)
- Selfridges (left)

CABBAGES & FROCKS

Although it's nothing to rival Camden Market or Borough Market, Marylebone's **Cabbages & Frocks Market** (www.cabbagesandfrocks.co.uk; St Marylebone Parish Church, Marylebone High St; 11am-5pm Sat) does sell some particularly fine designer frocks, along with arts and crafts and gourmet foodstuffs.

TRACEY NEULS *Shoes*

7935 0039; www.tn29.com; 29 Marylebone Lane; 11am-6.30pm Mon-Fri, noon-5pm Sat; Bond St

Boutique shoes, boots and sandals, beautifully displayed in a room full of knick-knacks. The footwear designed by Neuls is eye-catching, tasteful and timeless.

EAT

BUSABA EATHAI *Thai* £££

7518 8080; 8-13 Bird St; noon-11pm Mon-Thu, to 11.30pm Fri & Sat, to 10.30pm Sun; Bond St

Tucked away near Selfridges, this stylish Thai canteen offers more of the Alan Yau magic: fun, flavour-some, Asian cooking served at bench tables in minutes. There's a branch in Bloomsbury (p108).

CHAOPHRAYA *Thai* ££-£££

7486 0777; 22 St Christopher's Pl; noon-10.30pm; Baker St

This is one old-fashioned sit-down Thai restaurant that we can wholeheartedly recommend. Every dish we've eaten here has been full of complex flavours and subtle undertones. A top tip for a laid-back romantic dinner.

EAT & TWO VEG

Vegetarian ££

7258 8595; www.eatandtwoveg.com; 50 Marylebone High St; V ; Baker St

At last a place where carnivores and vegetarians can eat together without feeling like anyone is missing out. All the meat dishes are made with convincing meat substitutes.

GOLDEN HIND

Fish & Chips £

7486 3644; 73 Marylebone Lane; noon-3pm Mon-Fri, 6-10pm Mon-Sat; Bond St

Every traveller should try fish and chips at least once, and this 90-year-old chippy is a great place to start. The decor is as lost in time as the menu – cod and chips, mushy peas, pickled eggs, gherkins, and apple crumble for pudding.

LA FROMAGERIE

Deli-café ££

☎ 7935 0341; www.lafromagerie.co.uk; 2-6 Moxon St; ⌚ 10.30am-7.30pm Mon, 8am-7.30pm Tue-Fri, 9am-7pm Sat, 10am-6pm Sun; ⊖

Marylebone's answer to Neal's Yard Dairy, La Fromagerie is part cheese shop, part tasting café. West End ladies flock here at lunchtime for cheese and charcuterie plates and delicious ploughman's lunches.

LOCANDA LOCATELLI

Italian £££

☎ 7935 9088; www.locandalocatelli.com; 8 Seymour St; ⌚ noon-3pm daily, 6.45-11pm Mon-Thu, 6.45-11.30pm Fri & Sat, 6.45-10.15pm Sun; ⊖ Marble Arch

Chef Giorgio Locatelli is a member of the Michelin star club, but you won't see his face on reality TV shows. He's far too busy in the kitchen, making outstanding modern Italian food for Marylebone's glitterati. Book one calendar month in advance.

MAROUSH *Lebanese* ££

☎ 7723 0773; www.maroush.com; 21 Edgware Rd; ⌚ noon-2am; ⊖ Marble Arch

The original restaurant of the Maroush empire, this quaintly formal Lebanese restaurant serves meze dishes, pitta wraps, kebabs, felafel and salads, and, yes, there are belly dancers.

PROVIDORES & TAPA ROOM

Fusion ££

☎ 7935 6175; www.theprovidores.co.uk; 109 Marylebone High St; ⌚ noon-2.45pm & 6-10.30pm, breakfast from 9am Mon-Fri, from 10am Sat & Sun; ⊖ Baker St or Bond St

Tapas in London can be variable, but New Zealanders Peter Gordon and Anna Hansen don't seem to put an olive out of place. The imaginative menu of traditional and fusion tapas is backed up by a relaxed, European café vibe.

TEXTURE

Modern European £££

☎ 7224 0028; 34 Portman St; ⌚ noon-2.30pm & 6.30-11pm Tue-Sat; ⊖ Oxford Circus

Icelandic chef Agnar Sverisson is cooking up a treat using tricks he learnt while working at Raymond Blanc's Manoir Aux Quat' Saisons. The focus here is on mixing different textures, but you can rest easy – Iceland's famous *hakarl* (rotten shark) has yet to appear on the menu.

DRINK

DEVIGNE *Bar*

☎ 7935 3665; www.mandeville.co.uk; Mandeville Hotel, Mandeville Pl; ⊖ Baker St

Designer Stephen Ryan has filled the bar at the Mandeville Hotel with etched glass, Perspex and day-glo Regency upholstery. The result is simply outrageous, as are the prices.

THE VOLUNTEER *Pub*

☎ 7486 4091; 247 Baker St; ⊖ Baker St

Tucked around the corner from Baker St tube, the Volunteer pulls in a young, good-natured crowd who are always up for talking to strangers. Look out for hard-to-find beers and ciders at the bar.

PLAY

SCREEN ON BAKER STREET *Cinema*

☎ 7935 2772; www.everymancinemaclub.com; 96-98 Baker St; ⊖ Baker St

Sister-cinema to Islington's Screen on the Green (p165), this small theatre has two screens and a taste for small independent films.

>KNIGHTSBRIDGE, CHELSEA, SOUTH KENSINGTON & PIMLICO

The districts of Knightsbridge, Kensington, Chelsea and Pimlico contain some of the largest accumulations of wealth in the hemisphere. Residents of the area are often dubbed Sloane Rangers – after Sloane Sq in Chelsea – by people less well off than themselves (virtually everyone else).

The area has some of the best shops, hotels and restaurants in Europe, and it's shamelessly, incorrigibly glamorous. Kensington offers wonderful museums, Knightsbridge has chichi department stores and gorgeous Hyde Park and Kensington Gardens, and Chelsea boasts classy shops and eateries around Fulham Rd and the King's Rd.

Nearby Pimlico is best known as the location of Tate Britain and Victoria bus station, though it also has some extremely posh shops and restaurants. Be warned: everything in West London comes at a price, so budget accordingly!

KNIGHTSBRIDGE, CHELSEA, SOUTH KENSINGTON & PIMLICO

SEE

Albert Memorial 1 B2
Chelsea Physic Garden 2 D5
Diana, Princess of Wales Memorial Fountain 3 C2
Hyde Park 4 D1
Hyde Park Stables 5 C1
Kensington Gardens 6 B2
Kensington Palace 7 A2
Natural History Museum 8 B3
Royal Geographical Society 9 C3
Royal Hospital Chelsea 10 D5
Saatchi Gallery 11 D4
Science Museum 12 C3
Serpentine Boathouse 13 C2
Serpentine Gallery 14 C2
Tate Britain 15 G4
Victoria & Albert Museum 16 C3

SHOP

Daylesford Organic 17 E4
Harrods 18 D3
Harvey Nichols 19 D3
Peter Jones 20 D4
Shop at Bluebird 21 C5

EAT

Bibendum 22 C4
Boxwood Café 23 D2
Brompton Quarter Café 24 C3
Daquise 25 C4
Gordon Ramsay 26 D5
Hummingbird Bakery 27 C4
Itsu 28 C4
Ottolenghi 29 D3
Painted Heron 30 B6
Pétrus (see 23)
Tom Aikens 31 C4
Tom's Kitchen 32 C4
Vama 33 B5
Zuma 34 D3

DRINK

Botanist 35 D4
Nag's Head 36 D3
Thomas Cubitt 37 E4
Troubadour 38 A5

PLAY

Chelsea Cinema 39 C5
Chelsea Football Club 40 A6
Royal Albert Hall 41 B3
Royal Court Theatre 42 D4
Serpentine Lido 43 C2

Please see over for map

SEE

ALBERT MEMORIAL

☎ 7495 0916; Kensington Gardens, Kensington Gore; tours adult/concession £4.50/4; tours 2pm & 3pm 1st Sun of month Mar-Sep; ⊖ Knightsbridge or South Kensington

George Gilbert Scott's memorial to Albert, Queen Victoria's husband, is even more ostentatious than the memorial built for Victoria. Surrounding the central gilded statue of Albert are a series of allegorical sculptures representing Europe, Asia, the Americas and Africa. Book a place on the guided tour for a full explanation of all the carvings.

CHELSEA PHYSIC GARDEN

☎ 7352 5646; www.chelseaphysicgarden.co.uk; 66 Royal Hospital Rd; adult/concession £7/4; noon-5pm Wed-Fri, noon-6pm Sun, till 10pm Wed Jul-Aug; ; ⊖ Sloane Sq

A secret walled garden just off the Embankment, the Chelsea Physic Garden was founded in 1673 as an aid to trainee medics studying medicinal plants and healing. Today this delightful space brims with aromatic and medicinal plants from around the world.

HYDE PARK & KENSINGTON GARDENS

www.royalparks.gov.uk; ; ⊖ Hyde Park Corner, Marble Arch, Knightsbridge, Queensway, High St Kensington or Lancaster Gate

Hyde Park and Kensington Gardens were laid out separately, but now it's hard to tell where one starts and the other finishes. Londoners use this gorgeous green lung to its full potential, playing sports, swimming, picnicking, and going

GETTING THE MOST OUT OF HYDE PARK

Hyde Park is London's favourite picnic and promenade ground, but there's more to do here than just laze around in the sun. The track along the north bank of the Serpentine is used for rollerblading and there are miles of designated cycle tracks for pedallers. Equestrians can arrange horse-riding on the park bridleways through the **Hyde Park Stables** (☎ 7723 2813; www.hydeparkstables.com; 66 Bathurst Mews).

Our favourite park attraction is the **Serpentine Lido** (☎ 7706 3422; adult/child £4/1; 10am-6pm May-Sep), with a fenced-off swimming area in the cooling waters of the Serpentine. On the Serpentine, the **Serpentine Boathouse** (per hr adult/child £8/3) rents out pedalos and row boats from March to October.

As well as open-air concerts, fun runs and demonstrations, the park is also the setting for the chilly **Peter Pan Cup** swimming race in the lake on Christmas morning and the splash-tastic **Red Bull Flugtag** (p29) in June.

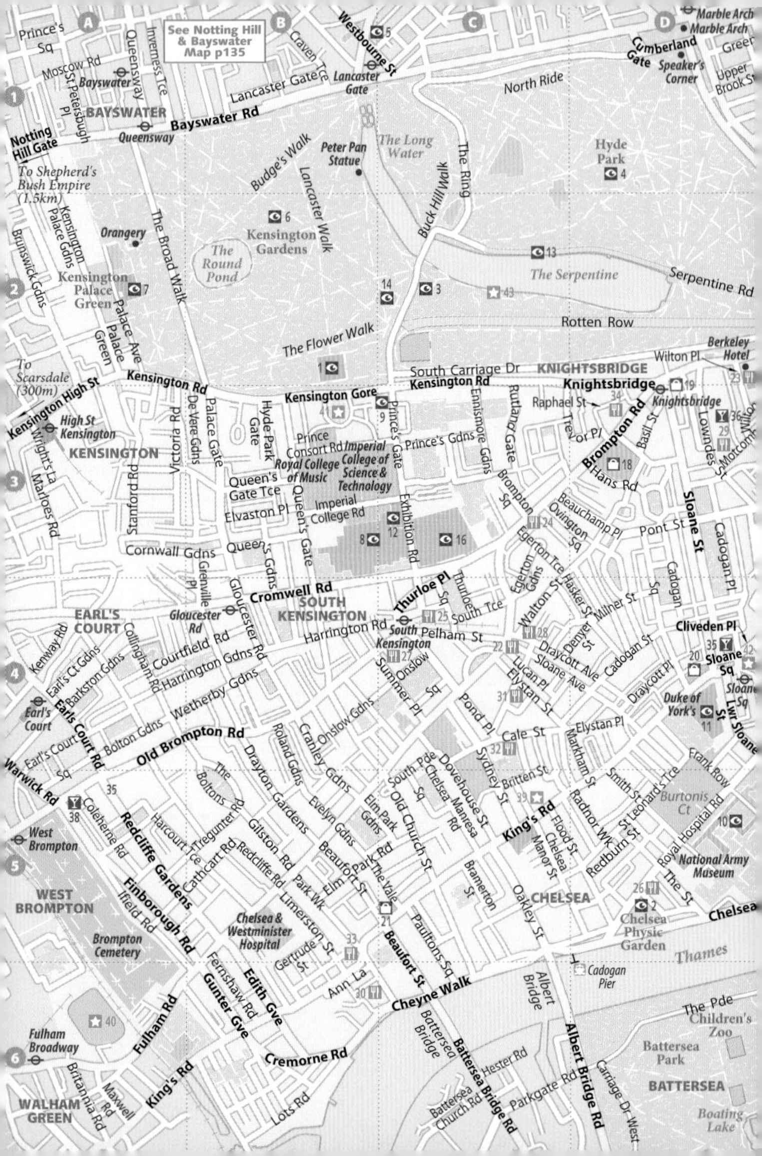

See Notting Hill & Bayswater Map p135
Prince's Sq
Moscow Rd
Bayswater
Queensway
Inverness Tce
St Petersburgh Pl
BAYSWATER
Bayswater Rd
Queensway
Notting Hill Gate
To Shepherd's Bush Empire (1.5km)
Craven Tce
Lancaster Gate
Westbourne St
Lancaster Gate
Marble Arch
Marble Arch
Cumberland Gate
Speaker's Corner
Upper Brook St
North Ride
Hyde Park
The Long Water
Peter Pan Statue
Budge's Walk
Lancaster Walk
The Ring
Buck Hill Walk
Kensington Palace Gdns
Brunswick Gdns
Orangery
The Broad Walk
Kensington Gardens
The Round Pond
Kensington Palace Green
Palace Green
Palace Ave
The Serpentine
Serpentine Rd
Rotten Row
The Flower Walk
To Scarsdale (300m)
Kensington Rd
South Carriage Dr
Kensington Rd
KNIGHTSBRIDGE
Wilton Pl
Berkeley Hotel
Knightsbridge
Knightsbridge
Kensington High St
High St Kensington
KENSINGTON
Wright's La
Marloes Rd
Victoria Rd
De Vere Gdns
Palace Gate
Hyde Park Gate
Kensington Gore
Prince's Gate
Prince Consort Rd
Royal College of Music
Imperial College of Science & Technology
Prince's Gdns
Ennismore Gdns
Rutland Gate
Raphael St
Trevor Pl
Brompton Rd
Basil St
Lowndes St
Hans Rd
Stanford Rd
Queen's Gate Tce
Queen's Gate
Elvaston Pl
Imperial College Rd
Exhibition Rd
Brompton Sq
Beauchamp Pl
Ovington Sq
Pont St
Sloane St
Cadogan Pl
Cornwall Gdns
Queen's Gdns
Grenville Pl
Gloucester Rd
Egerton Tce
Egerton Gdns
Hasker St
Cadogan Sq
Cromwell Rd
SOUTH KENSINGTON
Thurloe Pl
Thurloe Sq
Thurloe South Tce
Walton St
Milner St
EARL'S COURT
Gloucester Rd
Collingham Rd
Courtfield Rd
Harrington Rd
South Kensington
Pelham St
Denyer St
Cliveden Pl
Kenway Rd
Earl's Ct Gdns
Barkston Gdns
Harrington Gdns
Onslow Sq
Draycott Ave
Sloane Ave
Cadogan St
Sloane Sq
Sloane Sq
Draycott Pl
Earl's Court
Earls Court Rd
Wetherby Gdns
Sumner Pl
Lucan Pl
Elystan St
Pond Pl
Duke of York's
Lwr Sloane St
Bolton Gdns
Onslow Gdns
Old Brompton Rd
Roland Gdns
Cranley Gdns
Cale St
Elystan Pl
Earl's Court Sq
Warwick Rd
The Boltons
Drayton Gardens
South Pde
Chelsea Sq
Dovehouse St
Sydney St
Britten St
Markham St
Smith St
St Leonard's Tce
Frank Row
Burton's Ct
Coleherne Rd
Redcliffe Gardens
Harcourt Tce
Tregunter Rd
Evelyn Gdns
Elm Park Gdns
Old Church St
Manresa Rd
King's Rd
Flood St
Radnor Wk
Royal Hospital Rd
West Brompton
Cathcart Rd
Redcliffe Rd
Gilston Rd
Beaufort St
Elm Park Rd
Chelsea Manor St
Redburn St
National Army Museum
WEST BROMPTON
Finborough Rd
Ifield Rd
Limerston St
Park Wk
The Vale
Bramerton St
Oakley St
CHELSEA
The St
Chelsea
Chelsea & Westminister Hospital
Brompton Cemetery
Gertrude St
Beaufort St
Paultons Sq
Chelsea Physic Garden
Thames
Fernshaw Rd
Edith Gve
Gunter Gve
Ann La
Cheyne Walk
Albert Bridge
Cadogan Pier
Fulham Rd
Battersea Bridge
The Pde
Children's Zoo
Fulham Broadway
Cremorne Rd
Battersea Bridge Rd
Albert Bridge Rd
Hester Rd
Battersea Park
BATTERSEA
Britannia Rd
Maxwell Rd
King's Rd
WALHAM GREEN
Lots Rd
Battersea Church Rd
Parkgate Rd
Carriage Dr West
Boating Lake

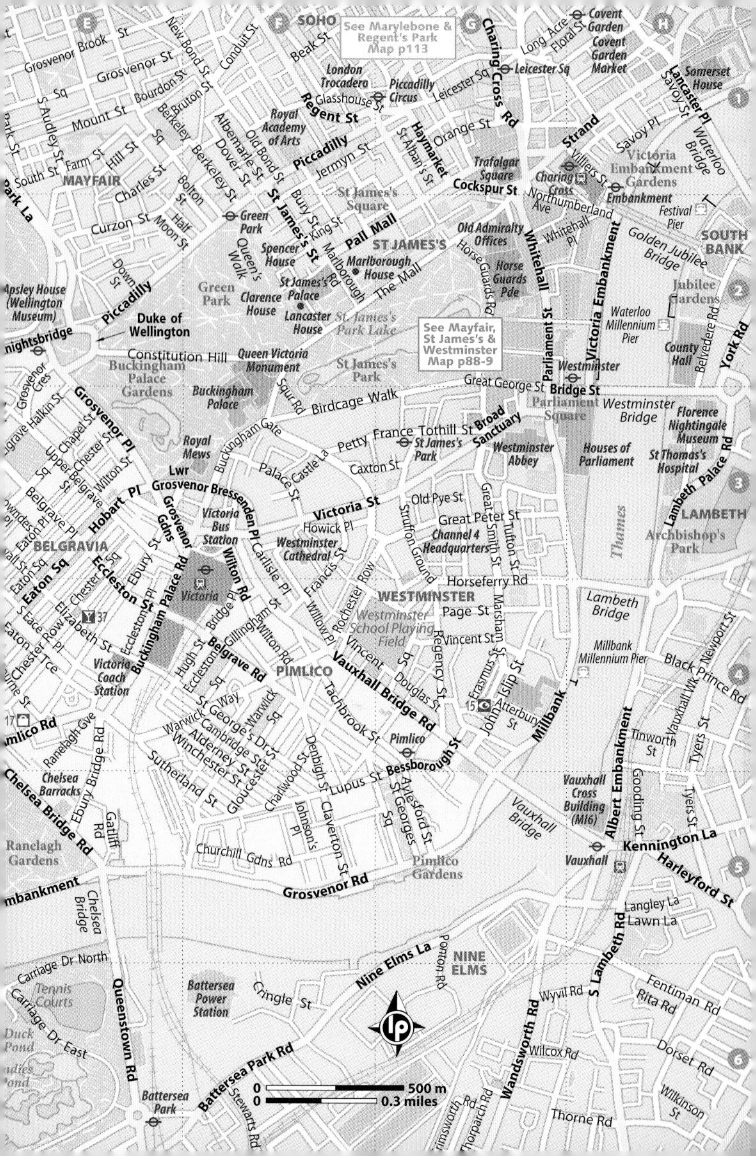

SOHO
See Marylebone & Regent's Park Map p113
Covent Garden
Covent Garden Market
Somerset House
Long Acre
Floral St
Leicester Sq
Charing Cross Rd
Grosvenor St
Brook St
New Bond St
Conduit St
Beak St
London Trocadero
Piccadilly Circus
Glasshouse St
Regent St
Royal Academy of Arts
Piccadilly
Jermyn St
Haymarket
St Alban's St
Orange St
Trafalgar Square
Strand
Savoy Pl
Savoy St
Lancaster Pl
Waterloo Bridge
Victoria Embankment Gardens
Charing Cross
Villiers St
Embankment
Cockspur St
Northumberland Ave
Festival Pier
Golden Jubilee Bridge
SOUTH BANK
Jubilee Gardens
Mount St
Bourdon St
Bruton St
Berkeley Sq
Albemarle St
Dover St
Old Bond St
S Audley St
Farm St
Hill St
South St
MAYFAIR
Charles St
Berkeley St
Bolton St
Half Moon St
Curzon St
Park La
Green Park
St James's St
Bury St
King St
St James's Square
Pall Mall
ST JAMES'S
Old Admiralty Offices
Whitehall
Whitehall Pl
Horse Guards Pde
Horse Guards Rd
Spencer House
Queen's Walk
Marlborough Rd
Marlborough House
The Mall
St James's Palace
Clarence House
Lancaster House
St James's Park Lake
Down St
Apsley House (Wellington Museum)
Duke of Wellington
Knightsbridge
Constitution Hill
Queen Victoria Monument
St James's Park
See Mayfair, St James's & Westminster Map p88-9
Parliament St
Victoria Embankment
Waterloo Millennium Pier
County Hall
Belvedere Rd
York Rd
Westminster
Bridge St
Great George St
Parliament Square
Westminster Bridge
Florence Nightingale Museum
St Thomas's Hospital
Lambeth Palace Rd
LAMBETH
Archbishop's Park
Thames
Buckingham Palace Gardens
Buckingham Palace
Birdcage Walk
Spur Rd
Buckingham Gate
Petty France
Tothill St
Broad Sanctuary
St James's Park
Westminster Abbey
Houses of Parliament
Caxton St
Castle La
Palace St
Royal Mews
Lwr Grosvenor Pl
Grosvenor Pl
Grosvenor Cres
Halkin St
Chapel St
Chester St
Upper Belgrave St
Wilton St
Belgrave Pl
Hobart Pl
Grosvenor Gdns
Bressenden Pl
Victoria Bus Station
Victoria St
Old Pye St
Great Peter St
Howick Pl
Strutton Ground
Great Smith St
Tufton St
Channel 4 Headquarters
Westminster Cathedral
BELGRAVIA
Eaton Pl
Eaton Sq
Eccleston St
Ebury St
Chester Sq
Buckingham Palace Rd
Victoria
Wilton Rd
Carlisle Pl
Francis St
Rochester Row
Horseferry Rd
WESTMINSTER
Westminster School Playing Field
Page St
Marsham St
Lambeth Bridge
Elizabeth St
Chester Row
Eaton Tce
Eccleston Pl
Bridge Pl
Gillingham St
Willow Pl
Vincent Sq
Vincent St
Regency St
Millbank Millennium Pier
Black Prince Rd
Newport St
Victoria Coach Station
Hugh St
Eccleston Sq
Belgrave Rd
PIMLICO
Vauxhall Bridge Rd
Douglas St
Erasmus St
John Islip St
Atterbury St
Millbank
Albert Embankment
Vauxhall Wk
Tyers St
Tinworth St
Pimlico Rd
Ranelagh Gve
Ebury Bridge Rd
Warwick Way
Warwick Sq
St George's Dr
Cambridge St
Alderney St
Winchester St
Gloucester St
Tachbrook St
Pimlico
Bessborough St
Sutherland St
Charlwood St
Denbigh St
Lupus St
Aylesford St
St Georges Sq
Chelsea Barracks
Chelsea Bridge Rd
Gatliff Rd
Johnson's Pl
Claverton St
Vauxhall Bridge
Vauxhall Cross Building (MI6)
Gooding St
Kennington La
Vauxhall
Harleyford St
Ranelagh Gardens
Churchill Gdns Rd
Grosvenor Rd
Pimlico Gardens
Embankment
Chelsea Bridge
Langley La
Lawn La
S Lambeth Rd
Carriage Dr North
Tennis Courts
Queenstown Rd
Battersea Power Station
Cringle St
Nine Elms La
Ponton Rd
NINE ELMS
Wyvil Rd
Wandsworth Rd
Fentiman Rd
Rita Rd
Carriage Dr East
Duck Pond
Battersea Park Rd
Wilcox Rd
Dorset Rd
Battersea Park
Stewarts Rd
Thorparch Rd
Thorne Rd
Wilkinson St
0 500 m
0 0.3 miles

A glimpse of Di's life at Kensington Palace

to rallies and open-air concerts throughout summer. Also here are the Serpentine Gallery (p126) and the pretty ring of babbling water that is the Diana Princess of Wales Memorial Fountain.

KENSINGTON PALACE

☎ 0844 482 7777; www.hrp.org.uk/kensingtonpalace; Kensington Gardens; adult/5-15yr/concession/family £12.30/6.15/10.75/34; 10am-6pm; Queensway, Notting Hill Gate or High St Kensington

Diana-philes flock to the royal palace in Kensington Gardens to gaze at the treasured belongings of the 'People's Princess'. A better reason to visit is to see the royal apartments, designed to provide royal newlyweds with intimacy and privacy, and to explore the beautifully laid-out Sunken Garden.

NATURAL HISTORY MUSEUM

☎ 7942 5000; www.nhm.ac.uk; Cromwell Rd; admission free; 10am-5.50pm; ; South Kensington

The Natural History Museum is one of the world's greats. As well as untold skeletons, fossils and stuffed animals, there are animatronic dinosaurs and a child-oriented simulation of the 1995 Kobe earthquake. Ask about special tours of the preserved 'wet' specimens in the expanding Darwin Centre. See p16 for further details.

ROYAL GEOGRAPHICAL SOCIETY

RGS; ☎ 7591 3000; www.rgs.org; 1 Kensington Gore, entrance on Exhibition Rd; admission free; 10am-5pm Mon-Fri; ; Knightsbridge

If you want to hear Sir Ranulph Feinnes tell his anecdote about removing his own frostbitten toes with a handsaw, check the list of upcoming speakers at the RGS. Every explorer worth their snowshoes speaks here sooner or later, and there are regular photo exhibitions, travel evenings and public debates, including spirited discussions from **Intelligence Squared** (www.intelligencesquared.com).

ROYAL HOSPITAL CHELSEA

☎ 7881 5303; www.chelsea-pensioners.co.uk; Royal Hospital Rd; admission free, fees apply for tours; 10am-noon & 2-4pm, closed Sun Oct-Mar; ; Sloane Sq

This stately Christopher Wren pile is famous as the setting for the **Chelsea Flower Show** (www.rhs.org.uk/chelsea) in May (see p29). It's also worth visiting at other times to tour the buildings and grounds with one of the liveried Chelsea Pensioners – war veterans who were granted leave to retire here by a law dating back to 1682. Call in advance to arrange a tour.

SAATCHI GALLERY

www.saatchi-gallery.co.uk; Duke of York's HQ, Sloane Sq; ; Sloane Sq

After losing more than 100 works of Britart in the 2004 Momart fire, ad-man and Britart aficionado Charles Saatchi has turned his attention to modern American painting in his gallery, which opened in the handsome Duke of York's buildings in 2008 – see the website for opening times and admission prices.

SCIENCE MUSEUM

☎ 0870 870 4868; www.sciencemuseum.org.uk; Exhibition Rd; admission free, separate charges for IMAX cinema & rides; 10am-6pm; ; South Kensington

The Science Museum will stimulate even the most technophobe

NIGHT AT THE MUSEUM

One a month the Science Museum (above) runs a special 'Science Night' for children and parents, with lots of mad-professor activities and a sleepover among the exhibits. The age limit for kids is eight to 11 years and tickets (book early) cost £30.

The British Museum (p104) runs its own museum sleepovers, open to youngsters who sign up as Young Friends of the Museum (annual fee £20). There are four sleepovers a year, with plenty of dressing up, games and torch-lit tours, but tickets should be booked well ahead.

Just so adults don't feel left out, the Victoria & Albert Museum (p126) runs special 'Friday Late' sessions on the last Friday of each month, with games and arty activities plus live DJs from 6.30pm to 10pm. See www.lates.org for more late-night activities in London.

of visitors. As well as science-based models with buttons to press and levers to pull, you can see such milestones of human endeavour as Stephenson's Rocket and the Command Module off the Apollo 10 rocket. See also p16.

SERPENTINE GALLERY

☎ 7402 6075; www.serpentinegallery.org; Kensington Gardens; admission free; 10am-6pm; ; Knightsbridge

The former tea pavilion at Hyde Park now serves as a gallery for cutting-edge art shows by the current darlings of the international art scene. It's worth coming here in summer to see the latest incarnation of the Serpentine summer pavilion, designed by a different architect every year (see boxed text, below).

TATE BRITAIN

☎ 7887 8000; www.tate.org.uk; Millbank; admission free, prices vary for temporary exhibitions, free tours; 10am-5.50pm, tours 11am, noon, 2pm, 3pm Mon-Fri, noon & 3pm Sat & Sun; ; Pimlico

The other, older half of the Tate, this huge Portland Stone edifice on the riverside focuses on paintings from the 16th to the 20th century, including works by Gainsborough, Turner, Hogarth, Constable and Francis Bacon. A special shuttle boat (one way adult £4) connects the Tate Britain to the Tate Modern (p79) every 40 minutes from 10.30am to 5.10pm.

VICTORIA & ALBERT MUSEUM

V&A; ☎ 7942 2000; www.vam.ac.uk; Cromwell Rd; admission free, £3 donation requested, prices vary for temporary exhibitions; 10am-5.45pm, to 10pm Fri; ; South Kensington

The V&A has the finest collection of decorative art and design ever assembled and galleries are being redeveloped and reinvented all

A PAVILION FOR ONE SEASON

Every year the Serpentine Gallery (above) invites one leading world architect to build a temporary pavilion to serve as the setting for the gallery's lavish summer party, as well as events for the general public. Past structures have included a flying saucer with a hat by Olafur Eliasson and Kjetil Thorsen, and a modernist concrete tent by Brasilia-creator Oscar Niemeyer. The 2008 pavilion – an angular wood and glass structure inspired by Leonardo Da Vinci's drawings of siege engines – was designed by Frank Gehry, of Guggenheim Bilbao fame.

ART ON THE RUN

Long seen as the classical (even stuffy) half of the Tate, Tate Britain (opposite) struck back in 2008 with a surprising piece of conceptual art – *Work No 850* by Martin Creed, who controversially won the Turner Prize for exhibiting a light bulb turning on and off in 2001. The installation was based on a team of runners sprinting the length of the gallery every 30 seconds. Art critics gushed indulgently about 'having to reassess their interaction with the space', while visitors just looked on in bemusement. With this new benchmark in silliness, many are wondering if conceptual art has anywhere left to go. If it has, the Tate Modern (p79) or Tate Britain will almost certainly put it on display.

the time. Visitors are particularly drawn to the fashion displays and the Islamic and Asian galleries with their carpets, ceramics, and ornate arms and armour. For more details, see p16.

SHOP

DAYLESFORD ORGANIC *Deli*

☎ 7881 8060; www.daylesfordorganic.com; 44B Pimlico Rd; 🕔 8am-8pm Mon-Sat, 10am-4pm Sun; ⊖ Sloane Sq

A Carluccio's (see boxed text, p53) for the Chelsea and Pimlico set, with a deli upstairs and a modernist downstairs café serving delicious light lunches.

HARRODS

Department Store

☎ 7730 1234; www.harrods.com; 87-135 Brompton Rd; 🕔 10am-9pm Mon-Sat, 11.30am-6pm Sun; ⊖ Knightsbridge

The ultimate department store, Harrods is famous for fashions, fragrances and its extravagant food court. Plenty of people will buy anything just to get hold of a Harrods bag. It's owned by Mohammed Al-Fayad, who waged a long but unsuccessful campaign to implicate the Queen and Prince Philip in the deaths of his son, Dodi, and Princess Diana in 1997.

HARVEY NICHOLS

Department Store

☎ 7235 5000; www.harveynichols.com; 109-125 Knightsbridge; 🕔 10am-8pm Mon-Sat, noon-6pm Sun; ⊖ Knightsbridge

Less ostentatious than Harrods, but arguably better stocked, Harvey Nichols is packed with the trappings of the high life – sleek brand-name fashions, expensive fragrances, and jewellery, shoes and bags to die for.

PETER JONES
Department Store

☎ 7730 3434; www.peterjones.co.uk; Sloane Sq; ⌚ 9.30am-7pm Mon-Sat, till 8pm Wed, 11am-5pm Sun; ⊖ Sloane Sq
This iconic 1960s Chelsea department store has been stylishly modernised – come for fashions, cosmetics and electricals in a funky, futuristic space.

SHOP AT BLUEBIRD *Clothing*

☎ 7351 3873; 350 King's Rd; ⌚ 10am-7pm Mon-Sat, noon-6pm Sun; ⊖ Sloane Sq, then bus 11, 19 or 22
This groovy boutique harks back to the way the King's Rd used to be in the era of Vivienne Westwood and the Sex Pistols. One-off pieces by chic designers are casually displayed around a room full of art books, DJ decks and papier-mâché sculptures.

EAT

BIBENDUM *French* £££

☎ 7581 5817; www.bibendum.co.uk; 81 Fulham Rd; ⌚ noon-2.30pm & 7-11pm, lunch from 12.30pm Sat & Sun; ⊖ South Kensington
Stained-glass Michelin Men smile down at you in this surreal French restaurant housed in the former British headquarters of the Michelin tyre company. The food is stunning, though – at these prices we would be outraged by anything less.

Shell out for a treat at Bibendum Oyster Bar

BROMPTON QUARTER CAFÉ
Mediterranean £-££

☎ 7225 2107; www.bromptonquarter cafe.com; 225 Brompton Rd; ⌚ 7.30am-11pm; ⊖ South Kensington or Knightsbridge
In case you can't afford the haute cuisine at Gordon Ramsay's, this bright modern café has some good inexpensive snack plates to share, including lots of salads

prepared with good, fresh deli ingredients.

DAQUISE *Polish* £-££

☎ 7589 6117; 20 Thurloe St; ⊖ South Kensington

This low-key Polish café has been around since the 1940s. Eastern-European expats come here for the tastes of home – *pierogi* (dumplings), *barszcz* (*bortsch*), *blinis* (pancakes) and Polish beef stroganoff.

GORDON RAMSAY *Modern European* £££

☎ 7352 4441; www.gordonramsay.com; 68-69 Royal Hospital Rd; set-price lunch/3 courses/7 courses £45/90/120; noon-2.30pm & 6.30-11pm Mon-Fri; ⊖ Sloane Sq

A gastronomic temple – the only restaurant with three Michelin stars. Gordon Ramsay's kitchen is still producing mouth-watering creations like lobster, langoustine and salmon ravioli, and sautéed fois gras for those lucky enough to secure a reservation. Bookings are taken exactly two months ahead to the day, and every slot is booked by mid-morning.

HUMMINGBIRD BAKERY *Café* £

☎ 7584 0055; www.hummingbirdbakery.com; 47 Old Brompton Rd; 10.30am-7pm; ⊖ South Kensington

Dainty cupcakes in a rainbow of pastel colours attract a largely female crowd to this pocket-sized cake shop. There's a branch in Notting Hill (p139).

ITSU *Japanese* £-££

☎ 7590 2400; www.itsu.co.uk; 118 Draycott Ave; noon-11pm, to 10pm Sun; ⊖ South Kensington

Board the sushi train at this fun Japanese restaurant close to South Ken tube. The colour-coded sushi plates cover all the bases (prices are calculated from the empty plates at the end of the meal), and there's a sleek upstairs bar where you can wait for a seat to come free.

OTTOLENGHI *Deli-café* £-££

☎ 7823 2707; www.ottolenghi.co.uk; 13 Motcomb St; 8am-8pm Mon-Fri, 8am-7pm Sat, 9am-6pm Sun; ⊖ Knightsbridge

The Belgravia branch of this small, classy chain offers similar deli treats to the flagship branch in Islington (p162).

PÉTRUS *Modern European* £££

☎ 7592 1609; www.gordonramsay.com; Berkeley Hotel, Wilton Pl; noon-2.30pm Mon-Fri, 6-11pm Mon-Sat; ⊖ Knightsbridge

Marcus Wareing came to Pétrus as a Gordon Ramsay protégé but the

MORE EXCLUSIVE EATS

If the cuisine at Ramsay or Pétrus amuses your bouche (and you have bottomless pockets), you might like to try the food at the following high-class establishments:

Boxwood Café (☎ 7235 1010; www.gordonramsay.com; Berkeley Hotel, Wilton Pl; ⊖ Knightsbridge) Renowned for its veal and foie gras burgers, Boxwood Café is another outing for the unstoppable Gordon Ramsay.

Painted Heron (☎ 7351 5232; www.thepaintedheron.com; 112 Cheyne Walk; ⊖ Sloane Sq) Starched white minimalism provides a backdrop for extravagant East-meets-West cuisine.

Tom Aikens (☎ 7584 2003; www.tomaikens.co.uk; 43 Elystan St; ⊖ South Kensington) The other Tom Aikens restaurant (see below), serving more exquisite Modern European taste sensations.

student has become the master. None of the critics seem to have a bad word to say about Wareing's stylish modern British food.

TOM'S KITCHEN
Modern European £££-£££

☎ 7349 0202; www.tomskitchen.co.uk; 27 Cale St; 🕐 7am-midnight Mon-Fri, 10am-midnight Sat & Sun; ⊖ South Kensington

Another Michelin-starred chef with a short temper, Tom Aikens has a reputation for using high-quality ingredients in simple but inspired combinations. This is the cheaper of the Aikens' restaurants (see boxed text, above), though the menu still includes a £56 steak.

VAMA *Indian* ££-£££

☎ 7565 8500; www.vama.co.uk; 438 King's Rd; 🕐 noon-3pm & 6.30-11.30pm; ⊖ Sloane Sq

Famous patrons of Vama have included Wesley Snipes and King Abdullah of Jordan – so it's touristy, but targeted at a special class of tourist. Don't let that put you off. Ordinary mortals come here, too, to sample some of the most imaginative North Indian cooking in town.

ZUMA *Japanese* £££

☎ 7584 1010; www.zumarestaurant.com; 5 Raphael St; 🕐 noon-2.15pm Mon-Fri, 12.30-3.15pm Sat & Sun, 6-11pm daily; ⊖ Knightsbridge

Zuma is about 20% less trendy than it used to be, which means that you can actually get a table these days. Chefs still produce the same high-quality Japanese food, including freshly rolled sushi and steaks cooked on the charcoal *robata* grill.

DRINK

As well as the following pubs, there are several popular gay bars and cafés on Old Brompton Rd.

BOTANIST *Bar*

☎ 7730 0077; www.thebotaniston sloanesquare.com; 7 Sloane Sq; ⊖ Sloane Sq

Named for the botanist Sir Hans Sloane, whose private collection formed the basis for the British Museum (p104), this sleek bar and restaurant exudes a chic, sexy, early 1950s elegance.

NAG'S HEAD *Pub*

☎ 7235 1135; 53 Kinnerton St; ⊖ Knightsbridge

A ban on mobile phones is one of several nice touches at this quaint local, tucked away on a quiet lane just east of Knightsbridge tube. Eccentric publican Kevin Moran has created a real escape from the Harrods-bound hordes.

THOMAS CUBITT *Gastropub*

☎ 7730 6060; www.thethomascubitt .co.uk; 44 Elizabeth St; ⊖ Sloane Sq or Victoria

This is how the Belgravia set like their gastropubs – classy Victorian décor, ales and premium imports behind the bar, and a brasserie serving good-quality modern European fare.

TROUBADOUR *Café*

☎ 7370 1434; www.troubadour.co.uk; 265 Old Brompton Rd; 9am-midnight; ⊖ Earl's Court

A proper café in the 1950s mould, Troubadour is genuinely eccentric – part coffee shop, part pub, with all sorts of salvaged bric-a-brac on the walls and a regular programme of gigs by unsigned bands and local musicians. A treat.

PLAY

CHELSEA CINEMA *Cinema*

☎ 0871 703 3990; www.curzoncinemas .com; 206 King's Rd; ⊖ Sloane Sq

The West London branch of this small, independent chain shows the same program of arty and thought-provoking films as the Curzons (see p55 and p101) in the West End.

CHELSEA FOOTBALL CLUB *Sport*

☎ information 0871 984 1955, tickets 7915 2900; www.chelseafc.com; Fulham Rd; tours adult/child £15/9; tours hourly 11am-3pm; ⊖ Fulham Broadway

Chelsea doesn't seem to be bringing quite so much silverware to

Stamford Bridge since the departure of José Mourinho, but Russian owner Roman Abramovich can still afford to load the team with quality. See www.chelseafctours.com to book a tour.

ROYAL ALBERT HALL

Live Music

☎ 7589 8212; www.royalalberthall.com; Kensington Gore; ⊖ South Kensington

The Royal Albert Hall is probably the most famous music venue in Britain. Jimi Hendrix, the Beatles and Abba all played here and it's still an important forum for music, circus and opera. The last night of the **BBC Proms** (www.bbc.co.uk/proms) has taken place here almost every year since 1942 – see p30.

The iconic shell of Battersea Power Station

THE SAGA OF BATTERSEA POWER STATION

People have been wondering what to do with Battersea Power Station (F6) ever since it stopped generating power in 1983. Propelled to fame by the cover of the Pink Floyd album *Animals* – with a giant floating pig tethered to one of the four smokestacks – this has always been the building that was too iconic to lose and too impractical to use. One flamboyant plan involved filling the building with looping rollercoasters, but this was shelved in 1996. Now something finally seems to be happening. Work is underway to stabilise the smokestacks and fill the shell with flats, shops, restaurants and entertainment facilities. In the meantime the site is being rented out to movie studios – the opening sequences for the Batman movie *Dark Knight* were filmed here in 2007.

ROYAL COURT THEATRE

Theatre

☎ 7565 5000; www.royalcourttheatre.com; Sloane Sq; ⊖ Sloane Sq

Chelsea's Royal Court Theatre is credited with launching modern British drama with the first performance of *Look Back in Anger* in 1956. It still stages groundbreaking work by new playwrights.

SHEPHERD'S BUSH EMPIRE

Live Music

☎ 8354 3300; www.shepherds-bush-empire.co.uk; Shepherd's Bush Green; ⊖ Shepherd's Bush or Goldhawk Rd

An iconic London gig venue, attracting big-name international artists. It's outside the main Kensington–Knightsbridge–Chelsea triangle, but worth the trip for its diverse line-up.

>NOTTING HILL & BAYSWATER

There's more to Notting Hill than that Hugh Grant movie and Portobello Rd. This fascinating neighbourhood played a vital role in the development of modern cosmopolitan London. Established by the Georgians to accommodate the St James's overspill in the 1840s, the area had become one of London's biggest slums by the time the first Caribbean immigrants arrived in the 1950s.

Over the next few decades, Notting Hill faced race riots – powerfully described in the book *Absolute Beginners* – then reconciliation as the Notting Hill Carnival bought the Afro-Caribbean community into the mainstream. The 1970s and '80s saw an influx of arty white liberals and trustafarians (bohemians with secret trust funds), who opened new boutiques, antique shops and trendy cafés.

Since then Notting Hill has become more prosperous, but also slightly blander as bankers and lawyers have displaced the old guard from their increasingly valuable houses. Nevertheless, it's a cracking place to shop and eat and there's an appealingly nonconformist feel to the area, particularly during the Saturday market on Portobello Rd.

Just east of Notting Hill, Bayswater is more mainstream but less snooty, with a string of good Asian restaurants along its main thoroughfare, Queensway.

NOTTING HILL & BAYSWATER

SHOP

Golborne Rd Market.......1 A2
Portobello Rd Market.....2 B4
Rellik..............................3 A2
Retro Clothing................4 C5
Retro Clothing................5 C5
Rough Trade...................6 B4
Travel Bookshop.............7 A4

EAT

202.................................8 B4
Bumpkin.........................9 B3
Cow...............................10 C3
E&O...............................11 A4
Four Seasons.................12 D5
Grocer on Elgin.............13 A4
Hummingbird Bakery.......................14 B4
Kiasu.............................15 D5
Ledbury.........................16 B4

DRINK

Churchill Arms..............17 C6
Crazy Homies................18 C3
Earl of Lonsdale............19 B4
Lisboa Patisserie...........20 A2
Lonsdale........................21 B4
Montgomery Place.......22 A4
Trailer Happiness..........23 B4

PLAY

Elbow Room..................24 C4
Electric Cinema.............25 B4
Notting Hill Arts Club....26 C5

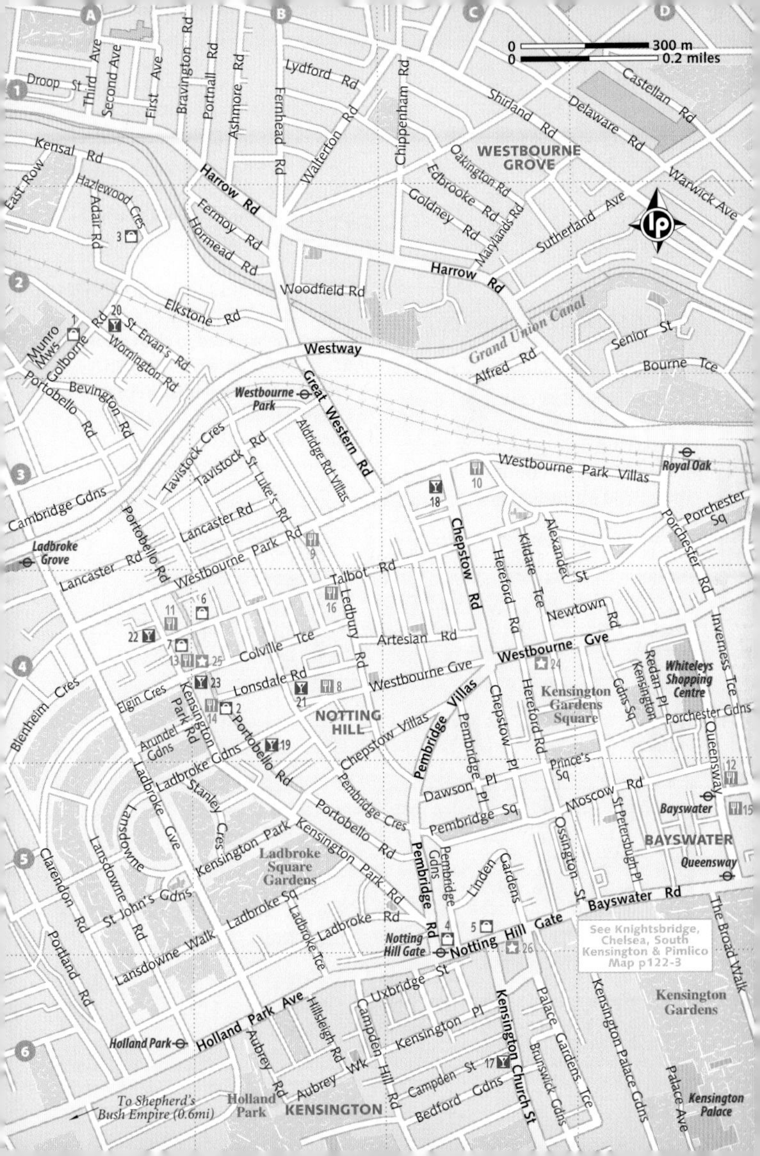

A
B
C
D
1
2
3
4
5
6
0 300 m
0 0.2 miles
WESTBOURNE GROVE
NOTTING HILL
BAYSWATER
KENSINGTON
Kensington Gardens Square
Ladbroke Square Gardens
Kensington Gardens
Holland Park
Kensington Palace
Whiteleys Shopping Centre
Grand Union Canal
Westbourne Park
Royal Oak
Ladbroke Grove
Bayswater
Queensway
Notting Hill Gate
Holland Park
To Shepherd's Bush Empire (0.6mi)
See Knightsbridge, Chelsea, South Kensington & Pimlico Map p122-3
Droop St
Third Ave
Second Ave
First Ave
Bravington Rd
Portnall Rd
Ashmore Rd
Lydford Rd
Fernhead Rd
Walterton Rd
Chippenham Rd
Shirland Rd
Delaware Rd
Castellan Rd
Warwick Ave
Kensal Rd
East Row
Hazlewood Cres
Adair Rd
Harrow Rd
Fermoy Rd
Hormead Rd
Oakington Rd
Edbrooke Rd
Goldney Rd
Marylands Rd
Sutherland Ave
Woodfield Rd
Elkstone Rd
Munro Mws
Golborne Rd
St Ervan's Rd
Wornington Rd
Portobello Rd
Bevington Rd
Westway
Senior St
Bourne Tce
Alfred Rd
Great Western Rd
Aldridge Rd Villas
Tavistock Cres
Tavistock Rd
St Luke's Rd
Cambridge Gdns
Lancaster Rd
Westbourne Park Rd
Westbourne Park Villas
Porchester Sq
Porchester Rd
Chepstow Rd
Hereford Rd
Kildare Tce
Alexander St
Talbot Rd
Ledbury Rd
Newtown Rd
Artesian Rd
Colville Tce
Westbourne Gve
Inverness Tce
Redan Pl
Kensington Gdns Sq
Porchester Gdns
Queensway
Lonsdale Rd
Elgin Cres
Kensington Park Rd
Arundel Gdns
Blenheim Cres
Chepstow Villas
Pembridge Villas
Pembridge Pl
Chepstow Pl
Prince's Sq
Dawson Pl
Moscow Rd
St Petersburgh Pl
Ossington St
Ladbroke Gdns
Stanley Cres
Ladbroke Gve
Lansdowne Cres
Lansdowne Rd
Pembridge Cres
Pembridge Sq
Pembridge Rd
Pembridge Gdns
Linden Gardens
Clarendon Rd
Kensington Park Gdns
St John's Gdns
Ladbroke Sq
Ladbroke Tce
Ladbroke Rd
Bayswater Rd
Notting Hill Gate
The Broad Walk
Portland Rd
Lansdowne Walk
Uxbridge St
Holland Park Ave
Hillsleigh Rd
Campden Hill Rd
Kensington Pl
Kensington Church St
Palace Gardens Tce
Kensington Palace Gdns
Aubrey Rd
Aubrey Wk
Campden St
Bedford Gdns
Brunswick Gdns
Palace Ave
1
2
3
4
5
6
7
8
9
10
11
12
13
14
15
16
17
18
19
20
21
22
23
24
25
26

SHOP

There are more good stores, including a funky open-plan grocery, in the Whiteleys shopping centre on Queensway (D4).

GOLBORNE RD MARKET

Market

Portobello Rd; 8am-6.30pm Fri & Sat; ⊖ Notting Hill Gate or Ladbroke Grove

During the week Golborne Rd is a busy shopping street for the local Moroccan community, but dealers in antiques, bric-a-brac and junk take over the entire street on Friday and Saturday – perfect if you need an empty picture frame or a mounted set of deer antlers.

PORTOBELLO RD MARKET

Market

www.portobelloroad.co.uk; Portobello Rd; 8am-6.30pm Mon-Wed, Fri & Sat, 9am-1pm Thu; ⊖ Notting Hill Gate or Ladbroke Grove

Sprawling most of the way from Notting Hill Gate to Westbourne Grove, this famous market

Browsing the lines at Portobello Rd Market

WEST LONDON RETRO

Although Shoreditch is the undisputed capital of retro fashion, there's a growing scene around Notting Hill. The big advantage of shopping at this end of town is that the clothes that end up in retro stores bear the names of people like Vivienne Westwood, Mary Quant and Zandra Rhodes. Try the following emporiums:

Rellik (☎ 8962 0089; www.relliklondon.co.uk; 8 Golborne Rd; 10am-6pm Tue-Sat; ⊖ Westbourne Park) Facing the towering concrete Trellick Tower, with neatly displayed racks of 1960's, '70's and '80's glad rags.

Retro Clothing (☎ 0845 644 1442; 16, 20, 28, 32-4 Pembridge Rd & 56 Notting Hill Gate; 10am-8pm; ⊖ Notting Hill Gate) An Aladdin's cave of retro, spread over five stores – this is how second-hand clothes were meant to be sold!

changes character several times as it winds through Notting Hill. The south end is given over to antique stalls and shops, but north of Westbourne Grove, the emphasis shifts to food, homewares, ornaments, and designer and vintage clothing. There's some good stuff on sale, but don't expect to find many bargains. Saturday is the main market day.

ROUGH TRADE *Music*

☎ 7229 8541; www.roughtrade.com; 130 Talbot Rd; 10am-6.30pm Mon-Sat, noon-5pm Sun; ⊖ Westbourne Park or Notting Hill Gate

The Notting Hill branch of this seminal record store is stacked with rare editions and obscure releases – it's indie paradise! There's also a branch in Shoreditch (p148).

TRAVEL BOOKSHOP *Books*

☎ 7229 5260; www.thetravelbookshop.co.uk; 13-15 Blenheim Cres; 10am-6pm Mon-Sat, noon-5pm Sun; ⊖ Ladbroke Grove

Hugh Grant doesn't really work here, though this neighbourhood bookshop was the inspiration for the bookshop in the soppy film *Notting Hill*. Ever since the movie came out, the owners have been trying to steer attention back to their strong selection of travel books.

EAT

202 *Bistro* £££

☎ 7727 2722; 202 Westbourne Grove; 10am-6pm Mon, 8.30am-6pm Tue-Sat, 10am-5pm Sun; ⊖ Westbourne Park or Ladbroke Grove

It's hard to tell where the shop displays end and the dining tables begin at 202. Chic but understated, this boutique-come-bistro is a firm favourite of the ladies who lunch, with intriguing deli salads and Asian-tinged main courses.

BUMPKIN

Gastropub £-££

7243 9818; www.bumpkinuk.com; 209 Westbourne Park Rd; noon-3pm & 6-11pm, lunch to 3.30pm Sat, to 4pm Sun; Westbourne Park or Royal Oak

One of a string of funky gastro-pubs along Westbourne Park Rd, Bumpkin styles itself as a 'country brasserie', which translates to rotisserie chicken, grilled salmon, out-sized sirloin steaks and roasts.

COW

Gastropub ££-£££

7221 0021; www.thecowlondon.co.uk; 89 Westbourne Park Rd; 7-11pm Mon-Fri, noon-3pm & 7-11pm Sat & Sun; Westbourne Park or Royal Oak

Tom Conran, son of design pioneer Sir Terence, has created one of London's finest eateries in this laid-back gastropub. The bar menu runs to winkles, whelks and rock oysters, while the smarter upstairs restaurant serves meaty English fare, done with panache.

E&O

Asian £££

7229 5454; www.rickerrestaurants.com/eando; 14 Blenheim Cres; 12.15-3pm & 6.15-11pm, lunch to 4pm Sat & Sun, Notting Hill Gate or Ladbroke Grove

E&O stands for Eastern & Oriental and this is exactly what you get on the menu, an eclectic mix of flavours and cooking styles from all over Asia. You can dine in the ultra-minimalist dining room or outside on the pavement, unless you foolishly forgot to make a reservation.

FOUR SEASONS

Chinese £-££

7229 4320; 84 Queensway; noon-11pm; Queensway

The huge pile of roast ducks in the window and the crowds of Chinese diners inside should alert you to the fact that something good is cooking at Four Seasons. The menu is straight-down-the-line Cantonese and there's often a queue for a table, even at lunchtimes.

GROCER ON ELGIN

Deli-café £

7437 7776; www.thegroceron.com; 6 Elgin Cres; 8.30am-5pm, shop to 8pm Mon-Fri & 6pm Sat & Sun; Ladbroke Grove

We defy you to enter this swish deli and café full of delicious Mediterranean smells and not buy something. As well as gourmet groceries, you can lunch on salads, freshly baked pizza and filled ciabatta rolls.

HUMMINGBIRD BAKERY

Café £

7229 6446; www.hummingbirdbakery.com; 133 Portobello Rd; 10.30am-5.30pm Tue-Sat, 11am-5pm Sun; Westbourne Park

The window displays at this American bakery look good enough to eat, and the fondant-covered cupcakes are positively dreamy; there's a branch in Knightsbridge (p129).

KIASU

Malaysian £

7727 8810; 48 Queensway; noon-11pm; Queensway

This informal Malacca Straits canteen is wowing locals with its home-style Peranakan cooking and budget prices. The laksa is as good as anything you'll find on the Malay peninsula.

LEDBURY *French* £££

7792 9090; www.theledbury.com; 127 Ledbury Rd; noon-2.30pm & 6.30-10.30pm; Westbourne Park or Notting Hill Gate

Michelin starred and swooningly elegant, Brett Graham's artful French restaurant attracts a well-heeled local clientele who wear jeans with designer jackets. To beat the credit crunch, go for the set-price menu.

DRINK

CRAZY HOMIES *Bar*

7727 6771; www.crazyhomieslondon.co.uk; 127 Westbourne Park Rd; 6-11pm Mon-Fri, noon-11pm Sat, noon-10.30pm Sun; Westbourne Park

Mexican Day of the Dead memorabilia and potted palms set the scene at this memorable bar and eatery. Everyone loves the food and the downstairs dive bar hosts such unlikely turntable stars as **DJ Wheelie Bag** (www.djwheeliebag.co.uk), with his shopping trolley sound system.

LISBOA PATISSERIE *Café*

8968 5242; 57 Golborne Rd; 8am-7.30pm; Westbourne Park or Ladbroke Grove

On market day there isn't room to flick a crumb in this phenomenally popular Portuguese patisserie on Golborne Rd. Regulars swear by the *pasteis de nata*

(Portuguese custard tarts), which are also available to takeaway.

LONSDALE *Bar*

☎ 7727 4080; www.thelonsdale.co.uk; 48 Lonsdale Rd; 6pm-midnight Mon-Thu, 6pm-1am Fri & Sat, 6-11.30pm Sun; ⊖ Notting Hill Gate or Westbourne Park

The Lonsdale looks like a groovy 1970's vision of what bars might look like in the future. The funky décor is backed up by quirky events and great cocktails made the old-fashioned way. Magic Wednesdays feature a magician performing parlour tricks at the tables.

MONTGOMERY PLACE *Bar*

☎ 7792 3921; www.montgomeryplace.co.uk; 31 Kensington Park Rd; 5pm-midnight Mon-Thu, 5pm-1am Fri, 2pm-1am Sat; ⊖ Ladbroke Grove

A sleek tribute to the glory days of the lounge bar, Montgomery Place combines savvy mixology with Rat Pack cool. The 1950's vibe extends to the cocktail list and the playlist.

TRAILER HAPPINESS *Bar*

☎ 7727 2700; www.trailerh.com; 177 Portobello Rd; 5-11pm Tue-Fri, 6-11pm Sat & Sun; ⊖ Notting Hill Gate

Trailer Happiness is styled after a 'mid-1960's California bachelor pad', which translates as dangling bead curtains, leather lounges, vintage Waikiki tat and that famous picture of the green Korean lady. It's good, tongue-in-cheek fun and the cocktail list is exemplary.

NOTTING HILL BOOZERS

If you feel the Portobello crowds starting to close in, duck into the quiet confines of one of the following public houses for a therapeutic pint of ale:

Churchill Arms (☎ 7727 4242; 119 Kensington Church St; ⊖ Notting Hill Gate) Covered in window boxes and hanging baskets, the cheerful Churchill was the 2007 winner of the Chelsea Flower Show 'Boozers in Bloom' competition.

Earl of Lonsdale (☎ 7727 6335; 277-81 Portobello Rd; ⊖ Notting Hill Gate or Westbourne Park) Part of the cost-effective Sam Smiths' chain, this convivial watering hole was restored to its original Victorian splendour in 2005.

Lucy Besant

Day manager at Lonsdale bar, Notting Hill

Best house cocktail The Lonsdale (opposite) specialises in the cocktails invented in London from the 1800s right up to the modern day. I love the Bloodhound – fresh raspberries, vermouth, gin and maraschino. **Best night out in Notting Hill** Here on Wednesday or Sunday nights – Wednesdays we have live table magic and there are free cocktails early on at our Soulful Sunday sessions. **Best tips for Notting Hill bar crawl** Well start at Lonsdale, obviously, then move on to Montgomery Place (opposite) and Trailer Happiness (opposite). It's a nice little circuit and you can continue to Crazy Homies (p139) if you like tequila. **Best eats before you party** E&O (p138) on Portobello Rd is good, and there are gorgeous little cakes at Hummingbird Bakery (p139). **Best Notting Hill shopping** Personally I like the retro shops around Notting Hill gate (see boxed text, p137).

PLAY

ELBOW ROOM *Pool Hall*

☎ 7221 5211; www.theelbowroom.co.uk; 103 Westbourne Grove; noon-11pm Mon-Sat, 1-10.30pm Sun; Bayswater

Pool sharks will love this slick modern pool and beer hall at the Notting Hill end of Ladbroke Grove. There's a branch in Shoreditch (p156).

ELECTRIC CINEMA *Cinema*

☎ 7908 9696; www.electriccinema.co.uk; 191 Portobello Rd; Ladbroke Grove or Notting Hill Gate

Built in the Edwardian era, this extravagant picture house has been kitted out with soft leather armchairs, two-seater sofas and footstools. There are tables for food and drink in the auditorium, and an upmarket brasserie for a pre-show dinner.

Caribbean culture and colours at the Notting Hill Carnival

CARNIVAL TIME

Notting Hill is busy at the best of times, but during the annual **Notting Hill Carnival** (p30) in August the crowds go into overdrive. Held yearly since 1965, this spirited celebration of Caribbean culture is the second-largest street festival in the world, attracting as many as two million visitors. The highlight of the carnival weekend is the Sunday costume parade, featuring some truly carnivalicious outfits and hordes of pretty young things in sequinned bikinis. Traditionally, the parade was accompanied by calypso and steel drum bands, but today the soundtrack is mainly provided by booming truck-mounted sound systems.

Despite its popularity, the carnival is not without its problems. The parade has been marred by gang violence on numerous occasions, and the cost of policing the event is estimated at £4 million each year. In 2000, former mayor Ken Livingstone launched a rival Caribbean festival over the same weekend at Hyde Park to reduce the crowds at Notting Hill, but the event was abandoned after two years amid allegations of financial impropriety. The debate rumbles on – see www.nottinghillcarnival.biz for the latest carnival news.

NOTTING HILL ARTS CLUB

Live Music

NHAC; ☎ 7460 4459; www.nottinghillartsclub.com; 21 Notting Hill Gate; Mon-Sat, event times vary; ⊖ Notting Hill Gate

Reached through an easy-to-miss doorway on Notting Hill Gate, the NHAC is more an underground club than an art centre, with a weekly line-up of DJs, bands and goofy 'craft nights' (see www.myspace.com/craftnight for details). Rough Trade Records provides a forum for up-and-coming bands from 4pm to 8pm on Saturday.

>HOXTON, SHOREDITCH & SPITALFIELDS

Twenty years ago Hoxton and Shoreditch were crumbling into ruin. Today these districts are the cutting edge of London cool. We should probably thank the Young British Artists – Tracey Emin, Damien Hirst et al – who put the area on the map by exhibiting at the White Cube Gallery. Since then the area has developed into a thriving entertainment district, with dozens of bars, restaurants and clubs. However, it's definitely a 'scene' and ironic haircuts á la Flock of Seagulls are all the rage. To get a handle on Hoxton, look for the free arts magazine *Vice*.

Until recently the action was centred on Spitalfields Market, but a corporate makeover has pushed the arty boutiques, cafés and bars east to the alleyways around Brick Lane. The south end of Brick Lane is something else again – the heart of London's Bangla-town and a popular spot for a curry.

HOXTON, SHOREDITCH & SPITALFIELDS

SEE

Brick Lane.......1 C6
Christ Church Spitalfields...2 B5
Dennis Severs' House.......3 B5
Geffrye Museum.......4 B2
White Cube Gallery.......5 A4
Whitechapel Art Gallery...6 C6

SHOP

A Gold.......7 B5
Absolute Vintage.......8 B5
Beyond Retro.......9 C4
Blondie.......10 B5
Brick Lane Market.......11 C4
Broadway Market.......12 D1
Columbia Road Flower Market.......13 B3
Hoxton Boutique.......14 A3
Laden Showrooms.......15 C5
Rough Trade.......(see 37)
Shop.......16 C4
Spitalfields Market.......17 B5
Start.......18 A4
Start.......19 A4
Start.......20 A4
Sunday Up Market.......21 C5
Tatty Devine.......22 C4
Westland London.......23 A4

EAT

Brick Lane Beigel Bake...24 C4
Buen Ayre.......25 D1
Canteen.......26 B5
Eyre Brothers.......27 A4
Giraffe.......(see 26)
Green & Red Cantina.......28 C4
Les Trois Garçons.......29 B4
Loong Kee.......30 B3
Mirch Masala.......31 D6
New Tayyab.......32 D6
Princess.......33 A4
Rochelle School Canteen.......34 B4
St John.......35 B5
Sông Qûe Café.......(see 30)
Viet Hoa.......36 B3

DRINK

Big Chill Bar.......37 C5
Commercial Tavern.......38 B5
Dove.......39 D1
Drunken Monkey.......40 B4
Favela Chic.......41 A4
Foundry.......42 A4
George & Dragon.......43 B3
Golden Heart.......44 B5
Hawksmoor.......45 B5
Loungelover.......46 B4
Ten Bells.......47 B5
Vibe Bar.......48 C5

PLAY

333.......49 A4
93 Feet East.......50 C5
Bethnal Green Working Men's Club.......51 D3
Cargo.......52 B4
Elbow Room.......53 A4
Old Blue Last.......54 B4
Rich Mix.......55 B4

DE BEAUVOIR TOWN
HOXTON
SHOREDITCH
SPITALFIELDS
WHITECHAPEL
CITY
Shoreditch Park
Regent's Canal
To Mangal II (500m)
To Bistrotheque (500m)
To V&A Museum of Childhood (400m)
To Fifteen (200m)
0 500 m
0 0.3 miles
Old Truman Brewery
Exchange Square
Liverpool St
Finsbury Circus
Petticoat La Market
Aldgate East
Whitechapel
Bethnal Green
Cemetery
See The City, Holborn & Clerkenwell Map p60-1
Kingsland Rd
Hackney Rd
Bethnal Green Rd
Old Bethnal Green Rd
Queensbridge Rd
Shoreditch High St
Commercial St
Whitechapel Rd
Whitechapel High St
Commercial Rd
Great Eastern St
Old St
Curtain Rd
Bishopsgate
Brick La

SEE

BRICK LANE

Aldgate East

Brick Lane has seen wave after wave of immigration over the centuries, from 17th-century Huguenots to the 19th-century Jewish diaspora. Today Brick Lane is London's Bangla-town, with dozens of sari shops and spice wholesalers and a famous strip of Bengali curry houses (see boxed text, p152). The Sunday markets around Brick Lane provide one of London's great days out (see boxed text, p150).

CHRIST CHURCH SPITALFIELDS

7859 3035; www.christchurchspitalfields.org; Commercial St; 11am-4pm Tue, 1-4pm Sun, 10am-4pm other weekdays if not in use; Liverpool St

Nicholas Hawksmoor, the architect behind London's most perfectly proportioned church, was reputedly a Dionysian. Pagan symbols are allegedly hidden in the architecture of this church; see what you think.

DENNIS SEVERS' HOUSE

7247 4013; www.dennisseverhouse.co.uk; 18 Folgate St; admission Sun/Mon £8/5, Mon evening £12; noon-4pm 1st & 3rd Sun of month, noon-2pm Mon following 1st & 3rd Sun of month, every Mon evening (times vary); Liverpool St

Inside this renovated Huguenot house, rooms are lit by flickering candles, meals are left unfinished on the table and clothes hang abandoned on the backs of chairs, creating an eerie sense that the 18th-century owners have just stepped out of the room. There's nowhere quite like it.

THE ORIGINAL GANGSTERS

If the movies are to be believed, the gangsters who ruled London's East End in the 1960s were loveable rogues. In truth, the real-life East End gangsters used violence and intimidation to control organised crime in the city.

During the '60s south London was controlled by Charlie Richardson, who became notorious for using torture to extort money from pub and club owners. In the East End, the Kray twins, Ronnie and Reggie, reigned supreme. The Krays' business interests covered everything from legitimate pubs and music halls to armed robbery and arson, but their empire fell when the pair were sentenced to life for the murder of Jack 'The Hat' McVitie in 1968.

Despite this, the East End hardman has become a movie cliché, spawning a whole genre of violent gangster films. Most portrayals of East End gangsters are hackneyed caricatures, but a few performers have pulled off the role with real aplomb – check out Sir Ben Kingsley's foul-mouthed character in *Sexy Beast*.

GUERRILLA ART

Look at the graffiti-covered walls of Shoreditch and you may notice some unusually artistic daubings among the tags and cuss words. Part vandal, part social commentator, the graffiti artist known as Banksy has created some of London's most eye-catching street art, stencilled onto the walls of shops, car parks and industrial buildings. Sadly, the authorities tend to have a different opinion – many famous Banksy murals have been deliberately painted over, though new work is appearing all the time. Rumours abound of the artist's real identity, but Banksy has refused to confirm or deny anything. For an interactive map listing locations of Banksy murals in London, see www.zeemaps.com/map.do?group=1571.

WHITE CUBE GALLERY

☎ 7930 5373; www.whitecube.com; 48 Hoxton Sq; admission free; 10am-6pm Tue-Sat; ; ⊖ Old St

The original Hoxton gallery, the White Cube is now an annexe to the larger White Cube (p93) in Mayfair. The gallery stills hosts some interesting exhibitions from time to time, among the conceptual tosh.

WHITECHAPEL ART GALLERY

☎ 7522 7888; www.whitechapel.org; 80-82 Whitechapel Rd; admission free, charges for some exhibitions; 11am-6pm Wed-Sun; ; ⊖ Aldgate East

A firm favourite of art students and the avant-garde cognoscenti, this gallery provides a less cliquey forum for modern art than the White Cube (above) in Hoxton. The gallery is currently under expansion, but should be offering its full exhibition programme by the time you read this.

SHOP

A GOLD *Food*

☎ 7247 2487; 42 Brushfield St; 9.30am-5.30pm Mon-Fri, 11am-6pm Sat, 10am-6pm Sun; ⊖ Liverpool St

Nostalgia alert – A Gold is a provisions store in the Victorian mould, full of old-fashioned sweeties, cakes, pickles and soda pop – remember dandelion and burdock anyone? The shop front has hardly changed in 100 years.

ABSOLUTE VINTAGE *Clothing*

☎ 7247 3883; www.absolutevintage.co.uk; 15 Hanbury St; noon-7pm Mon-Sat, 11am-7pm Sun; ⊖ Liverpool St

Finding it hard let go of the '80s? Well fear not, help is at hand. Absolute Vintage is stacked to the rafters with vintage shoes, clothes and bags covering four decades of fashion, all at reasonable prices.

HOXTON BOUTIQUE *Clothing*

7684 2083; www.hoxtonboutique.com; 2 Hoxton St; 10.30am-6.30pm Tue-Fri, 11am-6pm Sat, noon-5pm Sun; Liverpool St or Old St

This cute boutique is packed with the kind of labels that you have to be part of the scene to know about. We like the bright interior and spread-out displays.

LADEN SHOWROOMS *Clothing*

7247 2431; www.laden.co.uk; 103 Brick Lane; 11am-6.30pm Mon-Sat, 10.30am-6pm Sun; Liverpool St or Aldgate East

If you fancy shopping in the same place as Noel Gallagher, Pete Doherty and Sophie Ellis-Bextor, duck into this ubercool boutique on Brick Lane. All the clothes are by young, cool independent designers.

ROUGH TRADE *Music*

7392 7788; www.roughtrade.com; Dray Walk, Old Truman Brewery; 8am-9pm Mon-Thu, 8am-8pm Fri & Sat, 11am-7pm Sun; Liverpool St

Alternative rock fans will recognise Rough Trade as the record label behind the Smiths, Jarvis Cocker and the Butthole Surfers. There are now two excellent Rough Trade record stores – one here and one in Notting Hill (p137) – stocked up with hard-to-find limited releases.

START *Clothing*

7739 3334; www.start-london.com; 40, 42 & 59 Rivington St; 10.30am-6.30pm Mon-Fri, 11am-6pm Sat, 1-5pm Sun; Liverpool St

A mini Hoxton empire, this designer clothes store is spread over three buildings along Rivington St catering to local loft dwellers. Head to No 42 for womenswear, No 59 for men's off-the-peg and No 40 for bespoke tailoring.

RETRO IS THE NEW BLACK

The area around Hoxton and Spitalfields has become the epicentre of London's vintage clothing scene, with dozens of specialist shops offering racks of second-hand 1970s heels and scuffed-up leather jackets. As well as Absolute Vintage (p147), try the following:

Beyond Retro (7613 363; www.beyondretro.com; 112 Cheshire St; 10am-6pm Mon-Sat, 11am-7pm Sun; Bethnal Green) A warehouse of vintage wonders.

Blondie (7247 0050; 114-118 Commercial St; 10am-7pm; Liverpool St) Affiliated with Absolute Vintage, with the same taste in tasty second-hand shoes.

The Shop (7739 5631; 3 & 7 Cheshire St; 11am-6pm Mon-Sat, 9.30am-5pm Sun; Liverpool St) Two stores chock-full of retro fashions.

Deborah Pearson

Sales assistant at the Fred Bare hat shop on Columbia Rd

What makes Columbia Rd great The flowers, obviously, but the atmosphere is really buzzing on Sunday. There's a place where you can get old-fashioned cream teas at one end of the road, and there's a band with a cello and accordion that plays in the mornings in the courtyard. **Best shopping streets in the East End** Brick Lane (p146) is great, and there are cool small shops on Sclater St (Map p145, C4) and Cheshire St (Map p145, C4). **Best local secret** Broadway Market (p150) – it's not so much the shopping, it's just a really nice place to hang out. **Best East End pubs** The Commercial Tavern (see boxed text, p155) for the décor and the Dove (p157) for the delicious Belgian beers. **Best foodie tips** St John (p154) in Spitalfields does excellent cheap breakfasts, and the owner's wife runs the Rochelle School Canteen (p153).

EAST END MARKETS

The East End has been London's market ground since medieval times. Following is a quick guide to the top markets:

Brick Lane Market (cnr Brick Lane & Cheshire St; 9am-2pm Sun; Aldgate East) A huge flea market, selling everything from novelty T-shirts to broken record players; you could probably find real fleas here if you looked hard enough.

Broadway Market (www.broadwaymarket.co.uk; Broadway Market; 9am-5pm Sat; Bethnal Green or London Fields) Running south from London Fields to the Regent's Canal, with lots of world food, arts and crafts, and some good pubs and restaurants to relax in (see boxed text, p157).

Columbia Road Flower Market (www.columbia-flower-market.freespace.com; Columbia Rd; 8am-2pm Sun; Bethnal Green or Cambridge Heath) A riot of blooms – the main thoroughfare at is jammed with flowers and pot plants, while the shops behind sell bijou ornaments and accessories.

Spitalfields Market (www.spitalfields.co.uk; Brushfield St; 8am-11pm, from 9am Sat & Sun; Liverpool St) Less interesting since the corporate redevelopment, but worth a trip on Sundays for arty accessories and vintage fashions.

Sunday Up Market (www.sundayupmarket.co.uk; Ely's Yard, Old Truman Brewery; 10am-5pm Sun; Liverpool St) The Sunday market at the old Truman Brewery attracts more than 140 stalls, many of them refugees from the corporate refit of Spitalfields; good for knickknacks and retro gear.

TATTY DEVINE *Jewellery*

7739 9009; www.tattydevine.com; 236 Brick Lane; 11am-6pm; Liverpool St

The Shoreditch branch of this Soho boutique (p51) has more oddball jewellery and 1950s flavoured accessories. Look out for Perspex name chains and necklaces adorned with plastic moustaches and Pac Man ghosts.

WESTLAND LONDON

Interior Décor

7739 8094; www.westland.co.uk; St Michael's Church, Leonard St; 9am-6pm Mon-Fri, 10am-5pm Sat; Old St

If you're renovating, do not visit this treasure house of salvaged lamps, fireplaces and architectural elements. You'll want everything you see and you'll never be able to get it all home.

EAT

BISTROTHEQUE

British *££-£££*

8983 7900; www.bistrotheque.com; 23-27 Wadeson St; 6.30-10.30pm Mon-Thu, to 11pm Fri, 11am-4pm & 6.30-11pm Sat & Sun; Old St

You can't get much more urban than a bar, restaurant

and cabaret all hidden together inside a former garment factory. Diners make a night of it, eating fine modern British food at the upstairs restaurant before moving downstairs to the enjoyably camp cabaret.

BRICK LANE BEIGEL BAKE
Jewish £

☎ 7729 0616; 159 Brick Lane; 24hr; ⊖ Liverpool St

A clubbers' favourite, this old-fashioned Jewish bakery is open 24 hours for toasty hotel bagels stuffed with salt beef or salmon and cream cheese.

CANTEEN *British* £-££

☎ 0845 686 1122; www.canteen.co.uk; 2 Crispin Pl, Spitalfields Market; 8am-11pm Mon-Fri, 9am-11pm Sat & Sun; ⊖ Liverpool St

The best of the slick modern chain restaurants in the pedestrian arcade west of the redeveloped Spitalfields Market. Office workers descend on Canteen in a hungry tide at lunchtime for healthy British food and market-fresh salads.

Brick Lane is the beating heart of London's Bangla-town and a hot spot for a curry

EYRE BROTHERS

Spanish ££

7613 5346; www.eyreborthers.co.uk; 70 Leonard St; noon-3pm Mon-Fri, 6.30-10.45pm Mon-Sat; Old St

After the success of the Eagle in Clerkenwell (p67), the Eyre brothers turned their attention to this low-key tapas restaurant, serving dishes with a Mozambican twang, including great peri-peri prawns.

FIFTEEN *Italian* ££-£££

0871 330 1515; www.fifteenrestaurant.com; 15 Westland Pl; noon-3pm & 6.30-9.45pm; Old St

Regardless of whether you're a fan of Jamie Oliver's cheeky chappy TV persona, you have to admire the concept behind his Italian restaurant in Shoreditch. Fifteen is a registered charity, training disadvantaged youngsters up as career chefs. Take your pick from the posh dining room or the more relaxed and cheaper trattoria.

GIRAFFE *International* £-££

3116 2000; www.giraffe.net; 1 Crispin Pl, Spitalfields Market; 8am-11pm Mon-Fri, 9am-10.30pm Sat & Sun; ; Liverpool St

OK, so it's a chain, but kids love it, and hence so do parents. In any case, getting children eating healthy global food is better than stepping under the Golden Arches. There are numerous branches, including one in Bloomsbury (p108).

GREEN & RED CANTINA

Mexican ££

7749 9670; www.greenred.co.uk; 51 Bethnal Green Rd; 5.30pm-midnight Mon-Thu, 5.30pm-1am Fri & Sat, 5.30-10.30pm Sun; Liverpool St

THE KEBAB REVOLUTION

Brick Lane has long been the capital of Bangladeshi food in London, but the reputation of Brick Lane's Bengali curry houses has declined in direct proportion to the level of hard sell used by the touts working the restaurant doors. Many locals now skip Brick Lane entirely in favour of the Pakistani kebab houses further south around Commercial Rd. Here are our top picks (both are accessible by Whitechapel tube):

Mirch Masala (7377 0155; www.mirchmasalarestaurant.co.uk; 111-113 Commercial Rd; noon-midnight) Nothing about the modest-looking BYO will prepare you for the flavours coming out of the kitchen – the mixed grill is an almost religious experience.

New Tayyab (7247 9543; www.tayyabs.co.uk; 83 Fieldgate St; noon-midnight) The best of the subcontinent is served up at this phenomenally busy Pakistani food hall – with sizzling kebabs, meat pulaos (fried rice), *karahi* (steel pan) curries and authentic *mithai-wallah* sweets.

VIETNAMESE NORTH LONDON

Hackney is home to a small but thriving Vietnamese community and there are numerous low-key Vietnamese canteens along Kingsland Rd serving *pho* (beef noodle soup) and *banh cuon* (soft spring rolls). The following restaurants are all accessible by Liverpool St tube, then bus 149:

Loong Kee (☎ 7729 8344; 134g Kingsland Rd; ⌚ noon-3.30pm & 5.30-11.30pm Mon-Fri, noon-11pm Sat & Sun) Simple canteen with good pho and steamed spring rolls; handy for the Geffrye Museum.

Sông Qûe Café (☎ 7616 3222; 134 Kingsland Rd; ⌚ noon-3.30pm & 5.30-11.30pm Mon-Fri, noon-11pm Sat & Sun) Big and bright with a reputation as the best Vietnamese in town.

Viet Hoa (☎ 7729 8293; www.viethoarestaurant.com; 70-72 Kingsland Rd; ⌚ noon-3.30pm & 5.30-11.30pm Mon-Fri, noon-11pm Sat & Sun) One of the first Vietnamese restaurants to be 'discovered', this is still cheap and still good.

Serving authentic Mexican food, instead of the usual Tex-Mex gruel, Green & Red even looks like a proper cantina, with an extensive menu of tequilas and vintage propaganda posters on the walls.

LES TROIS GARÇONS

French *£££*

☎ 7613 1924; www.lestroisgarcons.com; 1 Club Row; ⌚ 7-10pm Mon-Sat; ⊖ Liverpool St

Crammed full of stuffed animals in tiaras, dangling handbags and a standing alligator with a sceptre, this is a piece of theatre as much as somewhere to eat. Flashy décor sometimes covers up for sins in the kitchen – we've enjoyed the creative French cooking here, but other people are of the opinion that it's overpriced.

PRINCESS

Gastropub *££*

☎ 7729 9270; 76 Paul St; ⌚ 12.30-3pm Mon-Fri, dinner 7-11pm Mon-Sat; ⊖ Old St

Meaty Mediterranean dishes and funky designer wallpaper are the order of the day at this Shoreditch gastropub in the tangle of lanes south of Old St. The restaurant is above the pub floor, reached via an old-fashioned spiral staircase.

ROCHELLE SCHOOL CANTEEN *British* *£-££*

☎ 7729 5667; www.arnoldandhenderson.com; Rochelle School, Arnold Circus; ⌚ 9.30am-4pm Mon-Fri; ⊖ Old St

A truly unique eatery, the canteen at the former Rochelle School ostensively exists to feed workers from the surrounding design studios, but passers-by are welcome to join in the fun. It's only open lunchtime, it's BYO, and the menu of top-class British food changes daily.

ST JOHN *British* *££-£££*

☎ 7251 0848; www.stjohnrestaurant.co.uk; 94-96 Commercial St; 9am-11pm Mon-Fri, 10am-10.30 Sat & Sun; ⊖ Liverpool St

The Shoreditch branch of this modern British specialist offers great breakfasts, as well as hearty lunches and dinners based on the flavoursome internal organs that used to be prevalent in British cooking before the arrival of prepackaged supermarket meat. The main branch is in Clerkenwell (p68).

DRINK

BIG CHILL BAR *Bar*

☎ 7392 9180; www.bigchill.net; Dray Walk, Old Truman Brewery; noon-midnight Sun-Thu, to 1am Fri & Sat; ⊖ Old St

Rocking the inside of the Old Truman Brewery, the Big Chill Bar is an extension of the Big Chill music festival (held every August in Herefordshire). Evenings are loud, brash and fun, and the front terrace is a great place to pass a sunny afternoon.

DRUNKEN MONKEY *Bar*

☎ 7490 7110; www.thedrunkenmonkey.co.uk; 222 Shoreditch High St; noon-midnight Mon-Fri, 6pm-midnight Sat, noon-11pm Sun; ⊖ Old St

What could go better with hip-hop DJs and bottled beers than dumplings? Red-cloth lanterns light the way at this throbbing Shoreditch bar and food lounge, where dim-sum favourites are served as an all-day bar snack.

FAVELA CHIC *Bar*

☎ 7613 5228; www.favelachic.com; 91-93 Great Eastern St; 5pm-1am Tue-Thu, 5pm-2am Fri & Sat; ⊖ Old St

Girls love Favela Chic, and where the girls go, the guys follow. Perhaps it's the Latin vibe; the place is decked out like a Brazilian beach hut and the DJs have a taste for South American beats.

FOUNDRY *Bar*

☎ 7739 6900; www.foundry.tv; 84-86 Great Eastern St; 4.30-11pm Tue-Fri, 2.30-11pm Sat & Sun; ⊖ Old St

This seminal down-and-dirty venue and beer hall exudes the 'whatever' nonchalance of a New York East Village grunge party. Check for out-of-left-field cultural happenings and events.

GEORGE & DRAGON *Bar*

☎ 7012 1100; 2 Hackney Rd; 6pm-midnight; ⊖ Old St

JUST SO THE PUB SET DON'T FEEL LEFT OUT...

Not everyone wants to drink surrounded by graphic designers with Kanye West sunglasses and 'ironic' haircuts. Fortunately, there are some decent pubs in the area, though none are entirely devoid of the sneaker pimp mentality. Seek out an honest pint at the following alehouses (all are accessible from Liverpool St tube):

Commercial Tavern (☎ 7247 1888; 142 Commercial St) Decked out with Victorian chintz, the Commercial has its tongue firmly in its cheek – it's trendy, but we like it.

Golden Heart (☎ 7247 2158; 110 Commercial St) Hoxton local made famous by the Chapman Bros and Tracey Emin, and its charmingly eccentric landlady Sandra.

Ten Bells (☎ 7366 1721; 84 Commercial St)This Victorian relic's Jack the Ripper links matter less to regulars than the agreeably decaying décor and chilled-out vibe.

An interesting bit of trivia – the author of this book once stayed at the George & Dragon when it was a derelict squat in the 1980s. How times change – today the George is Hoxton's hippest gay bar, if you didn't guess from the pink plastic flamingos and rhinestones.

HAWKSMOOR *Bar*

☎ 7247 7392; www.hawksmoor.co.uk; 157 Commercial St; noon-midnight Mon-Fri, 6pm-midnight Sat; ⊖ Liverpool St

Newspaper critics insist that Hawksmoor serves the best cocktails in London, and for once we're inclined to agree. The vast cocktail menu features no less than nine variations on the mint julep, along with martinis, manhattans, punches fizzes and sours. The attached restaurant serves thick, juicy steaks.

LOUNGELOVER *Bar*

☎ 7012 1234; www.loungelover.co.uk; 1 Whitby St; 6pm-midnight Sun-Thu, 6pm-1am Fri & Sat; ⊖ Liverpool St

Les Trois Garçons (p153) design team have worked their magic on this sublimely trendy cocktail bar just around the corner. So, the walls and ceiling drip with chandeliers, lanterns and lampshades. It's loud and camp but it's still a classy dame.

VIBE BAR *Bar*

☎ 0870 850 4989; www.vibe-bar.co.uk; Old Truman Brewery; 91-95 Brick Lane; 11am-11.30pm Sun-Thu, 11am-1am Fri & Sat; ⊖ Liverpool St

Champion of the 'retro new wave' Hoxton music scene, the Vibe is a popular end point for the Brick Lane bar crawl. The crowd is young and energetic, and the events line-up features everything from live bands and DJs to 'rave cabaret'.

PLAY

333 *Club*

☎ 7739 5949; www.333mother.com; 333 Old St; 10pm-4am Fri & Sat; Old St

The enigmatic 333 was an early pioneer of the Hoxton scene, but it's calmer these days. There's still plenty going on – techno and drum-and-bass sessions, indie nights, oddball theme events – but the queues are less absurd and the crowd less self-obsessed.

93 FEET EAST *Club*

☎ 7247 3293; www.93feeteast.co.uk; 150 Brick Lane; 5-11pm Mon-Thu, noon-1am Fri & Sat, noon-10.30pm Sun; Liverpool St or Aldgate East

Hip, happening and ear-splittingly loud, 93 Feet East takes up the other half of the old Truman Brewery. The popular Rock 'n' Roll Cinema events have sadly run their course, but you can still hear some thundering sounds on the dancefloor and recharge your dance batteries at the barbie in the courtyard.

BETHNAL GREEN WORKING MEN'S CLUB *Club*

☎ 7739 7170; www.workersplaytime.net; 44-46 Pollard Row; event times vary; Bethnal Green

Riding the wave of the swing-dance revival, this camped-up club feels deliciously underground. The live events fall somewhere between club nights and cabaret, with such diverse happenings as 'rockabilly burlesque' and 'Grind A Go Go' – a 1960s-style lounge evening with live go-go dancers. The vintage apparel dress code is only vigorously enforced for boogie woogie events.

CARGO *Club*

☎ 7739 3440; www.cargo-london.com; 83 Rivington St; noon-1am Mon-Thu, noon-3am Fri, 6pm-3am Sat, noon-midnight Sun; Old St

As well as respected disk-spinners, Cargo attracts edgy live acts of the Hot 8 Brass Band calibre. The music policy is broad and inclusive, and the chill-out spaces beside the railway arches are daubed with Banksy graffiti (see boxed text, p147).

ELBOW ROOM *Pool Hall*

☎ 7613 1316; www.theelbowroom.co.uk; 97-113 Curtain Rd; 5pm-2am Mon, noon-2am Tue-Sat, noon-midnight Sun; Liverpool St

A sleek American-style pool hall with plenty of tables; there's a branch in Notting Hill (p142).

WORTH A TRIP – HACKNEY & BETHNAL GREEN

North and east of Hoxton, Bethnal Green and Hackney took in the artists who moved when Shoreditch became too expensive. There's lots going on here, from the popular Broadway Market (see boxed text, p150) to the Vietnamese canteens around Kingsland Rd (see boxed text, p153). Following is a guide to the highlights:

Buen Ayre (☎ 7275 9900; www.buenayre.co.uk; 50 Broadway Market; 6.30-10.30pm Mon-Fri, noon-10.30pm Sat & Sun; London Fields) Huge slabs of meat and fine red wines in the best Argentinean tradition.

Dove (☎ 7275 7617; 24-28 Broadway Market; Bethnal Green or London Fields) Broadway Market's favourite boozer, with a full menu of full-flavoured Belgian beers.

Geffrye Museum (☎ 7739 9893; www.geffrye-museum.org.uk; 136 Kingsland Rd; admission free; 10am-5pm Tue-Sat, noon-5pm Sun; Liverpool St, then bus 149) A delightful exploration of English interior design, has mock-ups of rooms covering every period from the 17th to 21st centuries.

Mangal II (☎ 7254 7888; www.mangal2.com; 4 Stoke Newington Rd; noon-1am; Kingsland) The best of Hackney's Anatolian *ocakbasis* (kebab houses), with a huge charcoal grill piled high with skewers.

V&A Museum of Childhood (☎ 8980 2415; www.vam.ac.uk/moc; Cambridge Heath & Old Ford Rds; admission free; 10am-5.45pm; Bethnal Green) A brilliant collection of toys through the ages will bring childhood memories flooding back.

RICH MIX *Live Music*

☎ 7613 7498; www.richmix.org.uk; 35-47 Bethnal Green Rd; 9am-11pm Mon-Fri, 10am-11pm Sat & Sun, 10pm-4am Sun; Old St

Founded in 2006 in a converted garment factory, this modern cultural centre contains a three-screen cinema, a bar and a live-music venue. The movies are pretty mainstream, but the music events can be interesting.

THE OLD BLUE LAST *Live Music*

☎ 7739 7033; www.theoldbluelast.com; 38 Great Eastern St; noon-midnight Mon-Wed, noon-12.30am Thu & Sun, noon-1.30am Fri & Sat; Old St

NME recently described the Old Blue Last as the coolest pub in the world. That might be stretching it, but the nightly line-up of bands and DJs certainly rocks our world. We can't guarantee a secret show by Lily Allen or the Arctic Monkeys, but they do happen.

>KING'S CROSS & ISLINGTON

Long-neglected King's Cross seems poised on the verge of a renaissance with the opening of the Eurostar terminal at St Pancras International in 2007. High-speed trains now shuttle in daily from Paris, Brussels and Lille, marking another great step forward in the entente cordiale. Admittedly, there's still work to do – many streets around the gleaming St Pancras International need some tender loving care – but the outlook for the area seems brighter than it has done for decades.

Apart from the British Library and the glorious folly that is St Pancras Chambers, there isn't much to see or do in King's Cross, but it's only a short hop to Islington, the bower of the *Guardian*-reading middle classes. It's worth roaming this way to explore the small boutiques and restaurants along Upper St and take in a show at the charming Almeida Theatre. Islington also boasts a handful of good pubs, though most of the serious nightspots are closer to King's Cross (since the closure of the Cross, Egg and Scala dominate the scene).

A word of warning – King's Cross has ongoing problems with street prostitution, petty theft, drug abuse and general drunkenness. Keep your wits about you after dark.

KING'S CROSS & ISLINGTON

SEE
British Library................1 A4
St Pancras Chambers....................2 A4

SHOP
After Noah..................(see 10)
Camden Passage Antiques Market.........3 C3
Diverse (Men).................4 D2
Diverse (Women).......(see 11)
Eccentricities..................5 D3
Gill Wing.........................6 D2

EAT
Duke of Cambridge........7 D3
Gallipoli..........................8 D3
Gallipoli Again............(see 10)
Gallipoli Bazaar..............9 D3
Isarn.............................10 D2
Ottolenghi....................11 D2
S&M Café......................12 D3
Snazz Sichuan..............13 A4

DRINK
Elk in the Woods...........14 D3

PLAY
Almeida Theatre...........15 D2
Egg...............................16 A2
Scala.............................17 B4
Screen on the Green.....18 D3

A
B
C
D
1
2
3
4
5
6
0
500 m
0
0.3 miles
Caledonian Rd
To Arsenal Emirates Stadium (400m)
Highbury Corner
Highbury & Islington
St Paul's Rd
Mackenzie Rd
Lough Rd
Sheringham Rd
Westbourne St
Roman Way
Liverpool Rd
Furlong Rd
Bride St
Ellington St
Arundel Pl
Camden Park Rd
York Way
Market Rd
Caledonian Park
Brewery Rd
Blundell St
St Augustine's Rd
Marquis Rd
Pentonville Prison
BARNSBURY
Highbury Station Rd
Laycock St
Compton Ave
Compton Rd
Grange Gve
Alwyne Villas
Alwyne Pl
Alwyne Rd
Canonbury Rd
Caledonian Rd & Barnsbury
Agar Gve
Offord Rd
Belitha Villas
Huntingdon St
Thornhill Rd
Barnsbury Park
Bewdley St
Brooksby St
Islington Park St
College Cross
Upper St
Sebbon St
Halton Rd
Essex Rd
Vale Royal
Carnoustie Dr
Caledonian Rd
Thornhill Sq
Hemingford Rd
Lofting Rd
Barnsbury St
Ripplevale Gve
Pembroke St
Bingfield St
Almeida St
ISLINGTON
Cross St
Havelock St
Bemerton St
Matilda St
Richmond Ave
Barnsbury Rd
Cloudesley Rd
Cloudesley St
Gibson Sq
Barnard Park
Theberton St
Gaskin St
Popham St
Britannia Row
Packington St
Copenhagen St
KING'S CROSS
Carnegie St
Muriel St
Charlotte Tce
Copenhagen St
Cloudesley Pl
Camden Passage
Islington Green
Charlton Pl
St Peter's St
Rheidol Tce
Camley Street Natural Park
Camley St
St Pancras Gardens
Regent's Canal
Ritchie St
Tolpuddle St
Chapel Market
Parkfield St
Upper St
Duncan St
Colebrooke Row
Danbury St
Wynford Rd
Chapel Market
City Rd Basin
Wharfdale Rd
To St Aloysius Social Club (50m)
Pancras Rd
Cheney Rd
King's Cross
Balfe St
Northdown St
Killick St
Calshot St
Rodney St
Coller St
Donegal St
Panton St
White Lion St
Islington High St
Elia St
Graham St
Phoenix Rd
SOMERS TOWN
St Pancras International (Eurostar)
Midland Rd
Pentonville Rd
Penton Rise
PENTONVILLE
Angel
St John St
City Rd
To Wenlock Arms (500m)
King's Cross Rd
Britannia St
King's Cross St Pancras
Gt Percy St
Amwell St
Mydelton Sq
Chadwell St
Wakley St
Hall St
Charlton St
Euston Rd
Argyle St
Swinton St
Acton St
Wharton St
Lloyd Baker St
FINSBURY
Rawstorne St
Moreland St
Central St
Cromer St
King's Cross Rd
Hardwick St
Rosebery Ave
Myddelton St
Spencer St
City University
Goswell Rd
Hunter St
Cartwright Gdns
Sidmouth St
Cubitt St
Margery St
Lever St
Upr Woburn Pl
Burton St
ST PANCRAS
St George's Gardens
Eastman Dental Hospital
Mecklenburgh Sq
Calthorpe St
Mount Pleasant (Postal Sorting Office)
Exmouth Market
Skinner St
Spa Fields
Percival St
Tavistock Pl
Brunswick Centre
Marchmont St
Brunswick Sq
St Pancras Coram's Fields
Gray's Inn Rd
Phoenix Pl
Bowling Green La
Sekforde St
St John St
Compton St
Pear Tree St
Woburn Pl
Bernard St
Guildford St
Millman St
Gough St
See The City, Holborn & Clerkenwell Map p60-1
Aylesbury St
Gt Sutton St
Old St
BLOOMSBURY
University of London
Russell Sq
Hospital for Sick Children
Lamb's Conduit St
Roger St
Northington St
Clerkenwell Rd
Britton St
St John's La
St Bartholomew's Medical School
Aldersgate St
Russell Sq
Queen Sq
Gt Ormond St
CLERKENWELL
Farringdon Rd
Malet St
Leather La Market
Leather La
St Cross St
Farringdon
Cowcross St
Barbican
Southampton Row
Bedford Pl
Montague Pl
Montague St
British Museum
Theobald's Rd
Red Lion Sq
Red Lion St
Jockey's Fields
Gray's Inn Ct
Gray's Inn Ct
Brooke St
Hatton Garden
Charterhouse St
Long La
Smithfield Market
West Smithfield
Little Britain
Bloomsbury St
Bedford Ave
Great Russell St
Bloomsbury Way
Procter St
Eagle St
Chancery La
Holborn
Holborn Viaduct
Bartholomew's Hospital
Holborn
HOLBORN

SEE

BRITISH LIBRARY

☎ 0870 444 1500; www.bl.uk; 96 Euston Rd; admission free, tours adult/concession £8/6.50, free weekday mornings; 9.30am-6pm Mon-Fri, to 8pm Tue, 9.30am-5pm Sat, 11am-5pm Sun, tours 11am & 3pm Mon, Wed & Fri, 10.30am & 3pm Sat, 11.30am & 3pm Sun; ; King's Cross/St Pancras

Housed in a modernist building that still divides public opinion, the British Library preserves some of the nation's greatest literary treasures, including the original *Magna Carta* and the diaries of Captain Scott and Lewis Carroll. The stylish café is arranged around a glass tower containing the personal library of George III.

ST PANCRAS CHAMBERS

Euston Rd; King's Cross/St Pancras

Built in 1868 as the Midland Grand Hotel, this glorious Victorian Gothic brick tower is one of London's most distinctive buildings. It's currently being re-developed as a five-star hotel and apartment complex to serve the shiny new St Pancras International Eurostar terminal.

SHOP

AFTER NOAH *Antiques*

☎ 7359 4281; www.afternoah.com; 121 Upper St; 10am-6pm Mon-Sat, noon-5pm Sun; Angel

The kind of shop Islington does brilliantly – a quirky emporium of vintage furniture, retro lamps and ornaments and old-fashioned kids' toys. There's a branch on the King's Rd in Chelsea.

CAMDEN PASSAGE ANTIQUES MARKET *Antiques*

www.camdenpassageislington.co.uk; Camden Passage; 9am-5pm Wed & Sat; Angel

Twice a week, the alleyways around Camden Passage spill over with stalls selling bric-a-brac and serious antiques. It's like a

REGENT'S CANAL WALK

The towpath on the old Regent's Canal provides a handy and scenic shortcut between Limehouse, Hackney, Islington, King's Cross, Camden, Regent's Park, Marylebone and Little Venice near Paddington. The path is not continuous – you'll have to cross a few roads to get past the tunnels in Islington and Marylebone.

London's landmark of literature and learning – the British Library

less-touristy Portobello Rd, and there are some good cafés, pubs and restaurants where you can sit back and pick over your purchases.

DIVERSE *Clothing*

menswear 7359 0081, ladieswear 7359 8877; www.diverseclothing.com; 286 & 294 Upper St; 10.30am-6.30pm Mon-Sat, noon-5.30pm Sun; Angel or Highbury & Islington

You don't have to trek out to Mayfair for top designer clothing. This elegant boutique carries punchy gear from designers such as Paul Smith, Chloe and Stella McCartney. Head to No 286 for menswear or No 294 for womenswear.

ECCENTRICITIES *Interior Décor*

7359 5633; 46 Essex Rd; 10am-7pm Mon-Sat, 10am-6pm Sun; Angel

The name fits – the huge sprawling building is full of weird and wonderful furniture and bric-a-brac, some reproduction, some antique.

GILL WING *Accessories*

☎ 7226 5392; www.gillwing.co.uk; Upper St; 10am-6pm Mon-Sat, noon-5pm Sun; Angel or Highbury & Islington

A mini-empire of small boutiques selling everything an upwardly mobile Islingtonian might need. Go to No 182 for jewellery, No 190 for kitchenware, No 192 for shoes and No 194-5 for cute novelties.

EAT

DUKE OF CAMBRIDGE *Gastropub* £-££

☎ 7359 3066; www.dukeorganic.co.uk; 30 St Peter's St; noon-11pm, to 10.30pm Sun; Angel

Organic, before organic became cool, the Duke of Cambridge is Islington's favourite gastropub. There are plenty of organic wines and ales to accompany your organic British food.

GALLIPOLI *Turkish* £-££

☎ 7359 0630; www.cafegallipoli.com; 102 Upper St; 10.30am-11pm, to midnight Fri & Sat; Angel or Highbury & Islington

Islington's famous Gallipoli has been serving lip-smackingly authentic Anatolian kebabs for decades. As well as the original bistro at No 102, the owners run **Gallipoli Again** at No 120 and the small, souk-like **Gallipoli Bazaar** at No 107.

ISARN *Thai* ££

☎ 7424 5153; www.isarn.co.uk; 119 Upper St; noon-3pm & 6-11pm Mon-Fri, noon-11pm Sat, noon-10pm Sun; Angel

The flavours of freshly ground lemongrass, chilli, galangal and lime leaves sing out in the curry sauces and marinades at this small, stylish Thai restaurant. It's the perfect place for an informal dinner date.

OTTOLENGHI *Deli-café* £-££

☎ 7288 1454; www.ottolenghi.co.uk; 287 Upper St; 8am-11pm Mon-Sat, 9am-7pm Sun; Angel or Highbury & Islington

If the giant pink meringues in the window don't lure you inside this minimalist deli-café, you can't be human. Deli sandwiches, gourmet salads and stuffed pastries are available to eat in or take away. There are branches all over town, including one in Knightsbridge (p129).

S&M CAFÉ *British* *£-££*

☎ 7359 5361; www.sandmcafe.co.uk; 4-6 Essex Rd; 7.30am-11.30pm Mon-Thu, 7.30am-midnight Fri, 8.30am-midnight Sat, 8.30am-10.30pm Sun; ⊖ Angel

This sausage and mash specialist is set in the old Alfredo's Café, a chrome-filled former Mods and Rockers' joint. There are branches in Spitalfields and Notting Hill.

SNAZZ SICHUAN *Chinese* *££*

☎ 7388 0808; www.newchinaclub.co.uk; 37 Charlton St; 12.30-10.30pm; ⊖ King's Cross or Euston

One of the very few London restaurants brave enough to serve the fiery cuisine from Sichuan area in western China. Give your tastebuds a workout with the fire-exploded kidneys and *guai wei tu ding* (literally, 'strange-flavour rabbit').

DRINK

ELK IN THE WOODS *Pub*

☎ 7226 3535; www.the-elk-in-the-woods.co.uk; 39 Camden Passage; ⊖ Angel

Nudging towards gastropub status, the Elk is a favourite lunch stop for the 30-somethings who shop at Islington's boutiques. The interior has just the right amount of shabby chic furniture and exposed brickwork.

WENLOCK ARMS *Pub*

☎ 7608 3406; www.wenlock-arms.co.uk; 26 Wenlock Rd; ⊖ Angel

Regularly voted best pub in North London by the Campaign for Real Ale (CAMRA), the Wenlock Arms serves guest ales from all over the country. There are jazz evenings around the 'old Joanna' (piano) every Friday and Saturday.

SWINGING LONDON

The tradition of tea dances fizzled out in the 1950s, but swing dancing has seen a major revival in the last few years, thanks partly to cult events like **Viva Cake** (www.myspace.com/vivacakebitches), a jive-dancing, cake-baking, roller-skating celebration of all that was great about the '50s, held periodically at backstreet venues around London, including **Bloomsbury Bowling** (p111) and Islington's **St Aloysius Social Club** (☎ 7419 6891; 20 Phoenix Rd; event times vary; ⊖ Euston). Most jive, jitterbug and Lindy-hop events take place at small venues and community halls, and you may be expected to dress appropriately, down to the bouffant hairdo and crepe soles. For listings see www.swingdanceuk.com, www.swingland.com and www.jitterbugs.co.uk. One regular night that always draws a crowd is Stompin' Mondays at Oxford St's **100 Club** (p110).

FOOTBALL CRAZY

Londoners adopt an almost tribal identity when it comes to their favourite football clubs. Matches between arch-rivals like Arsenal and Tottenham Hotspurs are characterised by rousing crowd chants that recall medieval battles in their passion and intensity.

The two most famous Premiership clubs in London are **Arsenal Football Club** (below) and **Chelsea Football Club** (p131), but you can also see live Premiership action at **Tottenham Hotspurs Football Club** (www.tottenhamhotspur.com), **Fulham Football Club** (www.fulhamfc.com) and **West Ham United Football Club** (www.whufc.com).

Tickets go on general sale about a month in advance – it's often safer to sign up for club membership and buy your tickets online.

PLAY

ALMEIDA THEATRE *Theatre*

☎ 7359 4404; www.almeida.co.uk; Almeida St; ⊖ Angel

Who said big is better? This small, independent theatre offers the chance to see acting heavyweights of the Ralph Fiennes, Kevin Spacey, Imelda Staunton and Stephen Rea calibre, performing challenging contemporary plays in a small, intimate space.

ARSENAL EMIRATES STADIUM *Sports*

☎ box office 0870 3800 232, tours 7704 4504; www.arsenal.com; Ashburton Grove; tours adult/concession £12/6; ⊖ Arsenal or Holloway Rd

Despite being coached by a foreign coach, playing in a foreign-owned stadium and having almost no English players, Arsenal still has an obsessive local following. Tours of the stadium and dressing rooms run daily – see the website for times.

EGG *Club*

☎ 7609 8364; www.egglondon.net; 200 York Way; 9.30pm-6am Fri, 10pm-6am Sat, 5am-2pm Sun; ⊖ King's Cross/St Pancras

Cool, industrial and quintessentially urban, Egg booms till late with the latest electro, hip-hop and house. Saturday night revellers can continue through to Sunday lunchtime with Breakfast at Egg in the courtyard garden.

SCALA *Live Music*

☎ 7833 2022; www.scala-london.co.uk; 275 Pentonville Rd; 10pm-6am Fri & Sat, event times vary; ⊖ King's Cross/St Pancras

Scala almost went bust after losing a court case for illegally showing Stanley Kubrick's *A Clockwork Orange* in the 1990s. There's still an anti establishment feel to the club nights, gigs and special events.

SCREEN ON THE GREEN

Cinema

☎ 7226 3520; www.everymancinema club.com; 83 Upper St; ⊖ King's Cross/ St Pancras

Part of the small Everyman group, this cosy independent cinema screens art house imports and occasional blockbusters that pass the test of good film-making.

>CAMDEN, HAMPSTEAD & PRIMROSE HILL

You couldn't find three more different neighbours if you tried. Camden is London's capital of goth, grunge and all things alternative, while nearby Primrose Hill is almost impossibly genteel. Further north, Hampstead is another world, a village that never quite realised it was part of a capital city.

The attractions of this diverse corner of North London can be summed up in three words: markets, nightlife and the heath. Despite being nearly destroyed by fire in early 2008, Camden's sprawling markets support entire subcultures of music, art and fashion. Dotted around the market are cutting-edge bars and clubs that provide a vital training ground for up-and-coming musical talent.

If Camden sounds too noisy and down-with-the-kids, you'll find peace, quiet and splendid isolation on Hampstead Heath. Londoners flock to this vast open space on sunny days to swim, picnic, fly kites and get up to no good in the long grass. The views from Parliament Hill are so stunning they are protected by law, but if you don't feel up to the walk, there are more views from Primrose Hill, above Regent's Park.

CAMDEN, HAMPSTEAD & PRIMROSE HILL

SEE

Freud Museum 1 A5
Hampstead Heath 2 B2
Highgate Cemetery Entrance 3 D2
Keats House 4 B4
Kenwood House 5 B1
London Waterbus Company 6 D6

SHOP

Camden Lock Market 7 D5
Camden Market 8 D6
Inverness St Market 9 D6
Rosslyn Delicatessen.... 10 A4
Stables Market .... (see 19)

EAT

Gilgamesh .... (see 19)
Haché 11 D6
Lansdowne 12 C5
Mango Room 13 D6
Manna 14 C5
Marine Ices 15 C5
Masala Zone 16 D6
Trojka 17 C5
Wells 18 A3

DRINK

At Proud 19 C5
Hollybush 20 A3
Lock Tavern 21 D5

PLAY

Barfly 22 C5
Boogaloo 23 D1
Dublin Castle 24 D6
Jazz Cafe .... (see 16)
Koko 25 D6
Roundhouse 26 C5
Underworld 27 D6

1 km
0.5 miles
Hampstead Golf Course
HIGHGATE
Highgate
North Hill
Southwood La
North Rd
Archway Rd
Sheldon Ave
Denewood Rd
The Bishops Ave
Hampstead La
Wildwood Rd
North Wood
Ken Wood
Spaniard's Inn
Henry Moore Sculptures
Hampstead Heath Extension
Stock Pond
Kenwood Ladies' Pond
Bird Sanctuary
Highgate Men's Pond
Hampstead Heath
Spaniards Rd
North End Way
Fitzroy Park
Highgate West Hill
South Gve
Waterlow Park
Highgate Cemetery
Highgate Hill
Millfield La
Men's Bathing Pond
Parliament Hill
Highgate Ponds
Vale of Health Pond
West Heath
Mixed Bathing Pond
East Heath
Swains La
Dartmouth Park Hill
Highgate Rd
Heath St
Well Walk
East Heath Rd
Hampstead Ponds
Parliament Hill Fields
Parliament Hill Lido
Tufnell Park
Flask Wk
Hampstead
Downshire Hill
Keats Gve
South End Rd
Hampstead High st
HAMPSTEAD
Hampstead Heath
Gospel Oak
Gordon Ho Rd
GOSPEL OAK
Fortess Rd
Rosslyn Hill
Pond St
Fleet Rd
Mansfield Rd
Royal Free Hospital
Arkwright Rd
Southampton Rd
Grafton Rd
KENTISH TOWN
Kentish Town
Belsize Park
Spring Pl
Holmes Rd
Finchley Rd & Frognal
Maresfield Gardens
Fitzjohn's Ave
Daleham Gdns
Belsize La
Belsize Ave
Haverstock Hill
Maitland Park
Queen's Cres
Malden Rd
Kentish Town West
Kentish Town Rd
BELSIZE PARK
Finchley Rd
Belsize Park
Lancaster Gve
Prince of Wales Rd
Ferdinand St
Castle Rd
Chalk Farm
College Cres
Eton Ave
Eton Rd
Fellows Rd
Swiss Cottage
Adelaide Rd
CAMDEN
Chalk Farm Rd
Camden Rd
Erskine Rd
Hilgrove Rd
King Henry's Rd
Primrose Hill Rd
Belsize Rd
South Hampstead
Avenue Rd
Elsworthy Rd
Chalcot Rd
Regent's Canal
Gloucester Ave
Camden Town
Camden High St
Primrose Hill
Regent's Park Rd
Grand Union Canal
Oval Rd
Parkway
Finchley Rd
Queen's Gve
Norfolk Rd
Prince Albert Rd
ST JOHN'S WOOD
Abbey Rd
St John's Wood
Acacia Rd
Townshend Rd
Outer Circle
London Zoo
Regent's Park
See Marylebone & Regent's Park Map p113
Mornington Cres
A
B
C
D
1
2
3
4
5
6
1
2
3
4
5
6
7
8
9
10
11
12
13
14
15
16
17
18
19
20
21
22
23
24
25
26
27

SEE

FREUD MUSEUM

☎ 7435 2002; www.freud.org.uk; 20 Maresfield Gardens; adult/under 12yr/concession £5/free/3; noon-5pm Wed-Sun; ⊖ Finchley Rd

After fleeing Nazi-occupied Vienna, Sigmund Freud took refuge in this North London home until his death in 1939. As well as peering at Freud's personal effects, you can see the original psychoanalyst's chair where patients were encouraged to reveal intimate details of their relationship with their mother.

A pew with a view on Parliament Hill

HAMPSTEAD HEATH

☎ 7485 4491; ⊖ Hampstead, Gospel Oak or Hampstead Heath, 214 or C2 to Parliament Hill Fields

After the hectic pace of the rest of London, this enormous park feels like stepping into the wilds. Amazingly, this huge rolling blanket of meadows and lakes is just 4 miles from the City. On sunny days, thousands of Londoners come to stroll and admire the views. See p21 for more details.

HIGHGATE CEMETERY

☎ 8340 1834; www.highgate-cemetery.org; Swain's Lane; admission Eastern Cemetery £3, Western Cemetery tour £5; 10am-5pm Mon-Fri, 11am-5pm Sat & Sun, to 4pm Mon-Sun Nov-Mar, tours 2pm Mon-Fri & hourly 11am-4pm Sat & Sun, last tour 3pm Nov-Mar; ⊖ Highgate

This huge and overgrown Victorian cemetery is the final resting place of Karl Marx, among other London notables (see boxed text, opposite). You can explore the Eastern Cemetery under your own steam, but the atmospheric Western Cemetery is only accessible on an organised tour (call ahead to book; no bookings Sunday).

FAMOUS EX-LONDONERS

Karl Marx isn't the only famous name spending an eternity beneath the turf of **Highgate Cemetery** (opposite). Here are some other celebrity residents to look out for as you wander up and down the rows:

Western Cemetery

- Michael Faraday (1791–1867) Father of modern electromagnetism.
- Alexander Litvinenko (1962–2006) Russian dissident poisoned in London in 2006.
- Samuel Taylor Coleridge (1772–1834) Poet who wrote *Kubla Khan* in an opium-induced trance.

Eastern Cemetery

- Douglas Adams (1952–2001) Author of *Hitchhiker's Guide to the Galaxy*.
- William Foyle (1885–1963) Founder of **Foyles** (p49) bookshop on Charing Cross Rd.

KEATS HOUSE

☎ 7435 2062; www.cityoflondon.gov.uk/keats; Wentworth Pl, Keats Grove; ⊖ Hampstead

The great Romantic poet John Keats composed his famous *Ode to a Nightingale* in this house on Wentworth Pl, before falling in love with the girl next door and dying from consumption aged just 25. The house is now a museum, dedicated to his life and work.

KENWOOD HOUSE

☎ 8348 1286; www.english-heritage.org.uk; Hampstead Lane; admission free; 🕒 11am-4pm; ⊖ Archway or Golders Green, then bus 210

This magnificent neoclassical mansion makes an appearance in numerous period dramas, including the film version of Jane Austen's *Mansfield Park*. Inside are paintings by Gainsborough, Reynolds, Turner, Vermeer, Van Dyck and others, but plenty of people just come here for tea at the genteel Brew House Café. In summer classical concerts are performed on the lawn.

SHOP

CAMDEN MARKETS *Market*

www.camdenlockmarket.com; Camden High St/Chalk Farm Rd; 🕒 10am-6pm; ⊖ Camden Town

London's punks, goths and emos depend on this sprawling market for thigh-high wedge boots, message T-shirts, zip-leg trousers, 'healing' crystals and bondage leathers. This is alternative culture packaged for the masses – a bit like Japanese 'cosplay' but with a

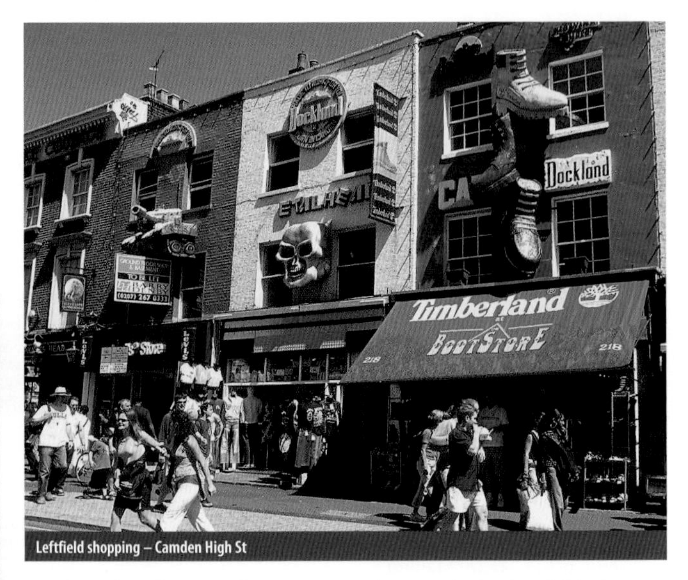

Leftfield shopping – Camden High St

London edge. There are four main areas: the original market on Camden High St (good for clothes), the street market on Inverness St (more clothes) and the Camden Lock and Stables markets on Chalk Farm Rd (antiques, crafts and ethnic food).

ROSSLYN DELICATESSEN

Deli

☎ 7794 9210; www.delirosslyn.co.uk; 56 Rosslyn Hill; 8.30am-8.30pm Mon-Sat, 8.30am-8pm Sun; ⊖ Hampstead or Belsize Park

This award-winning deli is the perfect pit stop for a few picnic provisions en route to the Heath – a definite step up from a prepackaged coffeeshop sandwich.

EAT

As well as the following restaurants, there are numerous small cafés along Regent's Park Rd (C6) in Primrose Hill, perfect for lunch after a stroll in Regent's Park.

GILGAMESH

Pan-Asian £££-£££

☎ 7482 5757; www.gilgameshbar.com; Stables Market, Chalk Farm Rd; noon-3pm & 6pm-midnight, bar to 2.30am Mon-Sat, to 1.30am Sun; Camden Town

Some accuse Gilgamesh of being overpriced and too self-conscious by half, but we've always enjoyed the dim sum and Southeast Asian curries and the Persian-themed décor is just outrageous.

HACHÉ *American* £-££

☎ 7485 9100; www.hacheburgers.com; 24 Inverness St; noon-10.30pm, to 10pm Sun; Camden Town

Forget McDonald's and Burger King, Haché has transformed the humble meat patty into a work of art. There are burgers here made from beef, duck, venison, chicken, lamb and veg. There's a branch on Fulham Rd in Chelsea.

LANSDOWNE

Gastropub ££

☎ 7722 0950; www.thelansdownepub.co.uk; 90 Gloucester Ave; Chalk Farm

Competing for trade with the Engineer down the road, the Lansdowne is a gastropub that still feels like a local. The menu features lots of locally sourced meat and fish, prepared with a Mediterranean twist.

MANGO ROOM

Caribbean ££

☎ 7482 5065; www.mangoroom.co.uk; 10-12 Kentish Town Rd; noon-11pm; Camden Town

This swish Camden eatery has taken Caribbean cooking uptown. Classic calypso adds just the right vibe while you enjoy island favourites like goat curry and saltfish and *ackees* (the national fruit of Jamaica).

CRUISING CAMDEN

Crazy property prices have driven many Londoners to take up residence in narrow boats on London's disused canals. You can get a taste of canal life on a narrow-boat tour with the **London Waterbus Company** (☎ 7482 2660; www.londonwaterbus.com; 2 Middle Yard, Camden Lock; Camden Town). The quaintly painted covered barges drift along the Regent's Canal from Camden Lock to Little Venice, passing through Regent's Park and London Zoo. Services run hourly (every 30 minutes on Sunday) from April to September and adult tickets cost £6.50/9 (one way/return).

MANNA *Vegetarian* *££-£££*

☎ 7722 8028; www.manna-veg.com; 4 Erskine Rd; 6.30-11pm Mon, noon-3pm & 6.30-11pm Tue-Sun; V; ⊖ Chalk Farm

One of the first veggie restaurants in Britain, Manna tempts Primrose Hill veggies with a global menu and lots of locally sourced, organic and GM-free ingredients. You won't save any money by foregoing meat, but the food is worth the splurge.

MARINE ICES

Italian *£*

☎ 7482 9003; www.marineices.co.uk; 8 Haverstock Hill; noon-3pm & 6-11pm Tue-Fri, noon-11pm Sat, noon-10pm Sun; ⊖ Chalk Farm

Marine Ices has been serving delicious Italian gelato since 1931 and locals still stop by after work for sweet scoops of pistachio ice-cream and mango sorbet.

MASALA ZONE

Indian *£-££*

☎ 7267 4422; www.masalazone.com; 25 Parkway; 12.30-3pm & 5.30-11pm Mon-Fri, 12.30-11pm Sat, 12.30-10.30pm Sun; ⊖ Chalk Farm

Indian food done the Wagamama (p109) way. Purists may find the spices a bit toned down, but the *thalis* (plate meals) go down a treat. There are branches all over town – see the website for details.

TROJKA

Eastern European *£-££*

☎ 7483 3765; www.trojka.co.uk; 101 Regent's Park Rd; 9am-10.30pm ⊖ Chalk Farm

A little slice of Russia in Primrose Hill, this charming tea house serves delicious Russian tea with raspberry syrup and hearty Russian, Polish and Ukrainian meals, like *blinis* (pancakes) with lumpfish caviar.

WELLS *Gastropub* *££-£££*

☎ 7794 3785; www.thewellshampstead.co.uk; 30 Well Walk; ⊖ Hampstead

A gastropub for the posh set, this 18th century public house attracts hordes of dressed-up locals at weekends. The chefs prepare top quality traditional British meals with real flair and imagination.

DRINK

AT PROUD *Bar*

☎ 7482 3867; www.atproud.net; Stables Market, Chalk Farm Rd; 11am-1am Tue-Wed, 11am-2am Thu-Sat, 11am-12.30am Sun; wi-fi; ⊖ Camden Town

Built for the nags that used to pull barges along the Regent's Canal, the old horse hospital is now a very cool bar. Pull up a deck chair on the terrace or sink a cold one surrounded by soft upholstery in one of the old wooden stables.

HOLLYBUSH *Pub*

☎ 7435 2892; www.hollybushpub.com; 22 Holly Mount; noon-11pm Mon-Sat, noon-10.30pm Sun; ⊖ Hampstead
Hidden away on an alley off Heath St, this Hampstead local is like all pubs ought to be. The interlinked rooms have a comforting aged patina, recalling the days when City-slickers dodged highwaymen to enjoy the rarefied country air.

LOCK TAVERN *Pub*

☎ 7482 7163; www.lock-tavern.co.uk; 35 Chalk Farm Rd; noon-midnight Mon-Thu, noon-1am Fri & Sat, noon-11pm Sun; ⊖ Camden Town or Chalk Farm
Subdued colours, worn leather sofas, live bands and a rooftop terrace lend just the right amount of urban cool to this revamped public house at the north end of Camden.

PLAY

BARFLY *Live Music*

☎ 0844 847 2424; www.barflyclub.com; 49 Chalk Farm Rd; from 7.30pm ⊖ Chalk Farm or Camden Town
If you want to find out who is going to be headlining Glastonbury in two years' time, look no further than this boisterous indie club at the Chalk Farm end of Camden. The line-up is spectacular and the volume cranks up to 11.

BOOGALOO *Live Music*

☎ 8340 2928; www.theboogaloo.co.uk; 312 Archway Rd; 6pm-late Mon-Fri, from 2pm Sat & Sun; ⊖ Highgate
It's worth hauling yourself out to the Archway end of Highgate to

ROCKING THE SUBURBS

Rock music and Camden go together like Sid Vicious and Dr Martens. A whole generation of rockers got their first big breaks at the Electric Ballroom and the Music Machine. Live music is still at the heart of the Camden scene, with venues like **The Underworld** (p175), **Koko** (p174), Dingwalls and the **Roundhouse** (p174) providing a platform for a new generation of musical rebels.

Every April Camden hosts the **Camden Crawl** (www.thecamdencrawl.com), a two-night rockathon, with more than 130 gigs spread out over 25 different venues. For details, see boxed text, p29. The borough goes live again in October for the **Electric Proms** (www.bbc.co.uk/electricproms), five days of live music at the Roundhouse, **Barfly** (above), Koko and other venues, covering every imaginable genre; see p31.

check out the DJ nights and gigs at this streetwise pub and venue. Stars like Pete Doherty and Shane MacGowan periodically turn up to revamp the track list on the jukebox.

DUBLIN CASTLE *Live Music*

☎ 7485 1773; www.myspace.com/thedublincastle; 94 Parkway; from 7.30pm ⊖ Camden Town

A very grungy pub with a very impressive pedigree. Madness and Travis were just two of the bands who got their first big breaks on the Dublin's tiny stage. Check www.bugbearbookings.com for listings.

TOP LONDON SONGS

Any playlist of songs about London will be extremely subjective, but here are a few suggestions for your iPod while you shuffle around the capital.

> 'Baker Street', Gerry Rafferty
> 'Down in the Tube Station at Midnight', The Jam
> 'Electric Avenue', Eddy Grant
> 'I Don't Want to Go to Chelsea', Elvis Costello
> 'LDN', Lily Allen
> 'London Calling', The Clash
> 'Mile End', Pulp
> 'Play with Fire', The Rolling Stones
> 'Up the Junction', Squeeze
> 'Warwick Avenue', Duffy
> 'Waterloo Sunset', The Kinks
> 'West End Girls', Pet Shop Boys
> '(White Man) in Hammersmith Palais', The Clash

JAZZ CAFÉ *Live Music*

☎ reservations 0870 060 3777, restaurant 7688 8899; www.jazzcafelive.co.uk; 5 Parkway; from 7pm; ⊖ Camden Town

Artists as diverse as De La Soul, Eddy Grant and Martha and the Vandellas have set the crowds grooving at this upbeat jazz club near Camden Town tube. The line-up is more mobo (music of black origin) than bebop, with lots of jazz-funk to R&B and soul.

KOKO *Live Music*

☎ 0870 432 5527; www.koko.uk.com; 1a Camden High St; ⊖ Mornington Cres

This Victorian theatre rocked through the 1970s as the Music Machine and thrashed out the '80s and '90s as the Camden Palace, before one last change of identity to Koko in 2004. It's still one of London's best music venues, not least because the building was actually designed with the audience in mind.

ROUNDHOUSE *Live Music*

☎ 0844 482 8008; www.roundhouse.org.uk; Chalk Farm Rd; ⊖ Chalk Farm

In the 1960s and '70s the Roundhouse was at the cutting edge of live entertainment – it hosted the only UK concert by The Doors, and the deeply weird plays of the Living Theatre of New York. Now the cutting-edge performances are back, with circus troupes, world music and big-name bands.

THE UNDERWORLD

Live Music

☎ 7482 1932; www.theunderworldcamden.co.uk; 174 Camden High St; ⊖ Camden Town

A forbidding set of steps leads down beneath the World's End pub to this seminal goth and rock venue. Bands who play here have names like Aortic Dissection and Monday Massacre and punters are obsessed with the colour black. Be afraid, be very afraid…

>GREENWICH & DOCKLANDS

With Britain's fortunes dependent on its empire and navy, the maritime precincts at Greenwich were almost as important to the nation as Westminster and St James's. Tucked below a looping bend in the River Thames, Greenwich is dominated by the Old Royal Naval College, built by Christopher Wren as the Royal Hospital for Seamen in 1692, and the Royal Observatory, whose aristocratic astronomers were beaten to the technique for calculating longitude by a Yorkshire carpenter in 1765.

Aside from being the Prime Meridian for time and longitude, Greenwich is notable for its grand neoclassical colonnades, its gorgeous setting beside the river, and the sweeping views from the top of Greenwich Park towards the Isle of Dogs and the skyscrapers of Docklands. Centred on distinctive One Canada Square (244m), these towering office blocks are Britain's tallest buildings, and they stand above a warren of underground shopping centres.

The Docklands Light Railway (DLR) and the boats of Thames River Services (see boxed text, p222) provide easy access to Docklands and Greenwich, so there's no reason not to see both on the same trip. On weekends Greenwich is crammed to capacity, while Docklands, emptied of its office workers, is almost surreally quiet.

GREENWICH & DOCKLANDS

SEE

Canary Wharf.................... 1 B2
Cutty Sark.......................... 2 C5
Greenwich Foot Tunnel.... 3 C5
Museum in Docklands...... 4 B1
National Maritime Museum......................... 5 D5
Old Royal Naval College ... 6 C5
Peter Harrison Planetarium.................. 7 D6
Queen's House.................. 8 D5
Royal Naval College Chapel 9 D5
Royal Naval College Painted Hall................. 10 C5
Royal Observatory.......... 11 D6

SHOP

Beehive.................. (see 12)
Emporium....................... 12 C5
Greenwich Market.......... 13 C5
Joy 14 C5
Meet Bernard 15 C5

EAT

Inside.............................. 16 C6
Pavement Restaurants.... 17 B1
Plateau 18 B1
Royal Teas....................... 19 C6
SE10 Restaurant & Bar............................... 20 C5

DRINK

Greenwich Union 21 C6
Gun................................. 22 C2
Trafalgar Tavern 23 D5

PLAY

The 02............................. 24 D2

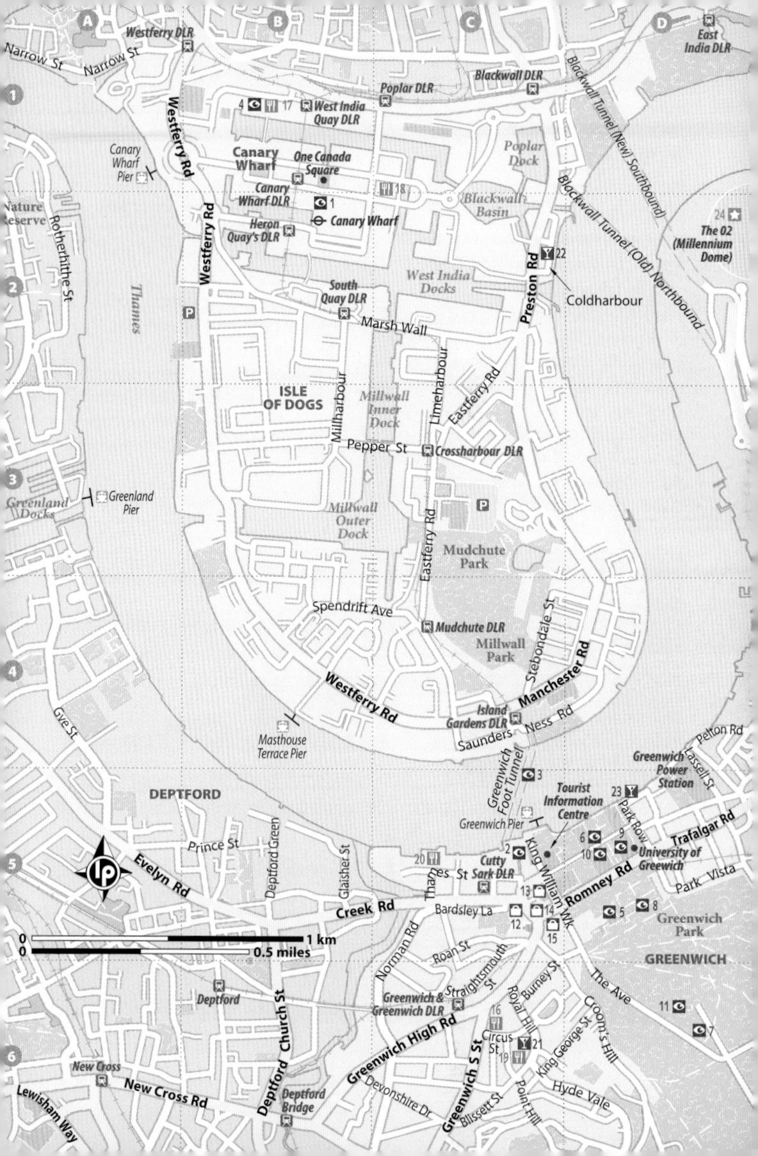

Westferry DLR
Narrow St
Westferry Rd
West India Quay DLR
Poplar DLR
Blackwall DLR
East India DLR
Canary Wharf Pier
Canary Wharf
One Canada Square
Canary Wharf DLR
Heron Quay's DLR
Canary Wharf
Poplar Dock
Blackwall Basin
Blackwall Tunnel (New) Southbound
Blackwall Tunnel (Old) Northbound
The O2 (Millennium Dome)
Rotherhithe St
Thames
West India Docks
Preston Rd
Coldharbour
South Quay DLR
Marsh Wall
ISLE OF DOGS
Millharbour
Millwall Inner Dock
Limeharbour
Eastferry Rd
Pepper St
Crossharbour DLR
Greenland Docks
Greenland Pier
Millwall Outer Dock
Mudchute Park
Spendrift Ave
Mudchute DLR
Millwall Park
Stebondale St
Manchester Rd
Westferry Rd
Island Gardens DLR
Saunders Ness Rd
Masthouse Terrace Pier
Gve St
Greenwich Foot Tunnel
Tourist Information Centre
Greenwich Power Station
Pelton Rd
Lassell St
Park Row
DEPTFORD
Greenwich Pier
Trafalgar Rd
Prince St
Deptford Green
Glaisher St
King William Wk
University of Greenwich
Evelyn Rd
Cutty Sark DLR
Thames St
Romney Rd
Park Vista
Creek Rd
Bardsley La
Greenwich Park
0 1 km
0 0.5 miles
Norman Rd
Roan St
GREENWICH
Straightsmouth St
Burney St
The Ave
Deptford
Deptford Church St
Greenwich & Greenwich DLR
Royal Hill
Croom's Hill
Greenwich High Rd
Circus St
King George St
New Cross
New Cross Rd
Devonshire Dr
Greenwich S St
Point Hill
Hyde Vale
Lewisham Way
Deptford Bridge
Blissett St
Nature Reserve

SEE

CANARY WHARF

www.mycanarywharf.com; 1 Canada Square, Docklands; admission free; ; Canary Wharf

Britain's tallest skyscraper, 244m-high One Canada Square, is actually more impressive from a distance. Nevertheless, it's worth breaking your journey at Canary Wharf DLR to experience the futuristic Docklands development and browse the underworld of subterranean malls.

CUTTY SARK

8858 2698; www.cuttysark.org.uk; Cutty Sark Gardens; Cutty Sark

Britain's last working tea clipper, the beautiful *Cutty Sark* was gutted by fire in 2007. All was not lost, though – many of the timbers had already been removed for conservation. A massive project is now underway to restore the ruined ship to its former glory – see the website for the latest information on when it will reopen.

Canary Wharf bathed in twilight

GREENWICH FOOT TUNNEL

Cutty Sark Gardens; admission free; via stairs 24hr, lifts 7am-7pm Mon-Sat, 10am-5.30pm Sun; Cutty Sark
Reached via glass-topped domes on either side of the river, this Victorian-era pedestrian tunnel provides an atmospheric back route from Greenwich to the Isle of Dogs. Wheelchair users can use the tunnel when the lifts are in operation.

MUSEUM IN DOCKLANDS

0870 444 3851; www.museumindocklands.org.uk; Warehouse No 1, West India Quay; adult/student or under 16yr/concession £5/3/free; 10am-5.30pm; ; West India Quay
This entertaining museum focuses on, predictably enough, the history of trade along the Thames. There are excellent exhibits on slavery and the Blitz, but the highlight is a re-creation of a Victorian wharf in Wapping, complete with animal traders, nautical stores and authentic smells.

NATIONAL MARITIME MUSEUM

8312 6565; www.nmm.ac.uk; Romney Rd; admission free, prices vary for special exhibitions; 10am-5pm; ; Cutty Sark
This rip-roaring seafaring museum uses dioramas and real boats to breathe life into a subject that might otherwise be stuffy. As well as ship models, paintings, swords and sextants, you can walk through a re-creation of a 1940's steamship, and see Nelson's blood-stained military uniform. Paintings are displayed in the attached **Queen's House**, constructed by Inigo Jones (whose work across the river was largely destroyed in the Great Fire of 1666).

THE MERIDIAN

There's a certain thrill to standing on the Prime Meridian. Global time and all points on the east–west access of the earth are measured from the line of 0° longitude at the Royal Observatory in Greenwich. Originally preserved on wind-up clocks, international time is now kept on an atomic clock that is accurate to within one second every million years. As the Prime Meridian is a concept rather then a physical entity, the Greenwich Observatory has made the line visible with a green laser beam that lights up the night sky and is visible from 9 miles away.

OLD ROYAL NAVAL COLLEGE

8269 4799; www.oldroyalnavalcollege.org; King William Walk; admission free; 10am-5pm Mon-Sat, grounds only 8am-6pm Sun; ; Cutty Sark
Now used mainly by the University of Greenwich, the former Naval

College is another grand civic building constructed by Christopher Wren. The highlights here are the chapel, with its ornate carved roof, and the Painted Hall, with its stunning Baroque murals by Sir James Thornhill. Parts of the films *Eastern Promises* and the *Golden Compass* were filmed in the precinct.

ROYAL OBSERVATORY

☎ 0870 781 5168; www.nmm.ac.uk; Greenwich Park; admission free, fees apply for planetarium shows; 10am-5pm Oct-May, courtyard to 8pm May-Aug; ; Cutty Sark

Come rain or shine, there's always a queue of people waiting to be photographed straddling the Prime Meridian – the line that divides the eastern and western hemispheres. More interesting sights include the futuristic **Peter Harrison Planetarium**, the 28in refracting telescope and four of John Harrison's amazing marine chronometers that solved the problem of longitude.

SHOP

BEEHIVE *Clothing*

☎ 8858 1964; www.flying-duck.com; 320-322 Creek Rd; 10.30am-6pm Mon-Fri, 10.30am-6.30pm Sat & Sun; Cutty Sark

Part café, part second-hand record store, part cathedral to kitsch, Beehive is leading the bohemian retro wave sweeping through Greenwich. Pop in for a cappuccino and leave with a mod-print dress and a vintage Bakelite telephone.

EMPORIUM *Clothing*

☎ 8305 1670; 330-332 Creek Rd; 10.30-6pm Wed-Sun; Cutty Sark

Austin Powers would love Emporium. The collection of vintage designer gear is replete with frilly cuffed shirts and brilliant-blue smoking jackets. However, famous-name clothes come at almost new prices.

GREENWICH MARKET *Market*

☎ 8293 3110; www.greenwichmarket.net; College Approach; 9.30am-5.30pm Fri-Sun, from 7.30am Thu; Cutty Sark

London has half a dozen markets like Greenwich Market, selling tourist-oriented arts and crafts to out of towners. The Greenwich version is worth a browse, but don't expect anything particularly earth-shattering.

JOY *Clothing*

☎ 8293 7979; www.joythestore.com; 9 Nelson Rd; 10am-7.30pm; Cutty Sark

Despite being a chain, Joy exudes just the right amount of quirkiness. Inside you'll find trendy menswear and womenswear, odd books, saucy homewares and trinkets that make good stocking fillers for kids at Christmas.

MEET BERNARD *Clothing*

☎ 8858 4047; www.meetbernard.com; 23 Nelson Rd; 10am-6pm; Cutty Sark

Who is Bernard? We're not sure. But his shop in Nelson Rd is crammed full of carefully selected designer clothes for hip, young men about town.

EAT

The Greenwich gyratory is full of unremarkable chain restaurants and Docklands' West India Quay is lined with corporate pavement restaurants – the following backstreet eateries are a better bet.

INSIDE

Modern European ££

☎ 8265 5060; www.insiderestaurant.co.uk; 19 Greenwich South St; noon-2.30pm Tue-Sun, 6.30-11pm Tue-Sat; Cutty Sark

An unexpected gourmet find in the more mainstream part of

Inside Inside – understated dining room, superior cooking

Greenwich. Enter the understated dining room and enjoy superior modern European cooking, with lots of seafood and meat sourced from small local suppliers.

PLATEAU *British* £££

☎ 7719 7800; www.danddlondon.com; Canada Place; ☎ noon-3pm Mon-Fri, 6-10.30pm Sat; Canary Wharf

Probably the slickest place to eat in Docklands, Terence Conran's Plateau serves modern British haute cuisine in a futuristic dining room full of monochrome design furniture. It's on the 4th floor, above the giant Waitrose.

ROYAL TEAS *Vegetarian* £

☎ 8691 7240; www.royalteascafe.co.uk; 76 Royal Hill; 9.30-5.30pm Mon-Fri, 10am-6pm Sat, 10.30am-6pm Sun; V; Cutty Sark

This delightful small café specialises in coffees, big breakfasts, superb cakes, stuffed baguettes and warming bowls of soup. It's a welcome retreat from the Greenwich crowds.

SE10 RESTAURANT & BAR *British* ££

☎ 8858 9764; www.se10restaurant.co.uk; 62 Thames St; 12.30-3.30pm & 7-10pm Mon-Sat, 10.30am-4pm Sun; Cutty Sark

Hidden away behind a housing complex just west of the Cutty Sark, this converted pub offers a laid-back interpretation of fine dining. The menu is modern British via Asia and the Mediterranean.

DRINK

GREENWICH UNION *Pub*

☎ 8692 6258; www.greenwichunion.com; 56 Royal Hill; Greenwich

A SURPRISING LOCATION

The Isle of Dogs has cropped up in a surprising variety of films during its transition from industrial wasteland to futuristic financial hub. The docks were so run-down in 1987 that Stanley Kubrick chose them as a stand in for war-torn Vietnam in *Full Metal Jacket*. The boom years of Thatcher's Britain saw the docks converted into office buildings and flats for 'young urban professionals'. The entrepreneurism and greed of the era was brilliantly sent up in the 1980 gangster thriller *The Long Good Friday*. One of the most entertaining movie chases ever set in London was the speedboat chase to the Millennium Dome in the 1999 Bond movie *The World is Not Enough*. More recently Canary Wharf served as the last hope of a virus-infected London in the apocalyptic *28 Weeks Later*.

Skip the touristy pubs around the park for this genteel local on Royal Hill. In-house brewery Meantime produces some lip-smacking beers, including a solid house stout and Belgian-style fruit beers.

THE GUN *Gastropub*

☎ 7515 5222; www.thegundocklands.com; 27 Coldharbour; Blackwall or Canary Wharf

Docklands is chain-pub central, but you'll find an honest pint – and some cracking British gastropub fare – at this restored dockers' pub on the riverside, facing the Millennium Dome.

TRAFALGAR TAVERN *Pub*

☎ 8858 2909; www.trafalgartavern.co.uk; Park Row; Cutty Sark

You pay a premium for the view at this stately Victorian tavern by the river, but what a view! Tables inside and out look across the Thames to the soaring towers of Docklands.

PLAY

THE O2 *Live Music*

☎ 8463 2000; www.theo2.co.uk; Drawdock Rd, Greenwich; North Greenwich, or boat to QEII pier

The former Millennium Dome has been reinvented as 'The O2', a massive leisure complex with a multiplex cinema, the O2 Bubble exhibition space, and the Indigo2 and O2 Arena concert halls, hosting famous names like James Blunt and Tina Turner.

>SNAPSHOTS

London has more facets to its character than the Cullinan diamond. Whether you are interested in art and history, music and theatre, shopping and fashion, monuments and markets, bars and dining, live bands and clubbing or cocktails and cabaret, you should find something here to pique your interest.

A bite-sized break from the hustle and bustle at Broadway Market (p150)

ACCOMMODATION

Accommodation in London isn't cheap. The average price of a hotel room in 2007 was £119 per night, and while five-star hotels push up the average, you'll still pay more for any class of room than you would in most of Europe.

The good news is that chintzy upholstery and tired velvet drapes have gone out with the recycling, to be replaced by ethnic fabrics and minimalist designer furniture. You can still find the cholesterol-infused monster that is the English fried breakfast, but you'll also get a real continental breakfast and freshly squeezed juice. Even the grande dames like Browns and the Dorchester have adopted a more contemporary approach in the face of growing competition from boutique hotels.

Once accommodation was thinly spread across the city, but today hotels and guesthouses are more evenly spaced out, though well-established places to stay like Earl's Court (p120), Russell Sq (p102), Bloomsbury (p102) and Mayfair (p86) still dominate the scene.

The most exciting things are happening in fringe districts like Hoxton (p144), Notting Hill (p134) and Clerkenwell (p58), which have some very cool hotels to go with their very cool nightlife. Top recommendations include Clerkenwell's Zetter, Shoreditch's Hoxton Hotel and Notting Hill's Miller's Residence.

The accommodation revolution has also permeated into traditional enclaves. If you want to be close to the Kensington–Knightsbridge–Chelsea triangle (p120), Earl's Court boasts trendy lodgings like the Mayflower and Base2Stay, as well as the budget easyHotel, and B+B Belgravia leads the way around Victoria. If you'd prefer to be in the thick of things in the West End, Bloomsbury (p102) has the myHotel, Fitzrovia (p102) has the Charlotte Street Hotel and Soho (p42) has the Soho Hotel.

Finding budget hotels that you'd actually want to stay in is still an issue, though some of the hotel booking sites on below can bring mid-range places into the budget category.

WEB RESOURCES

There are many generic online booking websites, but you'll find unbiased reviews at Lonely Planet's own accommodation website: lonelyplanet.com/hotels. Local booking engines like www.hotelconnect.co.uk, www.superbreak.com and www.laterooms.com can also be useful, while the official tourist website, www.visitlondon.com, has some good last-minute deals and www.londontown.com offers discounted hotel bookings.

BEST GRAND HOTELS

> Brown's (www.brownshotel.com)
> Claridge's (www.claridges.co.uk)
> Dorchester (www.dorchesterhotel.com)

BEST HOTELS FOR FOODIES

> Berkeley (www.the-berkeley.co.uk)
> Halkin (halkin.como.bz)

BEST DESIGNER DENS

> Charlotte Street Hotel (www.firmdale.com)
> Sanderson (www.sandersonlondon.com)
> Mandeville (www.mandeville.co.uk)
> Myhotel Bloomsbury (www.myhotels.com)
> Number Sixteen (www.firmdale.com)
> Soho Hotel (www.firmdale.com)
> Zetter (www.thezetter.com)

BEST BUDGET CHAINS

> Express by Holiday Inn (www.hiexpress.co.uk)
> Premier Inn (www.premierinn.com)
> Travelodge (www.travelodge.co.uk)

BEST ALADDIN'S CAVE OF ANTIQUES

> Miller's Residence (www.millershotel.com)

BEST STYLE ON A BUDGET

> Base2Stay (www.base2stay.com)
> B+B Belgravia (www.bb-belgravia.com)
> Hoxton Hotel (www.hoxtonhotels.com)
> Mayflower (www.mayflowerhotel.co.uk)

BEST HOSTEL

> Generator (www.the-generator.co.uk)

MOST RIDICULOUSLY ORANGE ROOMS

> easyHotel (www.easyhotel.com)

FOOD

As discussed on p23, dining in London is no longer restricted to stuffy formal dinners with silver service or plates of chips in greasy spoon cafés. In recent years, modern British cuisine has taken its place among the great cuisines of Europe, helped by the addition of rare ingredients and meat from traditional farm breeds. Even the Cordon Bleu crowd are looking nervously over their shoulders at the likes of Gordon Ramsay (see boxed text, p98) and Tom Aikens (see boxed text, p130).

There are two important things to note about dining in London compared to some cities around the world: first – vegetarians will have no problem finding good and varied food to eat (see the listings in the 'Eat' sections of this book), and second – the best food is not always served at the flashiest looking restaurants; many of the capital's finest restaurants are tiny local canteens in outlying ethnic neighbourhoods.

So, if you want top-notch Anatolian food, you should head to Islington (p162) or Hackney (p150), which also boasts London's best Vietnamese food (see boxed text, p153). Fans of Korean barbecues will find the best in town in the backstreets of Soho (p52) and Holborn (p66). Chinese restaurants dominate Gerrard and Lisle Sts in Chinatown (p52), but also Queensway in Bayswater (p137).

The best Indian – or, more accurately, Bangladeshi and Pakistani – food is found out east around Commercial and Whitechapel Rds (p150), while modern British restaurants are concentrated in Soho and within the Knightsbridge–Chelsea–Kensington triangle (p128). Edgware Rd (p117) is the undisputed Lebanese food capital of London.

Another place to turn for a superior meal is the nearest gastropub. Many London pubs have ditched the pie-and-chips menu for serious gourmet cuisine. The epicentre of the gastropub explosion is Clerkenwell (see boxed text, p67), but you'll now find upscale pub grub as far afield as Notting Hill (such as Cow, p138) and Docklands (The Gun, p183).

London's dining scene is always in flux. Long impressed by Baltic restaurants like Daquise (p129), Baltic (p80) and Trojka (p172), food pundits are now waiting to see what effect the recent influx of Polish and other Eastern European immigrants will have on the capital's cuisine.

BEST GASTROPUBS

> Anchor & Hope (p80)
> Eagle (p67)
> The Gun (p183)
> Wells (p172)

BEST FOR CELEB-SPOTTING

> Gordon Ramsay (p129)
> The Ivy (p54)
> Wolseley (p100)

BEST CHAINS

> Carluccio's (p53)
> Ottolenghi (p162)
> Wagamama (p109)

BEST ASIAN

> Painted Heron (p130)
> Hakkasan (p108)
> Nahm (p99)
> New Tayyab (p152)
> Nobu (p99)

Top left Putting the F into fine dining at Gordon Ramsay's Boxwood Café (p130) **Above** Coffee culture in the East End

ARCHITECTURE

Some cities have a unifying architectural style, but London is a glorious architectural mongrel, with a heritage forged over 2000 years of conflict, expansion and reinvention. If there was such a thing as a distinctive London building, it would have to be the tall Georgian townhouse, seen all over the West End and throughout West London. These handsome homes are a monument to a city on the up-and-up, when even residential housing was given that extra flourish.

Considering that more than one million bombs fell on London during WWII, it's remarkable that any ancient architecture survived, but London is awash with historical treasures. Traces of medieval London are hard to find thanks to the Great Fire of 1666, but several works by the celebrated Inigo Jones (1573–1652) have endured through the centuries, including Covent Garden Piazza (see boxed text, p49) and the gorgeous Queen's House at Greenwich (p179).

There are a few even older treasures – the Tower of London (p64) partly dates back to the 11th century, while Westminster Abbey (p92) and Temple Church (see boxed text, p59) are 12th- to 13th-century creations, and the gatehouse to St Bartholomew the Great church (p63) dates to 1595.

After the fire, renowned architect Sir Christopher Wren was commissioned to oversee reconstruction, but his grand scheme for a new city layout of broad, symmetrical avenues never made it past the planners. His legacy lives on, however, in the stunning St Paul's Cathedral (p63), the maritime precincts at Greenwich (p176) and the many churches dotted around the City (see p70).

Wren-protégé Nicholas Hawksmoor joined contemporary James Gibb in creating a new style known as English Baroque, which found its greatest expression in Spitalfields' Christ Church (p146) and St Martin-in-the-Fields on Trafalgar Sq (p47). However, remnants of Inigo Jones' classicism endured, morphing into neo-Palladianism in the Georgian era.

Like Wren before him, Georgian architect John Nash aimed to impose some symmetry on unruly London, and was slightly more successful in achieving this through grand creations like Trafalgar Sq and the curving arcade of Regent St. Built in similar style, the surrounding squares of St James's (p86) are still some of the finest public spaces in London – little wonder then that Queen Victoria decided to move into the recently vacated Buckingham Palace (p87) in 1837.

Pragmatism replaced grand vision with the arrival of the Victorians, who wanted ornate civic buildings that reflected the glory of empire but were open to the masses, not just the privileged few. The turrets, towers and arches of the Victorian neo-Gothic style are best exemplified by the Kensington Museums (see p16) and St Pancras Chambers (p160). The Victorians were also responsible for creating huge slums of cheap terraced houses that are now worth around £400,000 to £900,000 each.

A flirtation with Art Deco before WWII was followed by a functional, even brutal, modernism afterward, as the city rushed to build new housing to replace the terraces lost in the Blitz. The next big wave of development came in the derelict wasteland of the former London docks (p176), which were razed of their working-class terraces and warehouses and rebuilt as towering skyscrapers and 'loft' apartments for Thatcher's yuppies.

There was then a lull in new construction until 2000, when the London Eye (see p10) was raised on the south bank of the Thames. It was quickly followed by Norman Foster's wonderful 30 St Mary Axe (see boxed text, p65) – the 'Gherkin' to those who love it – marking the start of a new wave of skyscraper construction. The enthusiasm for new developments is causing nervous jitters among custodians of London's architectural heritage, including the UN World Heritage committee – for more information, see boxed text, p65.

BEST HISTORICAL GEMS

> Buckingham Palace (p87)
> Houses of Parliament (p90)
> St Paul's Cathedral (p63)
> Tower of London (p64)
> Westminster Abbey (p92)

BEST MODERN ICONS

> City Hall (p73)
> Gherkin (p65)
> Millennium Bridge (p76)
> Tate Modern (p79)

Above Cityscape view – complete with extra Gherkin (p65) – from Vertigo 42 (p69)

MUSEUMS & GALLERIES

Ever since the days of the Georgian gentleman collector, Britain has punched above its weight when it comes to museums. Even great treasure houses like the British Museum (p104) were originally based on the collection of a single individual: Hans Sloane, the physician and philanthropist who invented drinking chocolate. From small acorns grow mighty oaks – the British Museum has expanded to a vast collection of 13 million treasures from across the globe, including such iconic objects as the Parthenon Marbles and the Rosetta Stone. There are still some small offbeat museums dotted around the capital that recall the glory days of the eccentric private collector, most notably Sir John Soane's Museum (p62), the Wellcome Collection (p104) and the Hunterian Museum (p59).

Probably the best known museums in all of London are the Kensington Museums – the Natural History Museum (p124), the Science Museum (p125) and the Victoria & Albert Museum (V&A; p126). Commendably, all three are making their vast collections more accessible to the public. The Science Museum has come up with innovative ways to present scientific ideas, including the excellent Dan Dare exhibition in 2008, based on the comic-book hero. The Natural History Museum has opened up its 'wet' specimens collection to the public in the expanding Darwin Centre, and the once-stolid V&A looks brighter and bolder than it has done for decades.

Don't overlook London's smaller museums. The Museum of London (p62) and Museum in Docklands (p179) are becoming increasingly recognised for presenting stuffy events of history in an engaging modern way, and it's life-affirming that quirky museums like Pollock's Toy Museum (p104), the Old Operating Theatre Museum (p78) and the Wallace Collection (p115) exist and are open to the public.

London's galleries are something else again. New York and Paris may have more individual 'famous artworks', but London has made an artform of presenting art to the masses. The stunning Tate Modern (p79) is set to expand into a futuristic new wing by 2010, and even the staid Tate Britain (p126) has been breaking the mould by staging exhibitions of conceptual art (see boxed text, p127).

Elsewhere, the White Cube (p93 and p147) is still proving its worth as a talking point by indulging the pretensions of the Young British Artists, while former Britart collector Charles Saatchi has turned his attention to post-9/11 US painting in his new gallery (p125). For our money, some of the most engaging art can be seen outside of the traditional gallery setting – we loved graffiti-artist Banksy's satirical 'One Nation under CCTV' mural in Newham St, created in a hit-and-run raid in full view of the CCTV cameras (see also boxed text, p147).

Meanwhile, London's big art galleries have moved on from their love affair with blockbuster exhibitions of famous-name painters and delved a little deeper into the global art pool. Tate Britain recently proved it was still 'down with the kids' with shows from the Chapman brothers and a Britart-tastic Turner Prize retrospective, while the Royal Academy of Arts (p91) turned its attention to Cranach and fine French and Russian paintings from Moscow and St Petersburg. Showing its new sense of fun, the V&A displayed the extravagant costumes of Diana Ross and the Supremes.

A recent phenomenon has been the staging of high-profile exhibitions away from the established museums and galleries. The treasures of Tutankhamun went on show in The O2 (p183), while the works of Dali were given very commercial treatment in the County Hall gallery (p73). Our latest top exhibitions were the Chinese terracotta warriors at the British Museum and the Wellcome Collection's display of 26 ancient human skeletons unearthed around London.

For current exhibition news, check the following websites: www.visitlondon.com, www.thisislondon.co.uk and www.guardian.co.uk.

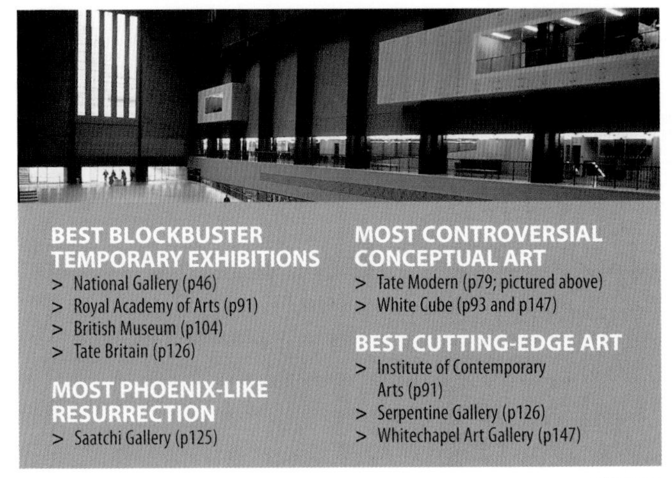

BEST BLOCKBUSTER TEMPORARY EXHIBITIONS
> National Gallery (p46)
> Royal Academy of Arts (p91)
> British Museum (p104)
> Tate Britain (p126)

MOST PHOENIX-LIKE RESURRECTION
> Saatchi Gallery (p125)

MOST CONTROVERSIAL CONCEPTUAL ART
> Tate Modern (p79; pictured above)
> White Cube (p93 and p147)

BEST CUTTING-EDGE ART
> Institute of Contemporary Arts (p91)
> Serpentine Gallery (p126)
> Whitechapel Art Gallery (p147)

VILLAGE LONDON

To understand how London fits together today, it pays to look at how London evolved. There was never a coherent plan for the city – London just surged out from the Roman walls of the City, absorbing parks, rivers, farms and villages as it struggled to accommodate its burgeoning population.

Though they found themselves absorbed into the urban sprawl, many of London's villages managed to preserve their quaint country atmosphere, with large public parks and quaint high streets full of locally-owned shops. Today these are among some of London's most desirable residences, with a bohemian air that marks them out from the rest of the metropolis.

Perhaps the most famous London village is Greenwich (p176), with its winding lanes, bijou shops and cafés, country-style pubs and sprawling Greenwich Park. North Londoners feel a similar affection for Hampstead (p168), which still feels only vaguely connected to the rest of the city. During the 18th century, residents of the City braved dirt roads and villainous highwaymen to reach Hampstead's rolling heath and genteel public houses.

You don't have to go as far as the suburbs to find villages in London. Primrose Hill (p166) is a classic London village, with a row of cultured cafés and small shops rising up the hill to Regent's Park (p115). Islington (p158) is also quintessentially villagey, with quirky restaurants, cafés and shops spread out along Upper St and around Islington Green. You can find the village vibe right in the city centre – Marylebone (p112) for example, with its wholefood groceries and pocket-sized boutiques. Even Bermondsey (p84) on the South Bank approaches village status with its foodie high street and tiny village park.

BEST VILLAGE GREENS

> Hampstead Heath (p168)
> Greenwich Park (p176)
> Regent's Park, Marylebone (p115)

BEST VILLAGE EATS

> Inside (p181), Greenwich
> Locanda Locatelli (p118), Marylebone
> Ottolenghi (p162), Islington
> Village East (p84), Bermondsey
> Trojka (p172), Primrose Hill

SHOPPING

Many people visit London just to shop, lured here by the tremendous range of things on sale, rather than bargain prices. While big chain stores are everywhere, London is notable for its small independent shops. Even world-famous designers like Stella McCartney and Matthew Williamson have one-off boutiques where ordinary mortals can browse the latest in high fashion away from the snobbery of the big couture houses.

Fashion is probably London's biggest retail commodity. The city has been setting trends ever since Mary Quant dreamt up miniskirts, and big designers' ideas are translated into high-street fashion faster than you can say Yves Saint Laurent. While Mayfair (p93) mainly caters to the high end of the market, Primark (p96) offers catwalk style on a shoestring.

The leading edge of the fashion scene is represented by cutting-edge Hoxton (p147), where trends seem to come and go in the time it takes to sign a credit card slip. At the other end of the spectrum are London's posh department stores – Harrods (p127), Selfridges (p116), Fortnum & Mason (p94) et al – if you're planning a splurge, time your visit to coincide with the sales in June and July.

And London is shaking off its reputation as a foodie desert with some excellent food markets – Borough Market, in particular (p79) – and dozens of delis stocked with fine cheeses, charcuterie and artisan ingredients. Don't overlook London's other markets – the best are in the East End (see boxed text, p150), Notting Hill (p136) and Camden (p169).

BEST FOR MARKET LOVERS

- > Borough Market (p79)
- > Columbia Road Flower Market (see boxed text, p150)
- > Portobello Rd Market (p136)
- > Camden Markets (p169)

BEST FOR STREETWISE CATS

- > Absolute Vintage (p147)
- > Retro Clothing (see boxed text, p137)
- > Laden Showrooms (p148)
- > Start (p148)

THEATRE

Along with Broadway in New York, London is the most important stage in the English-speaking world for drama and musical theatre. Some visitors plan a trip to London for the sole purpose of taking in a musical in London's West End (see p24). Performances range from the sublime – Michael Bourne's all-male *Swan Lake* – to the rousing – *Les Miserables,* now the longest-running London musical of all time – and the over-rated – *Spamalot,* with Eric Idle milking the Monty Python franchise for another few million.

Despite the high profile of musical theatre, more exciting things are happening in the fields of dance and drama. Kevin Spacey finally seems to be earning his keep at the Old Vic (p84), and Nicholas Hytner and David Lan have attracted some sterling new talent to the National Theatre (p83) and Young Vic (p85) respectively. Meanwhile, Sadler's Wells (p70) continues to lead the way in the field of dance, providing a forum for rare foreign artforms like Japanese *kabuki*.

Even the West End seems to be taking a few risks – critics are still gushing with praise for Damon Albarn and Jamie Hewlett's recent Gorillaz-style pop opera *Monkey: Journey to the West* at the Royal Opera House (p57). But all the old favourites remain – *Phantom of the Opera* continues to pack them in at Her Majesty's Theatre and you can still experience Shakespeare the bawdy way at the entertaining Globe Theatre (p84). For comprehensive listings, visit www.officiallondontheatre.co.uk.

BEST FOR THE CLASSICS

> National Theatre (p83)
> Shakespeare's Globe (p84; pictured right)

BEST FOR EDGY PERFORMANCES

> Almeida Theatre (p164)
> Donmar Warehouse (p55)
> Old Vic (p84)
> Royal Court Theatre (p133)
> Young Vic (p85)

GAY & LESBIAN

The pink pound certainly flexes its muscle in London. Homosexuality only became legal in 1967, but you wouldn't know that it had ever been proscribed from the plentiful gay venues dotted around the city. The London scene is out and proud – discrimination is rare and harassment even rarer, and gay and straight people mix in bars and clubs all over town.

The obvious first port of call for gay visitors to London is Soho's Old Compton St (p42). Night and day, a carnival atmosphere pervades, and the pavement cafés are prime spots for watching the beautiful people preen and promenade. The whole street is lined with gay bars, pubs and shops, where you can pick up freesheets like *Boyz* and *QX,* and magazines like *Attitude, Gay Times* or *Diva* for tips about the scene.

London's most famous gay club is Heaven (see boxed text, p56), under the railway arches at Charing Cross, but there are many more clubs in the alleys of Soho and around the West End. There's another cluster of essential gay clubs south of the river in the so-called 'Vauxhall Village' – see boxed text, p82 – and another low-key scene out in Kensington on Old Brompton Rd (p131).

Although the scene is criticised for being too commercial, gay London is a broad church. It doesn't just cater for muscled-up fans of mainstream techno – connoisseurs of indie, alternative and even heavy metal can rock out at nights like the famous Popstarz at SIN (see boxed text, p56).

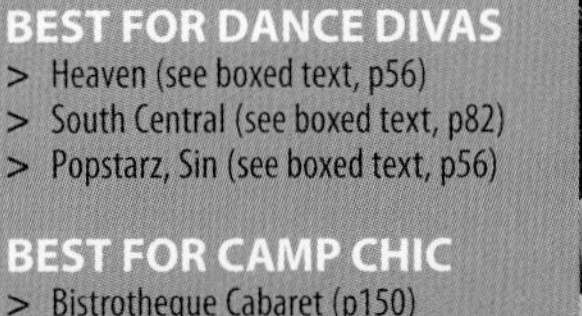

BEST FOR DANCE DIVAS

> Heaven (see boxed text, p56)
> South Central (see boxed text, p82)
> Popstarz, Sin (see boxed text, p56)

BEST FOR CAMP CHIC

> Bistrotheque Cabaret (p150)
> Royal Vauxhall Tavern (see boxed text, p82)
> Circus, Soho Revue Bar (see boxed text, p56)

CLUBBING

London is one of Europe's clubbing capitals, not just for the number of venues booming basslines out into the night, but for the sheer, dazzling variety of musical forms on offer. You'll find everything from vintage disco and soul to old-school house, indie, funk, punk, R&B, drum and bass, hip-hop, electro, techno and everything in between. Hell, there are even a few classic Russian pop nights in town.

The West End has the highest concentration of clubs and bars, but it also suffers the less pleasant consequences of binge-drinking and a lack of public toilets (you get the picture). Better nights out can be had away from the city centre in the bars and clubs of Clerkenwell (p70) and Notting Hill (p139). Probably the most innovative scene is around Hoxton (p154), though not everyone will go for to the cooler-than-thou attitude.

Rockers head for Camden, where venues like Koko (p174) and Barfly (p173) host rocking DJ nights as well as live indie bands. Fans of the urban dance scene favour gritty, streetwise venues like the End (p56) and Egg (p164), which also hosts London's favourite Saturday night and Sunday morning, Breakfast at Egg.

The dress code at London clubs varies from formal to smart casual via anything goes. As a general tip, tailor what you wear to match the venue crowd – if you turn up looking too different to the other punters, you may not make it past the style gurus on the door. For events listings, check out www.dontstayin.com or the weekly listings magazine *Time Out* (£2.99).

BEST FOR TURNTABLE FANATICS

> Egg (p164)
> End (p56)
> Fabric (p70)
> Ministry of Sound (p83)

BEST FOR BROAD MUSICAL MINDS

> Barfly (p173)
> Big Chill Bar (p154)
> Cargo (p156)
> Koko (p174)

MUSIC

There was a band called Londonbeat, but they came from the Netherlands – no real London band would ever call themselves something so predictable. Artists like Madness, the Sex Pistols and the Pogues immediately leap to mind when talking about the London music scene, but the city has been producing cutting-edge music for centuries – George Frideric Handel wrote some of his most famous oratorios from his home in Mayfair between 1723 and 1759.

Probably the most famous event in the modern-day classical calendar is the annual Proms (p30), which has expanded its repertoire in recent years to appeal to a broader audience – the 2008 concerts featured a Doctor Who show at the Royal Albert Hall (p132), complete with Daleks. The Royal Festival Hall (see p85) also turned a few heads by choosing Motorhead for its grand reopening concert in 2007.

If there's one music genre that rocks London's world, it's...well...rock. Famous venues like the Barfly (p173), Dublin Castle (p174) and The Old Blue Last (p157) are still providing a forum for teens with guitars and dreams of world domination. London's rock scene has always flirted with other genres, however, so there's more on offer than three-chord rock.

Then there's the jazz scene, still happily tootling away in clubs like Ronnie Scott's (p57) and the 100 Club (p110). Over in Camden, the Jazz Café (p174) has expanded the definition to include fusion acts like Finley Quaye and De La Soul.

BEST BIG-NAME VENUES

> Brixton Academy (p83)
> Koko (p174)
> Ronnie Scott's (p57)
> Royal Albert Hall (p132)
> Shepherd's Bush Empire (p133)

BEST GRUNGE DENS

> Barfly (p173)
> Boogaloo (p173)
> Dublin Castle (p174)
> The Old Blue Last (p157)

RIVERSIDE LONDON

Without the Thames, there would be no London. The Romans chose this location as a crossing point over the river, and the City depended on trade along it until the modern day, when attention shifted to moving vast sums of money between bank accounts. It would have been quite something to see the tall ships lined up along the docks at the turn of the 20th century.

A hundred years later and Londoners have almost forgotten how to use their river. Although there are regular commuter boats, most of the two million people who take to the Thames every year are sightseers. There are now plans to dramatically expand ferry services (see boxed text, p222) on the river in time for the 2012 Olympics.

With the decline of the port of London from the 1960s, the docks fell into ruin and even the art centres built along the South Bank (p85) in the 1950s started to look gloomy and grey. Things began to pick up from the 1990s, as the yuppie generation ploughed money into the redevelopment of Docklands (p176) as office towers and apartment buildings.

The raising of the London Eye (p76) in 2000 marked the beginning of a renaissance along the river. It was soon joined by the Millennium Bridge (p76), Tate Modern (p79) and the space-age City Hall (p73). Despite all this the most charming parts of the river are still the least developed – the riverfront by the Old Royal Naval College (p179) at Greenwich has hardly changed in 300 years.

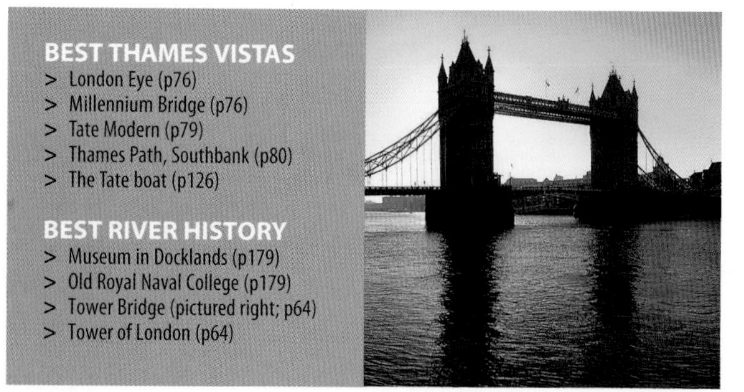

BEST THAMES VISTAS

> London Eye (p76)
> Millennium Bridge (p76)
> Tate Modern (p79)
> Thames Path, Southbank (p80)
> The Tate boat (p126)

BEST RIVER HISTORY

> Museum in Docklands (p179)
> Old Royal Naval College (p179)
> Tower Bridge (pictured right; p64)
> Tower of London (p64)

KIDS

The idea that children should be seen and not heard went out with the Victorians. London is becoming more child-friendly every year, with new activities and special events for kids adding interest at museums and monuments. Remember, however, that children are easily exhausted by the crowds and long walks involved in a day out in London – don't plan too much and build in some rest time in London's public parks (see p15).

An excellent activity for children is a sleepover at the Science Museum or Natural History Museum (see boxed text, p125), though you'll need to book months ahead. The Museum of London (p62) and Museum in Docklands (p179) have great activities during school holidays, and the Natural History Museum (p124) offers fantastic free explorer backpacks for under sevens. Most attractions offer family tickets and discounted entry for kids under 15 or 16 years (children under five are usually free).

When it comes to getting around, buses are much more child-friendly than the tube, not least because they're big and red and they offer views of all the sights. Open-topped bus tours are always fun for families (p224). Children aged five to 10 years travel free on public transport if accompanied by an adult, and children aged 11 to 15 years travel free on the buses with an Oyster photocard (follow the Oyster link at www.tfl.gov.uk). Child discounts also apply on London's riverboats.

BEST FOR INQUISITIVE YOUNG MINDS

- British Museum (p104)
- Natural History Museum (p124)
- Pollock's Toy Museum (p104)
- Science Museum (p125)
- V&A Museum of Childhood (p157)

BEST PLACES TO RECHARGE YOUNG BATTERIES

- Giraffe (pictured right; p152)
- Hyde Park & Kensington Gardens (p121)
- London Zoo (p114)
- St James's Park (p91)

SPORTING LONDON

To fully appreciate how Londoners feel about sport, you need to play Chas & Dave's *Ossie's Dream* at full volume and watch footage of Tottenham Hotspurs' victory in the 1981 FA Cup. Londoners are football crazy, and matches at the home grounds of the five London teams in the Barclays Premiership are truly passionate affairs.

There's a neighbourhood politics to football in London. Tottenham Hotspurs are fierce rivals of Arsenal (see p164), whose ground is just 2km away, and every team pitches themselves against Chelsea (see p131) because of the perception that Roman Abramovich has circumvented the spirit of the game by spending more than anyone else on new players. For more information, see boxed text, p164.

Cricket is London's other big passion. Lord's (p114) and the Oval (p85) are the venues where foreign teams often achieve their objective of thrashing the home side. For more on capital cricket, see boxed text, p115.

There's really only one thing we need to say about tennis: Wimbledon (see p29), the top event of the summer sporting season. As an interesting sidenote, modern tennis is derived from the medieval game of 'real tennis', which is still played at Hampton Court Palace (see boxed text, p92).

The other team sport of note in London is rugby union – the official headquarters of the Rugby Football Union (www.rfu.com) is Twickenham Stadium in the southwest of the city, the setting for high-octane England games in the Six Nations series in March.

BEST FOR FANS OF THE BEAUTIFUL GAME

> Arsenal Football Club (p164)
> Chelsea Football Club (p131)
> The rest of London's Premiership clubs (see boxed text, p164)

BEST FOR ENGLISH SPORTING PROWESS

> Lord's (p114)
> The Oval (p85)
> Twickenham Stadium (above)
> Wimbledon (p29)

>BACKGROUND

Two Beefeaters chew the cud

BACKGROUND

HISTORY

THE BIRTH OF A CAPITAL

What did the Romans ever do for us? Well, for one thing, they founded London. The walled town of Londinium was established in AD 43 in the area now covered by the City to protect a vital crossing over the Thames. The Romans filled their new metropolis with grand civic buildings and built a massive stone wall to protect its inhabitants – still visible in places in the City (see p62) – but Londinium was abandoned in AD 410 as the Roman Empire crumbled.

EXIT ROME, ENTER THE VIKINGS

You can't keep a good city down. Sometime in the 6th century the Roman city was reclaimed by Saxon tribes from Germany, who settled around today's Aldwych and Charing Cross. Over the following centuries the Saxons faced strong competition from Danish Vikings, who staged daring raids along the Thames, eventually conquering England in 1015 under the leadership of King Canute. The most important king of the era was Edward the Confessor, who passed control of the City of London to the merchants and founded a new political capital nearby at Westminster in 1042.

1066 & ALL THAT...

In 1066, Anglo-Saxon England fell to Norman troops led by William the Conqueror, adding the French to the list of successful invaders of the British Isles – though, in fact, the Normans were actually the descendents of 'Norse Men' from Scandinavia. William's distrust of the 'fierce populace' led him to build several strongholds, including the White Tower, at the heart of the Tower of London (p64).

COD PIECES & COLONIES

During the Middle Ages London grew into a major centre for international trade. Powerful guilds were founded to protect the interests of tradesmen, and English galleons set sail for distant lands to trade for silk, spices and gold. However, England remained an economic backwater compared to mainland Europe until the rise of the Tudors in the 16th century.

The most famous Tudor king – and arguably the most famous English ruler of all time – was Henry VIII (r 1509–1547). To facilitate a divorce from his first wife, Catherine of Aragon, King Henry created the Church of England in 1534, ushering in the Reformation. As well as chopping off wives' heads, Henry dissolved many of London's Catholic monasteries, seizing church land for use as royal hunting grounds that later formed the basis for Hyde Park (p121) and Regent's Park (p115).

When Henry's daughter, Elizabeth I (r 1558–1603), came to the throne she instructed the English navy to found new trading colonies in Asia and the Americas to rival the empires of Holland, Portugal, France and Spain, triggering numerous conflicts with her Catholic European neighbours. The 'Virgin Queen' became notorious for locking friends and enemies up in the Tower of London and she died without a heir. Elizabeth was succeeded by James I, the son of her Catholic cousin Mary, Queen of Scots, and the weak rule of James I and Charles I prepared the ground for the English Civil War.

ALL CHANGE IN THE CAPITAL

In the early years of the 17th century, wealthy aristocrats adopted the habit of spending the spring season in London, creating a demand for new venues for social interaction. Ballrooms, public gardens, coffee shops, ale houses and bordellos mushroomed all over the capital and theatres, such as the Rose and the Globe (p84), were built to stage the works of playwrights such as William Shakespeare.

This kind of aristocratic high living took a knock-back during the English Civil War (1642–51), when Oliver Cromwell's Roundheads (a coalition of Puritans, Parliamentarians, merchants and ordinary Londoners) wrested control of the country from the monarchy. Despite protestations that he was 'above the law', Charles I was executed at Whitehall (see p86) in 1649. Parliament restored the monarchy in 1660, but by then people power was firmly entrenched in the popular psyche.

TESTING TIMES FOR THE CITY

London was lucky to survive the rest of the 17th century. The Great Plague of 1665, carried to London by rats on the ships of City traders, killed one in five Londoners, and a year later the Great Fire of London destroyed 80 per cent of buildings inside the City walls. The diarist Samuel Pepys recorded a scene of total devastation from the steeple of All Hallows by the Tower (see p70), but miraculously, only nine Londoners lost their lives in the inferno.

LONDON THE PHOENIX

A less resilient city might have given up the fight, but London was made of stronger stuff. Citizens rebuilt their vanished wooden houses in brick and stone and Sir Christopher Wren was commissioned to rebuild St Paul's Cathedral (p63), along with dozens of other churches and civic buildings in the City (see p70) and at Greenwich (see p176). Wren's memorial to the blaze, the *Monument* (p62), can still be seen near London Bridge.

The Great Fire had other unexpected benefits for the city. Hygiene was massively improved, reducing the risk of further plague epidemics, and migrant workers flooded into London to assist with the reconstruction effort. In the years after the fire, Charles II (r 1660–85) shifted his palace from Westminster to St James's and members of the gentry built numerous squares and mansions in the surrounding area – including Buckingham Palace (p87), which become the official seat of the monarchy in 1837.

The 17th century also marked the period when London first became a truly international city. During the 1680s more than 13,500 French Protestants fled to London to escape persecution at the hands of Catholics in mainland Europe. These refugees were allowed to settle around Spitalfields (see boxed text, p150), Soho (see p42) and Clerkenwell (see p58), starting new industries such as watch-making, hat-making and silk weaving. Within a century, London had grown into the largest city in the Christian world, with half a million inhabitants.

MONEY, MADNESS & AMERICA

Britain's global empire expanded massively during the 18th century, with one minor setback – the loss of America, which proclaimed independence on 4 July 1776, during the reign of George III (1760–1820). Anger at the loss of America was widely blamed for the king going temporarily insane, though modern science has revealed that George actually suffered from porphyria, a disease possibly brought on by arsenic poisoning.

Perhaps the most distinctive feature of the Georgian period was the rise of the Regency and Neo-Classical styles of architecture, best expressed in the buildings of Sir John Soane (see p62), who constructed the Bank of England (p59), and John Nash, the genius behind Trafalgar Sq (p47) and Regent St (see p42).

LONDON LEADS THE EMPIRE

The Victorian era (1837–1901) was good for Britain and even better for London. The wealth created by the Industrial Revolution and the expansion of the British Empire flooded back to the capital and the population surged from under one million to more than 6.5 million. To accommodate these new arrivals, huge suburbs of neat terraced houses were built outside the city boundaries, bringing dozens of outlying villages into the urban sprawl.

Ground-breaking developments in the Victorian era included the introduction of gas lighting, the arrival of the steam railways and the massive redevelopment of the East End docks to cater to the ships of the East India Company. The Victorians were also partial to 'neo-Gothic' architecture – fine examples of this exuberant style include Kensington's Natural History Museum (p124) and Victoria & Albert Museum (V&A; p126) and Charles Barry's Houses of Parliament (p90), erected in 1840–60 on the site of the fire-gutted Westminster Palace

Not everyone shared in the wealth. As Dickens revealed, the Victorian era was also a time of terrible urban poverty. Debtors prisons were full to overflowing and thousands of destitute Londoners ended up toiling in workhouses. Victoria was succeeded on her death by her son Edward, whose reign saw the introduction of motorised buses, the expansion of the Victorian Underground train network and the first Olympics in London.

WAR COMES TO LONDON

Hundreds of thousands of Londoners fought in the Great War (WWI; 1914–18) and over 120,000 died on the battlefields of Europe. But even the losses of WWI were overshadowed by the carnage during WWII, when German bombers destroyed huge swathes of London during the Blitz (1940–41) and Winston Churchill's government had to coordinate the war from the Cabinet War Rooms (p90), hidden away below Whitehall.

By the time WWII ended in 1945, 30,000 had died in London, adding to the hundreds of thousands who were killed fighting overseas. After the war the government authorised the construction of huge areas of cheap modern housing to replace the homes lost in the Blitz, ushering in the era of the housing estate. The government also introduced policies to replace London's decimated workforce, inviting in thousands of immigrant workers from the Caribbean and South Asia, paving the way for the multi-ethnic London of today.

Perhaps the most obvious legacy of the post-war years is London's huge stock of brutal modernist architecture, characterised by concrete towers with little empathy for the practical needs of human beings. The Thames docks never really recovered from WWII; shipping moved east to

Tilbury and the Docklands were left to rot until they were redeveloped as a hyper-modern financial district in the 1980s and '90s.

Politically, London swung first left, under the Labour government of Clement Atlee and Aneurin Bevan, the 'father of the National Health Service', then right under a succession of Conservative governments. During Margaret Thatcher's 11 years at No 10 Downing St (p91), the City of London became the leading financial centre in the world, but these years also saw social deprivation, unemployment, strikes and soaring inflation. A failed attempt to introduce the Poll Tax led to riots in the streets of London and Thatcher was toppled from within her own party. Her successor John Major's occupation of No 10 was marked by political sleaze, however, and a Labour government romped in under Tony Blair in 1997.

LONDON IN THE NOUGHTIES

A spirit of optimism pervaded in the early years of the Labour government, helped by Blair's charisma and a strong economy, but things went rapidly downhill after British troops were committed to the Iraq War in 2003. Antiwar demonstrations in London were attended by more than a million people and Labour saw its majority slashed in the 2005 elections.

Anger at British foreign policy rumbled on. On 7 July 2005, just one day after London won the right to host the 2012 Olympics, Islamic terrorists carried out suicide bomb attacks on the London Tube and bus network,

DID YOU KNOW?

> London's population: circa 7.5 million
> Percentage of London residents born outside the UK: 33%
> London's share of UK GDP: 19%
> Average annual household income: £25,700
> Average London house price: £358,500
> Average house price in Kensington and Chelsea: £1,185,349
> Average hourly wage: £12.75
> Average hourly wage in the City: £28.20
> Time a Londoner must work to pay for a Big Mac: 11 minutes
> Unemployment: 6.8%
> Passengers travelling through Heathrow yearly: 67 million
> Overnight stays by tourists in 2005: 25.4 million
> Annual income from tourism: £15 billion

USEFUL WEBSITES

> www.cityoflondon.gov.uk – Read up on the latest from the City of London.
> www.corpoflondon.gov.uk – Get news from the Mayor of London.

killing 52 commuters. Two more attacks in 2005 and 2007 failed because the bombs didn't detonate.

Hoping to shake off the associations of the Iraq War, Labour replaced Tony Blair with Gordon Brown in 2007, but the new prime minister quickly floundered. Soaring oil prices and the global credit crisis wiped billions off the value of shares in the City, and Brown seemed powerless to stop the decline.

LOOKING FORWARDS TO THE OLYMPICS

Facing falling equity, rising knife crime and soaring fuel prices, Londoners ditched Labour mayor Ken Livingstone in favour of Conservative Boris Johnson in 2008, a political swing echoed in national opinion polls. However, flagging support for the London Olympics – already £7 billion over budget – received a boost with the impressive UK tally of medals at the 2008 Olympics in Beijing.

Meanwhile, the economy continued to slide towards recession, with the near collapse of two British banks, and several mergers and bankruptcies that repainted Docklands' landscape almost overnight. In an attempt to firm up the national economy the government moved to nationalise banks and curb risky investments by City brokers.

However, this is unlikely to help the thousands of Londoners who borrowed more money than they could afford to buy houses that were not worth the money they paid for them. London is going to have to be a lot more cautious with money in future, which may affect projects like the Olympics and the new skyscrapers planned for the City of London.

LIFE AS A LONDONER

Perhaps the most striking thing about London is the number of Londoners who originally come from somewhere else. The British capital has been attracting migrant workers, refugees and seekers of fame and fortune ever since it was founded in AD 43. Today the city is home to nearly 50 ethnic groups, speaking more than 300 languages, and one in every three residents was born in another country.

PAINTING LONDON

London has always been a rich subject for painters and artists. Some, like Sir James Thornhill (1675–1734), even painted *on* London – his murals in the Old Royal Naval College (p179) at Greenwich are still visible today. As the birthplace of the Royal Academy of Arts (p91), London gave birth to some of the finest artists in history, including Joshua Reynolds (1723–92), Thomas Gainsborough (1727–88), Joseph Turner (1775–1851) and John Constable (1776–1837), who painted grand portraits of members of London high society and dramatic landscapes of the city and river.

Foreign painters were also drawn to London by the rich subject material. Canaletto (1697–1768) painted some evocative images of Georgian London, with the new St Paul's Cathedral shining brilliantly in the background, and Claude Monet created a whole suite of impressionist paintings of the Houses of Parliament between 1900 and 1904.

Unconstrained by the rules of the Royal Academy, William Hogarth (1697–1764) was free to paint a rather different vision of London. Works like *Gin Lane, Beer Street* and the *Rake's Progress* were satirical propaganda pieces decrying the failing morals of London society.

For our money, the rightful successor to Hogarth is not one of the navel-gazing Young British Artists, but anti-establishment art-vandal Banksy (1974–), one of the few artists still passing comment on the failings of London society (see boxed text, p147).

London is often described as a melting pot, but different communities are concentrated in different areas, creating more of a plum pudding of cultures – locals often like to describe London as the world in a city. Even among new arrivals, there is a strong sense of geographic tribalism. Most visitors soon pick up on the (mostly good-natured) rivalry between North and South London, and the rivalry between west London and pretty much everywhere else. If you want to wind up South Londoners, show them the Tube map upside down (the south has been shamefully neglected by Transport for London).

One thing that unites Londoners is their enthusiasm for the city they call home. Locals may criticise London for its high prices, high crime rate, overcrowding and urban decay, but they are fiercely defensive of the city in the face of criticism from outsiders. As Londoners are keen to point out, few other places in the world offer more to do, more places to eat, more new experiences and more people to share those experiences with.

Internationally, Londoners have a reputation for being standoffish and unfriendly, and this is not entirely undeserved. If you go around smiling at passers-by and saying hello to strangers on the Tube, locals will probably assume you are deranged. On the other hand, Londoners are remarkably unfazed by the weird, eccentric or outlandish, and most people warm up quickly once the ice is broken. The biggest problem most visitors face is finding a local to ask for directions – quite a task in a city that attracts more than 20 million tourists a year.

GOVERNMENT & POLITICS

Since 2000, Greater London has been governed by the Greater London Authority (GLA), under the stewardship of a directly elected mayor, but the City of London largely remains under the control of the politically nonaligned Lord Mayor of London and the archaic Corporation of London, the richest local authority in the world, founded in the 14th century.

Based at City Hall (p73) on the South Bank, the GLA replaced the defunct Greater London Council (GLC), which was disbanded by the government of Margaret Thatcher in 1986 amid allegations of profligate spending and political bias. Ironically, in the 2000 mayoral elections, Londoners overwhelmingly voted for Labour candidate Ken Livingstone, the former head of the GLC.

Despite some major successes in his first two terms, including reducing traffic congestion, winning the 2012 Olympics and managing community tensions after the 7/7 terrorist attacks, Ken Livingstone was

DOS & DON'TS

There are a few simple rules that all visitors should follow to fit in with the London way of doing things.

DO

- Stand to the right – the left side of any set of stairs or escalators on the London transport system is the fast lane.
- Wear comfortable shoes – London pavements eat heels for breakfast.
- Check the bill to see if the tip is included – if not, add 10%.
- Be aware of the home nations – people of Scottish, Irish or Welsh descent can be British but never 'English'.

DON'T

- Jump to the front of the queue – to the English, this is worse than murder.
- Expect table service in the bar – go to the bar yourself, or go thirsty.
- Kiss strangers as a greeting – acceptable in Europe, Londoners are more reserved.
- Smile at people you don't know – unless you are deliberately flirting…

hounded by accusations of financial mismanagement and cronyism. His decision to replace the city's beloved Routemaster buses with 30m-long 'bendy buses' was also deeply unpopular with many Londoners.

In the 2008 elections voters ousted 'Red Ken' in favour of the eccentric Conservative candidate Boris Johnson. Among his many manifesto promises was a pledge to cut back overspending on the Olympics and bring back the Routemaster buses.

FURTHER READING

There's more to London literature than Charles Dickens. A veritable who's who of novelists have created their own unique visions of the city, including a number of Asian and Caribbean authors. London writers have always strived to explore the secret life of the city – the tensions and sexual intrigue behind the stiff upper lip. Here's a pick of the best:

253 (1998; Geoff Ryman) A deeply weird novel, with 253 words devoted to each of the 253 people on board an Underground train – buy the book or read it online at www.ryman-novel.com.

Absolute Beginners (1959; Colin MacInnes) A teenage photographer comes of age in 1950's London against a backdrop of sexual liberation and race riots.

Brave New World (1932; Aldous Huxley) Still the most original dystopian view of the future, set in a strange sex- and drug-obsessed London in AD 2540.

Brick Lane (2003; Monica Ali) The novel that launched Monica Ali, a tale of arranged marriage and escape from social constraints in London's Bangladeshi community.

Bridget Jones's Diary (1996; Helen Fielding) The novelisation of Helen Fielding's popular newspaper column about the life and loves of a London 'singleton'.

Great Apes (1997; Will Self) A surreal satire on London life, as a hedonistic Young British Artist wakes up to find himself the only human in a world of chimpanzees.

High Fidelity (1995; Nick Hornby) Hornby's classic novel of lost love, top fives and obsessive record collecting, set in the north London suburb of Holloway.

Last Orders (1997; Graham Swift) A funeral brings together four ageing East Enders in this moving story of unhealed wounds from WWII.

London: The Biography (2000; Peter Ackroyd) Despite gaps in the narrative, many say this highly accessible book is the definitive history of London.

London Fields (1989; Martin Amis) A dense but gripping study of London lowlife. 'Dickens plus swearing and sex, minus compassion', wrote one reviewer.

Neverwhere (2005; Neil Gaiman) Scary goings-on in a parallel London where Black Friars and Angel Islington are characters rather than districts, from the writer of the *Sandman* comics.

Oliver Twist (1837; Charles Dickens) A classic portrayal of Victorian London, seen through the eyes of a hapless orphan coerced into joining a criminal gang.

Small Island (2004; Andrew Levy) Warmth and humour characterise this prize-winning novel about the first wave of Caribbean immigration to London.

The Buddha of Suburbia (1991; Hanif Kureishi) Raunchy antics in the suburbs as a group of Asian immigrants struggle to assimilate in 1970s London.
The End of the Affair (1951; Graham Greene) Wartime attitudes come vividly to life in this tragic love story set against the background of the Blitz.
The Jeeves Omnibus (1931; PG Wodehouse) Gin and tonics abound in the comic tales of wealthy Mayfair resident Bertie Wooster and his long-suffering butler Jeeves.
White Teeth (2000; Zadie Smith) A vast rambling novel in the Gabriel Garcia Marquez mould, following the trials of three immigrant families in north London.

FILMS

Movies filmed in New York or Paris sparkle with optimism, but film-makers tend to see London through a lens darkly (it must be something to do with the British temperament, or the weather). Nevertheless, directors have produced some ground-breaking films here, covering every imaginable genre from rom-coms and crime capers to sci-fi and horror. The following list covers more than 50 years of film-making, exploring every aspect of life in London.

28 Days Later (2002) Superb scenes of a post-apocalyptic London, after a deadly virus transforms Londoners into flesh-eating zombies. Better than it sounds.
Alfie (1966) Forget the Jude Law remake, Michael Caine was the original Cockney Lothario, getting up to no good in 1960's London.
An American Werewolf in London (1981) Horror comedy that defined the genre, notable for an adrenaline-pumping chase on the London Underground.
Bridget Jones's Diary (2001) Renee Zellweger puts on a convincing London accent and big pants in the film adaptation of Helen Fielding's bestselling novel.
Children of Men (2006) A dystopian vision of the future, where human reproduction has failed and Battersea Power Station is an 'ark' for the world's artistic treasures.
Eastern Promises (2007) A flawed Russian agent infiltrates a London-based gang in David Cronenberg's violent exploration of people trafficking and ambition.
Frenzy (1972) A strangler stalks the streets of Covent Garden in Alfred Hitchcock's second-to-last shocker. Better than any of the Jack the Ripper spin-offs.
Lock, Stock and Two Smoking Barrels (1998) The iconic gangster flick that launched Guy Ritchie (later Mr Madonna) and a thousand clichés about 'shooters' and 'claret'.
Mona Lisa (1986) Michael Caine does a star turn as a villainous pimp while Bob Hoskins falls for the prostitute he drives around Soho and King's Cross.
My Beautiful Laundrette (1985) A taboo-breaking drama about a gay punk helping his Asian boyfriend open a music-filled laundrette.
Nil by Mouth (1997) Withering portrayal of life on a South London housing estate, written and directed by Gary Oldman.

Notting Hill (1999) Soppy blockbuster rom-com with Julia Roberts falling for bookshop-owner Hugh Grant in a conspicuously Rastafarian-free Notting Hill.
The Ipcress File (1965) Michael Caine plays it up as a working-class spy investigating a political conspiracy at a series of London locations.
The Krays (1990) Pop star brothers Gary and Martin Kemp from Spandau Ballet put on a convincing performance as the notorious Kray twins, gangland masters of 1960's London.
The Ladykillers (1955) The definitive Ealing comedy, with Alec Guinness as the buck-toothed criminal mastermind foiled by his gentle landlady in 1920's King's Cross.
The Long Good Friday (1980) The brooding thriller that launched Bob Hoskins to stardom, set against the greed and hedonism of Thatcher's London.
V for Vendetta (2005) Striking scenes of revolution in London, though the party political rhetoric weakens this adaptation of Alan Moore's dark comic masterpiece.
Vera Drake (2005) Realist film-maker Mike Leigh turns his camera to the post-war East End, where a backstreet abortionist pays the price for 'helping out' young girls.
Withnail & I (1986) Eminently quotable British comedy about a pair of louche 'resting' actors in 1960's Camden.

DIRECTORY

TRANSPORT

ARRIVAL & DEPARTURE

AIR

Just west of the city, Heathrow is still London's busiest airport, but flights also land at Gatwick to the south, Stansted to the north and City airport to the east.

Heathrow

The newly expanded **Heathrow** (LHR; ☎ 0870 000 0123; www.heathrowairport.com) is the world's busiest international airport. There now five terminals, reached by three different Underground stations. Always check the terminal for your flight before you travel to the airport.

Gatwick

London's second-biggest airport, **Gatwick** (LGW; ☎ 0870 000 2468; www.gatwickairport.com) has two terminals – north and south – linked by a monorail service. The train station is in the south terminal.

Stansted

London's fastest-growing airport, **Stansted** (STN; ☎ 0870 000 0303; www.stanstedairport.com) mainly caters to low-cost airlines, including Easyjet and Ryanair.

London City

Just east of the City, **London City Airport** (LCY; ☎ 7646 0088; www.londoncityairport.com) is used by commuter airlines serving cities around Europe.

Luton

Many charter and budget airlines use **Luton** (LTN; ☎ 01582-405100; www.london-luton.co.uk), roughly 35 miles north of central London.

CLIMATE CHANGE & TRAVEL

Travel – especially air travel – is a significant contributor to global climate change. At Lonely Planet, we believe that all who travel have a responsibility to limit their personal impact. As a result, we have teamed with Rough Guides and other concerned industry partners to support Climate Care, which allows people to offset the greenhouse gases they are responsible for with contributions to energy-saving projects and other climate-friendly initiatives in the developing world. Lonely Planet offsets all staff and author travel.

For more information, turn to the responsible travel pages on www.lonelyplanet.com. For details on offsetting your carbon emissions and a carbon calculator, go to www.climatecare.org.

Travel to/from Heathrow

	Tube, Piccadilly Line	Heathrow Express Train	National Express Bus	Taxi
Going to	Earls Court, Kensington, West End, Kings Cross	Paddington	Victoria coach station	central London
Duration	1hr to central London	15min	1hr (rush hour 1½hr)	1hr (rush hour 1½hr)
Cost	one way £4 (Oyster Card £2-3.50)	one way/return £14.50/28 (£17.50/31 from conductor)	one way/return from £4/8	black cab £55-70, service cab £35-45
Frequency	every 5-10min 5am-11.45pm Mon-Sat (5.45am-11.30pm Sun)	every 15min, from Heathrow 5.05am-11.45pm, from Paddington 5.10am-11.25pm, reduced service Sun	every 30-60min, from Heathrow 5.30am-9.30pm, from Victoria coach station 7.15am-11.30pm	
Contact	☎ 7222 1234; www.tfl.gov.uk	☎ 0845 600 1515; www.heathrowexpress.com	☎ 08717-818181; www.nationalexpress.co.uk	

Travel to/from Gatwick

	Gatwick Express Train	Southern Railway Train	Thameslink	National Express Bus	Taxi
Going to	Victoria rail station	Victoria rail station	London Bridge, City, King's Cross rail station	Victoria coach station	central London
Duration	30-35min	30-35min	1hr to King's Cross	1-1½hr	1½hr
Cost	one way/return from £17.90/30.80	one way/return from £9.50/18.10	one way/return from £8.90/17.80	one way/return from £6.60/12.20	black cab £90-95, service cab £55-65
Frequency	every 15min, from Gatwick 5.50am-12.35am, from Victoria 5am-11.45pm, some night services	every 15-30min, hourly midnight-4am	every 30min, 24hr	hourly, from Gatwick 5.15am-10.15pm, from Victoria coach station 3.30am-11.30pm	
Contact	☎ 0845 850 1530; www.gatwickexpress.com	☎ 0845 127 2920; www.southernrailway.com	☎ 0845 026 4700; www.firstcapitalconnect.co.uk	☎ 08717-818181; www.nationalexpress.com	

Travel to/from Stansted

	Stansted Express Train	National Express Bus	Terravision Bus	Taxi
Going to	Liverpool St station	Victoria coach station	Victoria coach station	central London
Duration	45-50min	1-1½hr	1-1½hr	1½hr
Cost	one way/return from £16/24	one way/return from £10/17	one way/return £8/12	black cab £105, service cab £50
Frequency	every 15min, from Stansted 5.30am-12.30pm, from Victoria coach station 4.10am-11.25pm, some additional night services	every 20min, 24hr, reduced night services	every 30min, from airport (coach bay 26) 7.15am-1am, from Victoria coach station 2.40am-11.10pm	
Contact	☎ 0845 600 7245; www.stansted express.com	☎ 08717-818181; www.national express.com	☎ 01279-662931; www.terravision.eu	

Travel to/from London City Airport

	DLR	Taxi
Going to	Bank tube/DLR station	City/King's Cross
Duration	22min	20/30min
Cost	£4	£20/30
Frequency	every 5-10min, 5.30am-12.30am Mon-Sat, 7am-11.30pm Sun	
Contact	☎ 7222 1234; www.tfl.gov.uk/dlr	

Travel to/from Luton

	Thameslink/Midland Mainline Train	Greenline Bus	Taxi
Going to	King's Cross/St Pancras station	Buckingham Palace Rd, Victoria	central London
Duration	21-25min	1hr	1hr
Cost	£9-11	one way/return £13/18	£95-100
Frequency	To/from Luton Airport Parkway every 6-15min, 7am-10pm, shuttle bus connects to airport	every 30min, 3.30am-11.30pm, hourly at other times	
Contact	☎ 7222 1234; www.tfl.gov.uk/dlr	☎ 0870 608 7261; www.greenline.co.uk	

BUS

Within the UK & to Europe

Long-distance and international buses arrive and depart from **Victoria coach station** (☎ information 7730 3466; 164 Buckingham Palace Rd; 🕐 booking office 7am-10pm), close to the Victoria tube and rail stations. The largest operator of buses around the UK is **National Express** (☎ 08717-818181; www.nationalexpress.com). The affiliated **Eurolines** (☎ 0870 514 3219; www.eurolines.com) handles international bus services.

TRAIN

Within the UK

Train services to the north tend to leave from King's Cross/St Pancras or Euston stations, services to the southwest leave from Waterloo, services to the south leave from London Bridge or Victoria, and services to the west leave from Paddington. Call **National Rail Enquiries** (☎ 0845 748 4950; www.rail.co.uk) for exact details.

Continental Europe

Eurostar (☎ 0870 518 6186; www.eurostar.com) links London's shimmering new **St Pancras International train station** (Map p159, A4; ☎ 7843 4250; www.stpancras.com) with the Gare du Nord in Paris (2¼ hours, 25 services daily), Brussels' international terminal (two hours, 10 services daily) and Lille (1½ hours, 10 services daily). For more details, contact **Rail Europe** (☎ 0870 584 8848; www.raileurope.com).

GETTING AROUND

Although it can get insanely busy, the Underground, or 'tube', is still the easiest way to get around London. In this book, the nearest tube station is noted after the ⊖ icon in each listing.

For timetable and fare information on the Underground, the bus network, suburban trains or the Docklands Light Railway (DLR), contact **Transport for London** (☎ 7222 1234; www.tfl.gov.uk). You can also get up-to-the-minute info on service or local trains from **Travelcheck** (☎ 7222 1200).

LET THE TRAIN TAKE THE STRAIN

Getting into central London from its major airports – London City excluded – can take as long as an entire flight from mainland Europe. There is an alternative, however, in the shape of the Eurostar, which will whisk you directly to the West End. The operators of Eurostar make much of the train's carbon-neutral status, but passengers are equally impressed by the convenience of the high-speed rail link to Paris, Brussels and Lille. The sparkling new St Pancras International train station is years ahead of London's airports, and you'll arrive unstressed in Zone 1 of the London transport network, within walking distance of the British Museum.

TRAVEL PASSES

The golden rule of getting around in London is buy an **Oyster Card** (www.tfl.gov.uk/oyster). This rechargeable pass will save you a fortune on the bus and Underground system and on some overland trains. You can buy the card at stations or online for £3 and charge it up with extra credit at stations and many convenience stores.

A single fare is deducted from the card for each journey you complete, but the total cost per day is capped at the value of a one-day Travelcard or bus pass, less 50p. The total will vary depending on the time of day and the number of zones you pass through – a one-day Travelcard for Zones 1 and 2 costs £6.80 (£5.30 on weekends and after 9.30am weekdays), while a daily bus pass costs £3.50. Paper tickets and passes are also available, but you'll pay more for single journeys.

You can also charge up the card with a weekly Travelcard, allowing unlimited travel on the entire system for a week. However, you should always swipe the card when boarding buses and when you enter and leave the Underground.

UNDERGROUND & DLR

Underground stations can be recognised by the distinctive red, white and blue circle logo outside. Trains generally run every five minutes or so from 5.30am (7am on Sunday) to around midnight – make a habit of checking the time of the last train home.

The Underground and overground train network is divided into six concentric zones. If you pay by cash, a single fare anywhere within Zones 1 to 6 is £4. Oyster Card fares are much cheaper at £1.50/2/2.50/3.50 for Zones 1/1-2/1-4/1-6. See www.tfl.gov.uk/oyster for details.

London's Docklands are served by the **Docklands Light Railway** (DLR; www.tfl.gov.uk/dlr), which connects with the tube network at Bank, Tower Hill, Bow Church, Stratford and Canary Wharf. The DLR also runs to City Airport. DLR trains run from 5.30am to 12.30am Monday to Saturday, and 7.30am to 11pm Sunday; fares are the same as the tube.

OVERGROUND TRAIN

Oyster Cards are valid on the suburban passenger trains operated by **London Overground** (☎ Transport for London 7222 1234; www.tfl.gov.uk), which mainly connect the suburbs to each other. Fares are the same as for the tube.

The useful **Thameslink** (☎ 0845 748 4950; www.firstcapitalconnect.co.uk) service run by First Capital Connect passes through Luton Airport, King's Cross, the City, London Bridge and Gatwick Airport,

ROUTEMASTER ROUTES

In 2008 London Mayor Boris Johnson announced a plan to reintroduce a modernised version of London's much-missed Routemaster bus by 2012. Until then you can ride the original Routemasters on two Heritage Routes from Trafalgar Sq. Route 9 runs west to Green Park and the Royal Albert Hall and Route 15 runs east to St Paul's Cathedral and Tower Hill. All the normal tickets are valid.

providing an alternative north–south transect across the city.

BUS

Daytime buses run regularly between 7am and midnight; less frequent night buses (prefixed with the letter 'N') take over between midnight and 7am. Some routes run 24 hours at the same intervals as daytime services. Those under 16 years with valid ID are entitled to free bus travel. If paying by Oyster Card, always swipe your card, or you could face a £20 penalty fare.

Recommended Modes of Transport

	Covent Garden	Piccadilly Circus	South Kensington	Westminster
Covent Garden	n/a	walk 10min	tube 10min	tube 15min, walk 25min
Piccadilly Circus	walk 10min	n/a	tube 8min	tube 8min, walk 20min
South Kensington	tube 10min	tube 8min	n/a	tube 8min
Westminster	tube 15min, walk 25min	tube 8min, walk 20min	tube 8min	n/a
Waterloo	tube 11min, walk 20min	tube 5min	tube 15min	tube 2min, walk 15min
London Bridge	tube 20min	tube 12min	tube 20min	tube 5min
Tower Hill	tube 20min	tube 15min	tube 20min	tube 10min
Liverpool St	tube 17min, walk 25min	tube 20min	tube 27min	tube 20min, bus 25min
Notting Hill	tube 20-25min	tube 15min	tube 8min	tube 15min

Single-journey bus tickets (valid for two hours) cost £2/0.90 cash/Oyster Card, and a daily bus pass (valid for unlimited travel on London buses only) costs £3.50/3 cash/Oyster Card. You must buy your ticket *before* boarding from the automatic machine (or use your Oyster Card) at any bus stop with a yellow sign. Trafalgar Sq, Tottenham Court Rd and Oxford Circus are the main terminals for night buses.

CAR

If you want to bring your car to London, note that all drivers must pay a £8 daily 'congestion charge' to enter the city centre or the western suburbs between 7am and 6pm Monday to Friday. See www.cclondon.com for more details.

TAXI

There are two kinds of cabs in London. The distinctive licensed black cabs are reliable and spacious and the knowledgeable drivers charge fares on the meter. Licensed minicabs (which can be any model of car) charge lower fares, which must be agreed before you start your journey.

Waterloo	London Bridge	Tower Hill	Liverpool St	Notting Hill
tube 11min, walk 20min	tube 20min	tube 20min	tube 17min, walk 25min	tube 20-25min
tube 5min	tube 12min	tube 15min	tube 20min	tube 15min
tube 15min	tube 20min	tube 20min	tube 27min	tube 8min
tube 2min walk 15min	tube 5min	tube 10min	tube 20min, bus 25min	tube 15min
n/a	tube 4min	tube 15min	tube 10min	tube 20min
tube 4min	n/a	tube 10min, bus 25min, walk 20min	tube 10min, bus 15min	tube 25min
tube 15min	bus 25min, tube 10min, walk 20min	n/a	tube 10min	tube 30min
tube 10min	tube 10min, bus 15min	tube 10min	n/a	tube 20min
tube 20min	tube 25min	tube 30min	tube 20min	n/a

LONDON BY RIVER

The Thames was a major thoroughfare in medieval times and several companies have revived the tradition using fast modern passenger boats. The **Thames Clippers** (☎ 0870 781 5049; www.thamesclippers.com) connect Waterloo, Embankment, London Bridge, the Tower, Docklands and Greenwich, and there's an additional service between Tate Britain and Tate Modern. Boats run from around 6am to 11pm (9am to 5.45pm on weekends) and fares start at £4, or you can buy an all-day roamer ticket for £8, valid from 10am to 5pm. There's a discount for Oyster Card holders.

There are also some useful daytime sightseeing services. **City Cruises** (☎ 7740 0400; www.citycruises.com) runs boats between Greenwich, the Tower, Waterloo and Westminster. Prices range from £6.40 for a single journey to £10.50 for an all-day rover ticket. **Thames River Services** (TRS; ☎ 7930 4097; www.thamesriverservices.co.uk) runs east from Westminster to Greenwich and the Thames Barrier, while **Thames River Boats** (☎ 7930 2062; wpsa.co.uk) runs west from Westminster to Richmond and Hampton Court Palace.

You can flag down black cabs in the street (empty cabs display a yellow light above the windscreen). To order a black cab by phone, for a surcharge, try **One-Number Taxi** (☎ 0871 871 8710). Minicabs can only be hired by phone or from a minicab office; every neighbourhood high street has one.

Be wary of unlicensed cabs – if in doubt, call **TFL** (☎ 7222 1234) or text the word 'Home' to ☎ 60835 to obtain the phone number of a licensed local minicab company (see www.cabwise.com).

PRACTICALITIES

BUSINESS HOURS

Unlike New York, London is a city that does sleep. Although licensing hours are getting later, few shops, bars, restaurants or entertainment venues are open 24 hours and most places close early on Sunday. Exact hours depend on the establishment, but the following hours are a useful rule of thumb:

Banks 9.30am to 5.30pm Monday to Friday
Bars 11am or 6pm to 1am or later; many bars close early on Sunday
Offices 9am to 5pm, 9.30am to 5.30pm or 10am to 6pm Monday to Friday
Pubs 11am to 11pm Monday to Thursday, 11am to midnight Friday and Saturday, 11am to 10.30pm Sunday
Restaurants Noon to 11pm, or noon to 3pm and 6pm to 11pm; many restaurants close Sunday or Monday
Shops 9am or 10am to 6pm Monday to Saturday; some shops also open noon to 6pm Sun; late-night shopping in central areas until 8pm Thursday

DISCOUNTS

Most museums and galleries are free, but those that charge generally offer discounts to children (various age limits apply), youth cardholders under 25, students with ISIC cards (age limits may apply), seniors (over 60 or 65), disabled visitors and families.

The **London Pass** (☎ 0870 242 9988; www.londonpass.com) allows free admission to over 55 attractions in the capital, including London Zoo, the Tower of London and most of the other famous sights. For adults the pass costs £38/49/60/82 for one/two/three/six days, or £43/62/79/120 with a Travelcard for the same number of days.

EMERGENCIES

London is remarkably safe considering its size. Although widely reported in the media, gun and knife crime rarely affects tourists. However, opportunistic crime is a problem, as in any large city. Keep an eye on your belongings in crowds and avoid dark empty streets at night.

Everyone gets emergency treatment and European Economic Area (EEA) nationals are entitled to free non-emergency treatment for conditions that arise while travelling in the UK. To qualify for free treatment, you must carry a European Health Insurance Card (EHIC) validated in your home country. There is usually a charge for prescription medicines. Reciprocal arrangements between the UK and some countries (including Australia) allow free medical treatment at hospitals and surgeries, and subsidised dental care.

There is a single number for the Ambulance Service, the Fire Brigade or the Police: ☎ 999.

Accident and Emergency (A&E) departments operating 24 hours:

Chelsea & Westminster Hospital (Map pp122-3, B5; ☎ 8746 8000; www.chelwest.nhs.uk; 369 Fulham Rd; ⊖ Earl's Ct or Fulham Broadway)

Royal Free Hospital (Map p167, B4; ☎ 7794 0500; www.royalfree.nhs.uk; Pond St; ⊖ Belsize Park)

Royal London Hospital (Map p145, D5; ☎ 7377 7000; www.bartsandthelondon.org.uk; Whitechapel Rd; ⊖ Whitechapel)

St Thomas' Hospital (Map pp122-3, H3; ☎ 7188 7188; www.guysandstthomas.nhs.uk; Westminster Bridge Rd; ⊖ Westminster or Waterloo)

University College Hospital (Map p103, B4; ☎ 0845 155 5000; www.uclh.nhs.uk; 253 Euston Rd; ⊖ Warren St or Euston Sq)

Late-opening chemists:

Boots (Map pp88-9, E2; ☎ 7734 6126; www.boots.com; 44-46 Regent St; 🕑 9am-midnight Mon-Fri, 9am-midnight Sat, noon-6pm Sun; ⊖ Piccadilly Circus)

Pharmacentre (Map p113, B5; ☎ 0808 108 5721; www.pharmacentre.com; 149 Edgware Rd; 🕑 24hr; ⊖ Edgware Rd or Marble Arch)

HOLIDAYS

New Year's Day 1 January
Good Friday Late March/April
Easter Monday Late March/April
May Day Bank Holiday First Monday in May
Spring Bank Holiday Last Monday in May
Summer Bank Holiday Last Monday in August
Christmas Day 25 December
Boxing Day 26 December

INTERNET

London is chock-a-block with cybercafés, including the widespread chain **easyInternetcafe** (www.easyeverything.com). Wi-fi internet access is available in many cafés, restaurants and bars. Useful websites include the following:
BBC London (www.bbc.co.uk/London)
Evening Standard (www.thisislondon.co.uk)
Streetmap (www.streetmap.co.uk)
Transport for London (www.tfl.gov.uk)
Walk It (www.walkit.com/london)

MONEY

London isn't cheap – you could easily find yourself spending several hundred pounds a day if you travel by taxi and eat in the best restaurants. As a rule, you should be able to live comfortably for £50 to £75 a day on top of your hotel bill, with enough left over for the occasional splurge. If you're on a tight budget, you might get away with £30 per day by sticking to public transport and avoiding expensive eateries.

London has hundreds of *bureaus de change* and most banks, post offices and travel agencies offer foreign exchange. London banks are usually open 9.30am to 5.30pm Monday to Friday; private exchange offices open seven days a week. Alternatively, there are ATMs on every street corner and almost all accept international credit and debit cards.

For currency exchange rates, see the inside front cover of this book.

ORGANISED TOURS

There are numerous organised tours of London by bus, boat, bicycle and foot, ranging from the tacky to the strange and obscure.

BUS TOURS

Although touristy, hop-on, hop-off open-roofed bus tours are a useful way to see how London fits together. The double-decker buses of the **Original London Sightseeing Tour** (☎ 8877 1722; www.theoriginaltour.com) and the **Big Bus** (☎ 7233 9533; www.bigbustours.com) cruise around the city centre daily; adult tickets cost £22 to £24 for unlimited journeys in one 24-hour period.

RIVER & CANAL CRUISES

As well as the scheduled riverboats (see boxed text, p222), you can float down the Thames in luxury on a lunch or dinner cruise, travel London's canals by barge or ride the river on a rubber inflatable boat. Reliable operators:

Bateaux London (☎ 7696 1800; www.bateauxlondon.com)
City Cruises (☎ 7740 0400; www.citycruises.com)
Crown River Cruises (☎ 7936 2033; www.crownriver.com)
London RIB Voyages (☎ 7928 8933; www.londonribvoyages.com)
London Waterbus Company (☎ 7482 2550; www.londonwaterbus.com)

WALKING TOURS

Numerous companies offer themed walks around the city, visiting ancient ruins, historic buildings, movie locations, public houses, haunted houses and even the sites of notorious murders. **London Walks** (☎ 7624 3978; www.walks.com) offers the best selection of tours; prices start at £7 per adult.

SPECIALIST TOURS

Based on D-Day landing vehicles, the yellow amphibious vehicles of **London Duck Tours** (☎ 7928 3132; www.londonducktours.co.uk; adult/child/concession/family £19/13/15/57.50) cruise the streets of central London before making a dramatic descent into the Thames. Tours depart from outside County Hall (p73).

TELEPHONE

The British GSM 900 mobile phone network is compatible with roaming-enabled phones from Continental Europe, Australia and New Zealand, but not with phones on the North American GSM 1900 or Japanese PDC/CDMA/WCDMA system. If your phone is unlocked, you can use a British prepaid SIM card, available from mobile phone shops and newsagents. Public phones, when they are working, accept cash, credit cards or phonecards sold by newsagents.

COUNTRY & CITY CODES

London (☎ 020)
UK (☎ 44)

USEFUL NUMBERS

Directory enquiries (☎ 118 118, 118 500)
International dialling code (☎ 00)
International directory (☎ 118 505) Premium rate.
Local & international operator (☎ 100)
Reverse-charge/collect calls (☎ 155)
Time (☎ 123)
Weathercall (☎ 0906 850 0401)

TIPPING

Tipping is customary at restaurants and other venues with table service. Follow these guidelines:

Porters Around £2 per bag.
Restaurants From 10% to 15% (usually included in the bill).
Taxis Round up to nearest pound.

TOURIST INFORMATION

The official website of the London tourism authorities is **Visit London** (☎ 0870 156 6366; www.visitlondon.com). As well as information on sights and entertainment, you can book accommodation online.

TRAVELLERS WITH DISABILITIES

Pavement ramps are commonplace but only a few stations on the Victorian-era Underground network have access for travellers with impaired movement. Most buses have automatic ramps for wheelchair access, except on Heritage Routes 9 and 15. London's black cabs are wheelchair accessible and most Thames boat services have facilities for mobility-impaired travellers. Larger museums and attractions have hearing loops, and some offer tours for the visually impaired.

For more details, contact **Transport for London** (☎ 7222 1234, minicom 7918 3015; www.tfl.gov.uk) or the **Royal Association for Disability and Rehabilitation** (RADAR; ☎ 7250 3222; www.radar.org.uk). For information on access to galleries and cultural venues, visit **Artsline** (www.artsline.org.uk).

>INDEX

See also separate subindexes for See (p234), Shop (p236), Eat (p237), Drink (p238) and Play (p239).

000 map pages

000 map pages

SEE

000 map pages

EAT